Classical Geometries in Modern Contexts

Walter Benz

Geometry of Real Inner Product Spaces

Birkhäuser Verlag
Basel · Boston · Berlin

Author:

Walter Benz
Fachbereich Mathematik
Universität Hamburg
Bundesstr. 55
20146 Hamburg
Germany
e-mail: benz@math.uni-hamburg.de

2000 Mathematical Subject Classification 39B52, 51B10, 51B25, 51M05, 51M10, 83C20

Math
QA
431
·B455
2005

A CIP catalogue record for this book is available from the
Library of Congress, Washington D.C., USA

Bibliographic information published by Die Deutsche Bibliothek
Die Deutsche Bibliothek lists this publication in the Deutsche Nationalbibliografie;
detailed bibliographic data is available in the Internet at <http://dnb.ddb.de>.

ISBN 3-7643-7371-7 Birkhäuser Verlag, Basel – Boston – Berlin

© 2005 Birkhäuser Verlag, P.O. Box 133, CH-4010 Basel, Switzerland
Part of Springer Science+Business Media
Cover design: Micha Lotrovsky, CH-4106 Therwil, Switzerland
Printed on acid-free paper produced from chlorine-free pulp. TCF ∞
Printed in Germany
ISBN-10: 3-7643-7371-7   e-ISBN: 3-7643-7432-2
ISBN-13: 978-3-7643-7371-9

9 8 7 6 5 4 3 2 1   www.birkhauser.ch

Contents

Preface

The basic structure playing the key role in this book is a *real inner product space* (X, δ), i.e. a real vector space X together with a mapping $\delta : X \times X \to \mathbb{R}$, a so-called *inner product*, satisfying rules (i), (ii), (iii), (iv) of section 1 of chapter 1. In order to avoid uninteresting cases from the point of view of geometry, we will assume throughout the whole book that there exist elements a, b in X which are linearly independent. But, on the other hand, we do *not* ask for the existence of a positive integer n such that every subset S of X containing exactly n elements is linearly dependent. In other words, we do *not* assume that X is a finite-dimensional vector space. So, when dealing in this book with different geometries like euclidean, hyperbolic, elliptic, spherical, Lorentz–Minkowskian geometry or Möbius (Lie) sphere geometry over a real inner product space (X, δ), the reader might think of $X = \mathbb{R}^2$ or $\mathbb{R}^3$, of X finite-dimensional, or of X infinite-dimensional. In fact, it plays no role, whatsoever, in our considerations whether the dimension of X is finite or infinite: the theory as presented does not depend on the dimension of X. In this sense we may say that our presentation in question is *dimension-free*.

The prerequisites for a fruitful reading of this book are essentially based on the sophomore level, especially after mastering basic linear algebra and basic geometry of $\mathbb{R}^2$ and $\mathbb{R}^3$. Of course, hyperspheres are defined via the inner product δ. At the same time we also define hyperplanes by this product, namely by $\{x \in X \mid \delta(a, x) = \alpha\}$, or, as we prefer to write $\{x \in X \mid ax = \alpha\}$, with $0 \neq a \in X$ and $\alpha \in \mathbb{R}$. This is a quite natural and simple definition and familiar to everybody who learned geometry, say, of the plane or of $\mathbb{R}^3$. For us it means that we do not need to speak about the existence of a basis of X (see, however, section 2.6 where we describe an example of a quasi-hyperplane which is not a hyperplane) and, furthermore, that we do not need to speak about (affine) hyperplanes as images under translations of maximal subspaces $\neq X$ of X (see R. Baer [1], p. 19), hence avoiding *transfinite* methods, which could be considered as somewhat strange in the context of geometries of Klein's Erlangen programme. This programme was published in 1872 by Felix Klein (1849–1925) under the title *Vergleichende Betrachtungen über neuere geometrische Forschungen, Programm zum Eintritt in die philosophische Facultät und den Senat der k. Friedrich-Alexander-Universität zu Erlangen* (Verlag von Andreas Deichert, Erlangen), and it gave rise to an ingenious and fundamental principle that allows distinguishing between different geometries (S, G) (see section 9 of chapter 1) on the basis of their groups G, their invariants and invariant notions (section 9). In connection with Klein's Erlangen programme compare also Julian Lowell Coolidge, A History of Geometrical Methods, Clarendon Press, Oxford, 1940, and, for instance, W. Benz [3], p. 38 f.

The papers [1] and [5] of E.M. Schröder must be considered as pioneer work for a dimension-free presentation of geometry. In [1], for instance, E.M. Schröder proved for arbitrary-dimensional $X, \dim X \geq 2$, that a mapping $f : X \to X$ satisfying $f(0) = 0$ and $\|x_1 - x_2\| = \|f(x_1) - f(x_2)\|$ for all $x_1, x_2 \in X$ with $\|x_1 - x_2\| = 1$ or 2 must be orthogonal. The methods of this result turned out to be important for certain other results of dimension-free geometry (see Theorem 4 of chapter 1 of the present book, see also W. Benz, H. Berens [1] or F. Radó, D. Andreescu, D. Válcan [1]).

The main result of chapter 1 is a common characterization of euclidean and hyperbolic geometry over (X, δ). With an implicit notion of a *(separable) translation group* T of X with axis $e \in X$ (see sections 7, 8 of chapter 1) the following theorem is proved (Theorem 7). Let d be a function, not identically zero, from $X \times X$ into the set $\mathbb{R}_{\geq 0}$ of all non-negative real numbers satisfying $d(x, y) = d\left(\varphi(x), \varphi(y)\right)$ and, moreover, $d(\beta e, 0) = d(0, \beta e) = d(0, \alpha e) + d(\alpha e, \beta e)$ for all $x, y \in X$, all $\varphi \in T \cup O(X)$ where $O(X)$ is the group of orthogonal bijections of X, and for all real α, β with $0 \leq \alpha \leq \beta$. Then, up to isomorphism, there exist exactly two geometries with distance function d in question, namely the euclidean or the hyperbolic geometry over (X, δ). We would like to stress the fact that this result, the proof of which covers several pages, is also dimension-free, i.e. that it characterizes classical euclidean and classical (non-euclidean) hyperbolic geometry without restriction on the (finite or infinite) dimension of X, provided $\dim X \geq 2$. Hyperbolic geometry of the plane was discovered by J. Bolyai (1802–1860), C.F. Gauß (1777–1855), and N. Lobachevski (1793–1856) by denying the euclidean parallel axiom. In our Theorem 7 in question it is not a weakened axiom of *parallelity*, but a weakened notion of *translation with a fixed axis* which leads inescapably to euclidean or hyperbolic geometry and this for all dimensions of X with $\dim X \geq 2$. The methods of the proof of Theorem 7 depend heavily on the theory of functional equations. However, all results which are needed with respect to functional equations are proved in the book. Concerning monographs on functional equations see J. Akzél [1] and J. Akzél–J. Dhombres [1].

In chapter 2 the two metric spaces (X, eucl) (*euclidean metric space*) and (X, hyp) (*hyperbolic metric space*) are introduced depending on the different distance functions $\text{eucl}(x, y)$, $\text{hyp}(x, y)$ $(x, y \in X)$ of euclidean, hyperbolic geometry, respectively. The lines of these metric spaces are characterized in three different ways, as lines of L.M. Blumenthal (section 2), as lines of Karl Menger (section 3), or as follows (section 4): for given $a \neq b$ of X collect as *line through a, b* all $p \in X$ such that the system $d(a, p) = d(a, x)$ and $d(b, p) = d(b, x)$ of two equations has only the solution $x = p$. Moreover, subspaces of the metric spaces in question are defined in chapter 2, as well as spherical subspaces, parallelism, orthogonality, angles, measures of angles and, furthermore, with respect to (X, hyp), equidistant surfaces, ends, horocycles, and angles of parallelism. As far as isometries of (X, hyp) are concerned, we would like to mention the following main result (Theorem 35, chapter 2) which corresponds to Theorem 4 in chapter 1. Let $\varrho > 0$ be a fixed real number and $N > 1$ be a fixed integer. If $f : X \to X$ satisfies hyp $\big(f(x), f(y)\big) \le \varrho$ for all $x, y \in X$ with hyp $(x, y) = \varrho$, and hyp $\big(f(x), f(y)\big) \ge N\varrho$ for all $x, y \in X$ with hyp $(x, y) = N\varrho$, then f must be an isometry of (X, hyp), i.e. satisfies hyp $\big(f(x), f(y)\big) = \text{hyp}(x, y)$ for all $x, y \in X$. If the dimension of X is finite, the theorem of B. Farrahi [1] and A.V. Kuz'minyh [1] holds true: let $\varrho > 0$ be a fixed real number and $f : X \to X$ a mapping satisfying hyp $\big(f(x), f(y)\big) = \varrho$ for all $x, y \in X$ with hyp $(x, y) = \varrho$. Then f must already be an isometry. In section 21 of chapter 2 an example shows that this cannot generally be carried over to the infinite-dimensional case.

A geometry $\Gamma = (S, G)$ is a set $S \ne \emptyset$ together with a group G of bijections of S with the usual multiplication $(fg)(x) = f\big(g(x)\big)$ for all $x \in S$ and $f, g \in G$. The geometer then studies invariants and invariant notions of (S, G) (see section 9 of chapter 1). If a geometry Γ is based on an arbitrary real inner product space $X, \dim X \ge 2$, then it is useful, as we already realized before, to understand by "Γ, dimension-free" a theory of Γ which applies to every described X, no matter whether finite- or infinite-dimensional, so, for instance, the same way to $\mathbb{R}^2$ as to $C[0, 1]$ with $fg = \int_0^1 t^2 f(t) g(t) \, dt$ for real-valued functions f, g defined and continuous in $[0, 1]$ (see section 2, chapter 1). In chapter 3 we develop the geometry of Möbius dimension-free, and also the sphere geometry of Sophus Lie. Even Poincaré's model of hyperbolic geometry can be established dimension-free (see section 8 of chapter 3). In order to stress the fact that those and other theories are developed dimension-free, we avoided drawings in the book: drawings, of course, often present properly geometrical situations, but not, for instance, convincingly the ball $B(c, 1)$ (see section 4 of chapter 2) of the above mentioned example with $X = C[0, 1]$ such that $c : [0, 1] \to \mathbb{R}$ is the function $c(\xi) = \xi^3$. The close connection between Lorentz transformations (see section 17 of chapter 3) and Lie transformations (section 12), more precisely Laguerre transformations (section 13), has been known for almost a hundred years: it was discovered by H. Bateman [1] and H.E. Timerding [1], of course, in the classical context of four dimensions (section 17). This close connection can also be established dimension-free, as shown in chapter 3. A fundamental theorem in Lorentz–Minkowski geometry (see section 17, chapter 3) of A.D. Alexandrov [1] must be mentioned here with respect to Lie sphere geometry: if $(2 \le) \dim X < \infty$, and if $\lambda : Z \to Z$, $Z := X \oplus \mathbb{R}$, is a bijection such that the Lorentz–Minkowski distance $l(x, y)$ (section 1 of chapter 4) is zero if, and only if $l\big(f(x), f(y)\big) = 0$ for all $x, y \in Z$, then f is a Lorentz transformation up to a dilatation. In fact, much more than this follows from Theorem 65 (section 17, chapter 3) which is a theorem of Lie (Laguerre) geometry: we obtain from Theorem 65 Alexandrov's theorem in the dimension-free version and this even in the Cacciafesta form (Cacciafesta [1]) (see Theorem 2 of chapter 4).

All Lorentz transformations of Lorentz–Minkowski geometry over (X, δ) are determined dimension-free in chapter 4, section 1, by Lorentz boosts (section 14, chapter 3), orthogonal mappings and translations. Also this result follows from a theorem (Theorem 61 in section 14, chapter 3) on Lie transformations. In Theorem 6 (section 2, chapter 4) we prove dimension-free a well-known theorem of Alexandrov–Ovchinnikova–Zeeman which these authors have shown under the assumption $\dim X < \infty$, and in which all causal automorphisms (section 2, chapter 4) of Lorentz–Minkowski geometry over (X, δ) are determined.

In sections 9, 10, 11 (chapter 4) Einstein's cylindrical world over (X, δ) is introduced and studied dimension-free; moreover, in sections 12, 13 we discuss de Sitter's world. Sections 14, 15, 16, 17, 18, 19 are devoted to elliptic and spherical geometry. They are studied dimension-free as well. In section 19 the classical lines of spherical, elliptic geometry, respectively, are characterized via functional equations. The notions of Lorentz boost and hyperbolic translation are closely connected: this will be proved and discussed in section 20, again dimension-free.

It is a pleasant task for an author to thank those who have helped him. I am deeply thankful to Alice Günther who provided me with many valuable suggestions on the preparation of this book. Furthermore, the manuscript was critically revised by my colleague Jens Schwaiger from the university of Graz, Austria. He supplied me with an extensive list of suggestions and corrections which led to substantial improvements in my exposition. It is with pleasure that I express my gratitude to him for all the time and energy he has spent on my work.

Waterloo, Ontario, Canada, June 2005   Walter Benz

Chapter 1

Translation Groups

1.1 Real inner product spaces

A *real inner product space* (X, δ) is a real vector space X together with a mapping $\delta : X \times X \to \mathbb{R}$ satisfying

(i) $\delta(x, y) = \delta(y, x)$,

(ii) $\delta(x + y, z) = \delta(x, z) + \delta(y, z)$,

(iii) $\delta(\lambda x, y) = \lambda \cdot \delta(x, y)$,

(iv) $\delta(x, x) > 0$ for $x \neq 0$

for all $x, y, z \in X$ and $\lambda \in \mathbb{R}$. Concerning the notation $\delta : X \times X \to \mathbb{R}$ and others we shall use later on, see the section *Notations and symbols* of this book. Instead of $\delta(x, y)$ we will write xy or, occasionally, $x \cdot y$. The laws above are then the following:

$$xy = yx, \quad (x + y)z = xz + yz, \quad (\lambda x) \cdot y = \lambda \cdot (xy)$$

for all $x, y, z \in X$, $\lambda \in \mathbb{R}$, and $x^2 := x \cdot x > 0$ for all $x \in X \backslash \{0\}$. Instead of (X, δ) we mostly will speak of X, hence tacitly assuming that X is equipped with a fixed *inner product*, i.e. with a fixed $\delta : X \times X \to \mathbb{R}$ satisfying rules (i), (ii), (iii), (iv).

Two real inner product spaces (X, δ), (X', δ') are called *isomorphic* provided (in the sense of *if, and only if*) there exists a bijection

$$\varphi : X \to X'$$

such that

$$\varphi(x + y) = \varphi(x) + \varphi(y), \quad \varphi(\lambda x) = \lambda \varphi(x), \quad \delta(x, y) = \delta'\big(\varphi(x), \varphi(y)\big)$$

hold true for all $x, y \in X$ and $\lambda \in \mathbb{R}$. The last of these equations can be replaced by the weaker one $\delta(x, x) = \delta'(\varphi(x), \varphi(x))$ for all $x \in X$, since

$$2\overline{\delta}(x, y) = \overline{\delta}(x + y, x + y) - \overline{\delta}(x, x) - \overline{\delta}(y, y)$$

holds true for all $x, y \in X$ for $\overline{\delta} = \delta$ as well as for all $x, y \in X'$ for $\overline{\delta} = \delta'$.

1.2   Examples

a)   Let $B \neq \emptyset$ be a set and define $X(B)$ to be the set of all $f : B \to \mathbb{R}$ such that $\{b \in B \mid f(b) \neq 0\}$ is finite. Put

$$(f + g)(b) := f(b) + g(b)$$

for $f, g \in X$ and $b \in B$, and

$$(\alpha f)(b) := \alpha f(b)$$

for $f \in X$, $\alpha \in \mathbb{R}$, $b \in B$. Finally set

$$fg := \sum_{b \in B} f(b) g(b)$$

for $f, g \in X$.

b)   Let $\alpha < \beta$ be real numbers and let X be the set of all continuous functions $f : [\alpha, \beta] \to \mathbb{R}$ with $[\alpha, \beta] := \{t \in \mathbb{R} \mid \alpha \leq t \leq \beta\}$. Define $f + g$, αf as in a) and put

$$fg := \int_{\alpha}^{\beta} h(t) f(t) g(t) \, dt$$

for a fixed $h \in X$ satisfying $h(t) > 0$ for all $t \in [\alpha, \beta] \setminus T$ where T is a finite subset of $[\alpha, \beta]$. This real inner product space will be denoted by $X([\alpha, \beta], h)$.

c) Suppose that X is the set of all sequences

$$(a_1, a_2, a_3, \ldots)$$

of real numbers $a_1, a_2, a_3, \ldots$ such that $\sum_{i=1}^{\infty} a_i^2$ exists. Define

$$
\begin{aligned}
(a_1, a_2, \ldots) + (b_1, b_2, \ldots) &:= (a_1 + b_1, a_2 + b_2, \ldots), \\
\lambda \cdot (a_1, a_2, \ldots) &:= (\lambda a_1, \lambda a_2, \ldots), \\
(a_1, a_2, \ldots) \cdot (b_1, b_2, \ldots) &:= \sum_{i=1}^{\infty} a_i b_i,
\end{aligned}
$$

by observing

$$(a_i + b_i)^2 = a_i^2 + b_i^2 + 2a_i b_i \leq a_i^2 + b_i^2 + a_i^2 + b_i^2$$

from $(a_i - b_i)^2 \geq 0$, i.e. by noticing

$$\sum_{i=1}^{n}(a_i + b_i)^2 \leq 2\sum_{i=1}^{n} a_i^2 + 2\sum_{i=1}^{n} b_i^2,$$

i.e. that $\sum_{i=1}^{\infty}(a_i + b_i)^2$ exists. Because of

$$4\sum_{i=1}^{n} a_i b_i = \sum_{i=1}^{n}(a_i + b_i)^2 - \sum_{i=1}^{n}(a_i - b_i)^2,$$

also $\sum_{i=1}^{\infty} a_i b_i$ exists.

1.3 Isomorphic, non-isomorphic spaces

Let n be a positive integer. The $\mathbb{R}^n$ consists of all ordered n-tuples

$$(x_1, x_2, \ldots, x_n)$$

of real numbers x_i, $i = 1, 2, \ldots, n$. It is a real inner product space with

$$
\begin{aligned}
(x_1, \ldots, x_n) + (y_1, \ldots, y_n) &:= (x_1 + y_1, \ldots, x_n + y_n), \\
\alpha \cdot (x_1, \ldots, x_n) &:= (\alpha x_1, \ldots, \alpha x_n), \\
(x_1, \ldots, x_n) \cdot (y_1, \ldots, y_n) &:= x_1 y_1 + \cdots + x_n y_n
\end{aligned}
$$

for $x_i, y_i, \alpha \in \mathbb{R}$, $i = 1, \ldots, n$.

Obviously, $\mathbb{R}^n$ and $X(\{1, 2, \ldots, n\})$ are isomorphic: define $\varphi(x_1, \ldots, x_n)$ to be the function $f : \{1, \ldots, n\} \to \mathbb{R}$ with $f(i) = x_i$, $i = 1, \ldots, n$.

Suppose that B_1, B_2 are non-empty sets. The real inner product spaces $X(B_1)$, $X(B_2)$ are isomorphic if, and only if, there exists a bijection $\gamma : B_1 \to B_2$ between B_1 and B_2. If there exists such a bijection, define $\varphi(f)$ for $f \in X(B_1)$ by

$$\varphi(f)(\gamma(b)) = f(b)$$

for all $b \in B_1$. Hence $\varphi : X(B_1) \to X(B_2)$ establishes an isomorphism. If $X(B_1)$, $X(B_2)$ are isomorphic, there exists a bijection

$$\varphi : X(B_1) \to X(B_2)$$

with $\varphi(x + y) = \varphi(x) + \varphi(y)$, $\varphi(\lambda x) = \lambda \varphi(x)$ for all $x, y \in X(B_1)$ and $\lambda \in \mathbb{R}$. We associate to $b \in B_1$ the element $\hat{b}$ of $X(B_1)$ defined by $\hat{b}(b) = 1$ and $\hat{b}(c) = 0$ for all $c \in B_1 \backslash \{b\}$. Then $\hat{B}_1 := \{\hat{b} \mid b \in B_1\}$ is a basis of $X(B_1)$, and $\hat{B}_2$ and $\varphi(\hat{B}_1)$ must be bases of $X(B_2)$. Since $\hat{B}_1$, $\varphi(\hat{B}_1)$ are of the same cardinality, and also $\hat{B}_2$, $\varphi(\hat{B}_1)$, we get the same cardinality for $\hat{B}_1$, $\hat{B}_2$, and hence also for B_1, B_2.

Suppose that $\alpha < \beta$ are real numbers and that $h : [\alpha, \beta] \to \mathbb{R}$ is continuous with $h(\eta) > 0$ in $[\alpha, \beta]$. Then the real inner product spaces $X([\alpha, \beta], h)$ and $X([0, 1], 1)$ are isomorphic. Here 1 designates the function $1(\xi) = 1$ for all $\xi \in [0, 1]$. In order to prove this statement, associate to the function $f : [0, 1] \to \mathbb{R}$, also written as $f(\xi)$, the function $\varphi(f) : [\alpha, \beta] \to \mathbb{R}$ defined by

$$\varphi(f)(\eta) := \sqrt{\frac{(\beta - \alpha)^{-1}}{h(\eta)}} \, f\left(\frac{\eta - \alpha}{\beta - \alpha}\right).$$

Obviously, $\varphi : X([0, 1], 1) \to X([\alpha, \beta], h)$ is a bijection. It satisfies

$$\varphi(f + g) = \varphi(f) + \varphi(g), \ \varphi(\lambda f) = \lambda \varphi(f)$$

for all $\lambda \in \mathbb{R}$ and $f, g \in X([0, 1], 1)$. Moreover, we obtain

$$\varphi(f) \cdot \varphi(g) = \int_\alpha^\beta h(\eta) \, \varphi(f)(\eta) \, \varphi(g)(\eta) \, d\eta = \int_0^1 f(\xi) g(\xi) \, d\xi = f \cdot g,$$

and hence that $X([0, 1], 1)$, $X([\alpha, \beta], h)$ are isomorphic.

Remark. There exist examples of (necessarily infinite-dimensional) real vector spaces X with mappings $\delta_\nu : X \times X \to \mathbb{R}$, $\nu = 1, 2$, satisfying rules (i),(ii), (iii), (iv) of section 1.1 such that (X, δ_1) and (X, δ_2) are not isomorphic (J. Rätz [1]).

1.4   Inequality of Cauchy–Schwarz

Inequality of Cauchy–Schwarz: If a, b are elements of X, then $(ab)^2 \leq a^2 b^2$ holds true.

Proof. Case $b = 0$. Observe, for $p \in X$,

$$pb = p \cdot 0 = p \cdot (0 + 0) = p \cdot 0 + p \cdot 0,$$

i.e. $pb = p \cdot 0 = 0$, i.e. $a \cdot b = 0$ and $b^2 = 0$.

Case $b \neq 0$. Hence $b^2 > 0$ and thus

$$0 \leq \left(a - \frac{ab}{b^2} b\right)^2 = a^2 - \frac{(ab)^2}{b^2}, \tag{1.1}$$

i.e. $(ab)^2 \leq a^2 b^2$. $\square$

Lemma 1. *If a, b are elements of X such that $(ab)^2 = a^2 b^2$ holds true, then a, b are linearly dependent.*

Proof. Case $b = 0$. Here $0 \cdot a + 1 \cdot b = 0$.

Case $b \neq 0$. Hence, by (1.1), $\left(a - \frac{ab}{b^2} \cdot b\right)^2 = 0$, i.e.

$$a - \frac{ab}{b^2} \cdot b = 0. \qquad \square$$

For $x \in X$, the real number $s \geq 0$ with $s^2 = x^2$ is said to be the *norm* of x, $s =: \|x\|$. Obviously, $\|\lambda x\| = |\lambda| \cdot \|x\|$ for $\lambda \in \mathbb{R}$ and $x \in X$. Moreover, $\|x\| = 0$ holds true for $x \in X$ if, and only if, $x = 0$. Observing $xy \leq |xy| \leq \|x\| \cdot \|y\|$ for $x, y \in X$, from the inequality of Cauchy–Schwarz, we obtain $(x + y)^2 \leq (\|x\| + \|y\|)^2$, i.e. we get the *triangle inequality*

$$\|x + y\| \leq \|x\| + \|y\| \text{ for all } x, y \in X. \tag{1.2}$$

1.5 Orthogonal mappings

Let X be a real inner product space. In order to avoid that the underlying real vector space of X is $\mathbb{R}$ or $\{0\}$, *we will assume throughout the whole book that there exist two elements in X which are linearly independent.* Under this assumption the following holds true: *if x, y are elements of X, there exists $w \in X$ with $w^2 = 1$ and $w \cdot (x - y) = 0$.* Since there are elements a, b in X, which are linearly independent, put $w = \frac{a}{\|a\|}$ in the case $x = y$. If $x \neq y$, there exists z in X such that $z \notin \mathbb{R} \cdot (x - y)$, because otherwise $a, b \in \mathbb{R} \cdot (x - y)$ would be linearly dependent. Hence

$$v := z - \frac{z(x - y)}{(x - y)^2}(x - y) \neq 0.$$

Thus $w := \frac{v}{\|v\|}$ satisfies $w^2 = 1$ and $w \cdot (x - y) = 0$.

A mapping $\omega : X \to X$ is called *orthogonal* if, and only if,

$$\omega(x + y) = \omega(x) + \omega(y), \ \omega(\lambda x) = \lambda \omega(x), \ xy = \omega(x)\omega(y)$$

hold true for all $x, y \in X$ and $\lambda \in \mathbb{R}$.

An orthogonal mapping ω of X must be injective, but it need not be surjective. Assume $\omega(x) = \omega(y)$ for the elements x, y of X. Because of

$$\omega(x - y) = \omega(x + [(-1)y]) = \omega(x) + (-1)\omega(y) = 0,$$

we obtain $(x - y)^2 = [\omega(x - y)]^2 = 0$, i.e. $x - y = 0$, i.e. $x = y$.

Define $B := \{1, 2, 3, \ldots\}$ and take the space $X = X(B)$ of type a). For $f \in X$ put

$$\omega(f)(1) = 0 \text{ and } \omega(f)(i) = f(i - 1), \ i = 2, 3, \ldots.$$

Since $g \in X$ with $g(1) = 1$, $g(i) = 0$ for $i = 2, 3, \ldots$, has no inverse image, ω cannot be surjective. But, trivially, ω is orthogonal.

A linear mapping $\omega : X \to X$ is orthogonal if, and only if, $\|x\| = \|\omega(x)\|$ holds true for all $x \in X$. From $x^2 = (\omega(x))^2$ we get, for all $a, b \in X$,

$$(a - b)^2 = (\omega(a - b))^2 = (\omega(a) - \omega(b))^2,$$

i.e. $ab = \omega(a)\omega(b)$, in view of $a^2 = (\omega(a))^2$, $b^2 = (\omega(b))^2$. On the other hand, if ω is orthogonal, then $x \cdot x = \omega(x)\omega(x)$ holds true, i.e. $\|x\| = \|\omega(x)\|$.

Lemma 2. *Suppose that* $a, b, m \in X$ *satisfy*

$$\|m - a\| = \|b - m\| = \frac{1}{2}\|b - a\|.$$

Then $m = \frac{1}{2}(a + b)$ *holds true.*

Proof. Put $\mu := \|m - a\|$, $a' := m - a$, $b' := b - m$. Hence

$$(b - a)^2 + (a' - b')^2 = (a' + b')^2 + (a' - b')^2 = 4\mu^2.$$

Thus $\|b - a\| = 2\mu$ implies $(a' - b')^2 = 0$, i.e. $a' = b'$, i.e. $m = \frac{1}{2}(a + b)$. □

Proposition 3. *A mapping* $f : X \to X$ *satisfying* $f(0) = 0$ *and* $\|x - y\| = \|f(x) - f(y)\|$ *for all* $x, y \in X$ *must be orthogonal.*

Proof. Obviously, for all $a, b \in X$,

$$\left\|\frac{a + b}{2} - a\right\| = \left\|b - \frac{a + b}{2}\right\| = \frac{1}{2}\|b - a\|.$$

This implies, by $\|x - y\| = \|f(x) - f(y)\|$,

$$\left\|f\left(\frac{a + b}{2}\right) - f(a)\right\| = \left\|f(b) - f\left(\frac{a + b}{2}\right)\right\| = \frac{1}{2}\|f(b) - f(a)\|.$$

Hence, by Lemma 2, we obtain $f\left(\frac{a+b}{2}\right) = \frac{1}{2}(f(a) + f(b))$, i.e. Jensen's functional equation. From $b = 0$ we get $f\left(\frac{a}{2}\right) = \frac{1}{2}f(a)$ for all $a \in X$. Thus

$$f(a + b) = f(a) + f(b)$$

for all $a, b \in X$. This implies $f(\lambda a) = \lambda f(a)$ for all rationals λ and all $a \in X$. Let now $\lambda \in \mathbb{R}$ be given, and let λ_n be a sequence of rational numbers with $\lim \lambda_n = \lambda$. By (1.2),

$$\|f(\lambda a) - \lambda f(a)\| \leq \|f(\lambda a) - f(\lambda_n a)\| + \|f(\lambda_n a) - \lambda f(a)\| =: R,$$

and with $\|x - y\| = \|f(x) - f(y)\|$,

$$R = \|\lambda a - \lambda_n a\| + \|\lambda_n f(a) - \lambda f(a)\| = |\lambda - \lambda_n| \cdot (\|a\| + \|f(a)\|)$$

for all $a \in X$. Hence $\|f(\lambda a) - \lambda f(a)\| = 0$, and thus f must be linear. Finally observe $\|x\| = \|f(x)\|$ for all $x \in X$ from $\|x - 0\| = \|f(x) - f(0)\|$ and $f(0) = 0$. $\qquad\square$

Of course, the set of all surjective orthogonal mappings $\omega : X \to X$ forms a group under the permutation product, the so-called *orthogonal group* $O(X)$ of X.

1.6   A characterization of orthogonal mappings

The following theorem characterizes the orthogonal mappings of a real inner product space under mild hypotheses, i.e. under especially weak assumptions.

Theorem 4. *Let $\varrho > 0$ be a fixed real number and $N > 1$ be a fixed integer. If the mapping $f : X \to X$ satisfies*

$$\forall_{x,y \in X} \ \|x - y\| = \varrho \ \Rightarrow \ \|f(x) - f(y)\| \le \varrho, \tag{1.3}$$

$$\forall_{x,y \in X} \ \|x - y\| = N\varrho \ \Rightarrow \ \|f(x) - f(y)\| \ge N\varrho, \tag{1.4}$$

it must be of the form

$$\forall_{x \in X} \ f(x) = \omega(x) + t, \tag{1.5}$$

where ω is an orthogonal mapping, and t a fixed element of X.

Proof. We will prove

$$\|f(x) - f(y)\| = \|x - y\| \tag{1.6}$$

for all $x, y \in X$. Then, by Proposition 3,

$$g(x) := f(x) - f(0) =: \omega(x)$$

must be orthogonal, and f is of the form (1.5).

a) $\|x - y\| \in \{\varrho, 2\varrho\}$ *implies* $\|x - y\| = \|f(x) - f(y)\|$.

If $x, y \in X$ are given with $\|x - y\| = \varrho$, define $z := 2y - x$. Hence $\|x - z\| = 2\varrho$. If $x, z \in X$ are given with $\|x - z\| = 2\varrho$, define $y := \frac{1}{2}(x + z)$, and we obtain $\|x - y\| = \varrho$. Put

$$p_\lambda := x + \frac{1}{2}\lambda(z - x) \text{ for } \lambda = 0, 1, \ldots, N.$$

Hence, by (1.4) and $\|p_0 - p_N\| = N\varrho$,

$$\|f(p_0) - f(p_N)\| \geq N\varrho.$$

By (1.3), for $\lambda = 0, \ldots, N - 1$,

$$\|f(p_\lambda) - f(p_{\lambda+1})\| \leq \varrho,$$

on account of $\|p_\lambda - p_{\lambda+1}\| = \varrho$. Thus

$$N\varrho \ \leq \|f(p_0) - f(p_N)\| \leq \|f(p_0) - f(p_2)\| + \sum_{\lambda=2}^{N-1} \|f(p_\lambda) - f(p_{\lambda+1})\|$$

$$\leq \|f(p_0) - f(p_1)\| + \|f(p_1) - f(p_2)\| + \sum_{\lambda=2}^{N-1} \|f(p_\lambda) - f(p_{\lambda+1})\| \leq N\varrho,$$

i.e. $\|f(p_\lambda) - f(p_{\lambda+1})\| = \varrho$ for $\lambda = 0, \ldots, N - 1$, and, moreover,

$$\|f(p_0) - f(p_2)\| = \|f(p_0) - f(p_1)\| + \|f(p_1) - f(p_2)\| = 2\varrho.$$

Putting $p_0 = x$, $p_1 = y, p_2 = z$, we obtain

$$\|f(x) - f(y)\| = \varrho \text{ and } \|f(x) - f(z)\| = 2\varrho.$$

b) *If $x, y \in X$ satisfy $\|x - y\| = \varrho$, then*

$$f(x + \lambda(y - x)) = f(x) + \lambda(f(y) - f(x)) \qquad (1.7)$$

holds true for $\lambda = 0, 1, 2, \ldots$.

This is clear for $\lambda = 0$ and $\lambda = 1$. Put

$$p_\lambda := x + \lambda(y - x) \text{ for } \lambda = 0, 1, 2, \ldots.$$

If $\lambda \in \{1, 2, 3, \ldots\}$, we obtain

$$\varrho = \|p_\lambda - p_{\lambda-1}\| = \|p_{\lambda+1} - p_\lambda\| = \frac{1}{2}\|p_{\lambda+1} - p_{\lambda-1}\|,$$

i.e. by a),

$$\varrho = \|f(p_\lambda) - f(p_{\lambda-1})\| = \|f(p_{\lambda+1} - f(p_\lambda)\| = \frac{1}{2}\|f(p_{\lambda+1}) - f(p_{\lambda-1})\|,$$

and hence, by Lemma 2,

$$f(p_\lambda) = \frac{1}{2}\left(f(p_{\lambda-1}) + f(p_{\lambda+1})\right) \text{ for } \lambda = 1, 2, 3, \ldots.$$

This equation implies (1.7), by induction.

c) *If $x, y \in X$ satisfy $\|x - y\| = \frac{\lambda}{\mu} \varrho$ with $\lambda, \mu \in \{1, 2, 3, \ldots\}$, then $\|x - y\| = \|f(x) - f(y)\|$.*

We assumed that there exist two elements in X which are linearly independent. This implies, as we already know, the existence of an element $w \in X$ satisfying $w \cdot (x - y) = 0$ and $\|w\| = 1$. Put

$$z := \frac{1}{2}(x + y) + \lambda \varrho \sqrt{1 - \frac{1}{4\mu^2}} \cdot w$$

and observe $\|z - x\| = \lambda \varrho = \|z - y\|$. Define a, b, x', y' by means of

$$x = z + \lambda(a - z), \quad y = z + \lambda(b - z),$$
$$x' = z + \mu(a - z), \quad y' = z + \mu(b - z).$$

Hence, by b) and $\|a - z\| = \varrho = \|b - z\|$,

$$f(x) = f(z) + \lambda\big(f(a) - f(z)\big), \quad f(y) = f(z) + \lambda\big(f(b) - f(z)\big),$$
$$f(x') = f(z) + \mu\big(f(a) - f(z)\big), \quad f(y') = f(z) + \mu\big(f(b) - f(z)\big),$$

i.e. $\lambda\big(f(x') - f(y')\big) = \mu\big(f(x) - f(y)\big)$. Thus, by a) and $\|x' - y'\| = \varrho$, we obtain

$$\|f(x) - f(y)\| = \frac{\lambda}{\mu}\varrho = \|x - y\|.$$

d) *Suppose that $t > 0$ is a rational number, and that for $x, y \in X$ we have $\|x - y\| < t\varrho$. Then $\|f(x) - f(y)\| \le t\varrho$.* As in step c) we take $w \in X$ with $w(x - y) = 0$ and $\|w\| = 1$. Put

$$z := \frac{1}{2}(x + y) + \frac{t\varrho}{2}\sqrt{1 - \left(\frac{\|x - y\|}{t\varrho}\right)^2} \, w,$$

and observe $\|z - x\| = \frac{1}{2}t\varrho = \|z - y\|$. Hence, by c),

$$\|f(z) - f(x)\| = \frac{1}{2}t\varrho = \|f(z) - f(y)\|,$$

and thus

$$\|f(x) - f(y)\| \le \|f(x) - f(z)\| + \|f(z) - f(y)\| = t\varrho.$$

e) *Let $r > 0$, $s > 0$ be rational numbers, and x, y be elements of X satisfying*

$$r\varrho < \|x - y\| < s\varrho.$$

Then $r\varrho \le \|f(x) - f(y)\| \le s\varrho$ holds true.
Put

$$p := x + \frac{s\varrho}{\|x - y\|}(y - x)$$

and observe

$$\|p - y\| = \left(\frac{s\varrho}{\|x - y\|} - 1\right)\|y - x\| = s\varrho - \|y - x\| < (s - r)\varrho.$$

Hence, by d), $\|f(p) - f(y)\| \le (s - r)\varrho$. Because of $\|p - xVert = s\varrho$ and step c), we obtain

$$\|f(p) - f(x)\| = s\varrho.$$

Thus

$$\|f(x) - f(y)\| \ge \|f(x) - f(p)\| - \|f(y) - f(p)\| \ge s\varrho - (s - r)\varrho.$$

Since, moreover, $\|x - y\| < s\varrho$ yields, by d), $\|f(x) - f(y)\| \le s\varrho$, e) is proved.

f) (1.6) *holds true for all* $x, y \in X$.

Assuming $x \ne y$, we will consider two sequences $r_\nu, s_\nu (\nu = 1, 2, 3, \ldots)$ of rational numbers with $\lim r_\nu = \frac{1}{\varrho}\|x - y\| = \lim s_\nu$, and such that

$$r_\nu \varrho < \|x - y\| < s_\nu \varrho$$

is satisfied for all $\nu = 1, 2, 3, \ldots$. Step e) implies

$$r_\nu \varrho \le \|f(x) - f(y)\| \le s_\nu \varrho.$$

Hence $\|x - y\| = \|f(x) - f(y)\|$.   □

Remark. Steps b), c), d), e) f) of the previous proof were given by E.M. Schröder in [1]. For generalizations of Theorem 4, or similar results, see F.S. Beckman, D.A. Quarles [1], W. Benz, H. Berens [1], W. Benz [8], K. Bezdek, R. Connelly [1], J.A. Lester [1], F. Radó, D. Andreescu, D. Valcán [1], E.M. Schröder [5], among others.

1.7   Translation groups, axis, kernel

Let X be a real inner product space such that there exist two linearly independent elements in X. By Perm X we designate the group of all permutations of X with the usual permutation product

$$(fg)(x) = f(g(x)), \text{ for all } x \in X,$$

for $f, g \in \text{Perm } X$. Let e be a fixed element of X with $e^2 = 1$. Put

$$H := e^\perp := \{x \in X \mid xe = 0\},$$

and we obtain $X = H \oplus \mathbb{R}e$: this means, by definition, that to every $x \in X$ there exist uniquely determined elements $\overline{x} \in H$ and $x_0 e \in \mathbb{R}e$ satisfying

$$x = \overline{x} + x_0 e,$$

and that $H, \mathbb{R}e$ are subspaces of X. In fact,

$$x = \left(x - (xe)\, e\right) + (xe)\, e$$

holds true with $x - (xe)\, e \in H$, $(xe)\, e \in \mathbb{R}e$, and

$$x = \bar{x} + x_0 e, \ \bar{x} \in H, \ x_0 \in \mathbb{R},$$

implies $xe = (\bar{x} + x_0 e)\, e = x_0$, i.e. $\bar{x} = x - x_0 e = x - (xe)\, e$. Observe $H \neq \{0\}$, since there exists $w \in X$ with $\|w\| = 1$ and $w \cdot (e - 0) = 0$ (see the beginning of section 1.5).

Remark. Occasionally, the following statement will be useful. If α, β are mappings from X into X such that $\alpha\beta = \mathrm{id} = \beta\alpha$ holds true, where id designates the identity element of Perm X, then α, and, of course, β as well, must be bijections of X. In fact, $\beta(x)$ is an inverse image of x, since $\alpha\left(\beta(x)\right) = x$, and if $\alpha(x)$ is equal to $\alpha(y)$, then

$$\mathrm{id}\,(x) = \beta\left(\alpha(x)\right) = \beta\left(\alpha(y)\right) = \mathrm{id}\,(y)$$

implies $x = y$.

A mapping $T : \mathbb{R} \to$ Perm X is called a translation group of X in the direction of e, or with axis e, if, and only if, the following properties hold true.

(T1) *$T_{t+s} = T_t \cdot T_s$ for all $t, s \in \mathbb{R}$,*

(T2) *For $x, y \in X$ satisfying $x - y \in \mathbb{R}e$ there exists exactly one $t \in \mathbb{R}$ with $T_t(x) = y$,*

(T3) *$T_t(x) - x \in \mathbb{R}_{\geq 0} e$ for all $x \in X$ and all real $t \geq 0$.*

Here T_t designates the image of $t \in \mathbb{R}$ under T, moreover, $T_t \cdot T_s$ the permutation product, and $T_t(x)$ the image of $x \in X$ under the permutation T_t of X. By $R_{\geq 0}$ we denote the set of all non-negative reals. (T1) is called *translation equation* in the theory of functional equations (J.Aczél [1]).

If we associate to $t \in \mathbb{R}$ the permutation

$$\forall_{x \in X}\, x \to x + te,$$

of X, we get an example of a translation group with axis e. Another important example is given by $t \to T_t$ with

$$\forall_{x \in X}\, T_t(x) = x + [(xe)(\cosh t - 1) + \sqrt{1 + x^2}\, \sinh t]\, e. \tag{1.8}$$

In order to prove that $t \to T_t$ is a translation group, observe that the elements of X can be written in the form

$$h + \sinh \tau \cdot \sqrt{1 + h^2}\, e$$

with $h \in H$ and $\tau \in \mathbb{R}$: for $x \in X$ put

$$h := \bar{x}, \; \sinh \tau := \frac{x_0}{\sqrt{1 + \bar{x}^2}},$$

and notice that (1.8) yields

$$T_t(h + \sinh \tau \cdot \sqrt{1 + h^2}\, e) = h + \sinh(\tau + t)\, \sqrt{1 + h^2}\, e. \tag{1.9}$$

From (1.9) we get at once $T_s T_t = T_{t+s} = T_{s+t}$, especially $T_{-t} T_t = $ id, id the identity permutation of X, so that T_t must be a bijection of X. If $x, y \in X$ satisfy $x - y \in \mathbb{R}e$, put $y = x + \alpha e$ for a suitable $\alpha \in \mathbb{R}$. Hence, by

$$h := \bar{x} \text{ and } x_0 =: \sinh \tau \cdot \sqrt{1 + h^2},$$

we obtain $y = h + (x_0 + \alpha)\, e =: h + \sinh(\tau + t') \cdot \sqrt{1 + h^2}\, e$, i.e. $T_t(x) = y$ has the uniquely determined solution $t = t'$. Property (T3), finally, follows from (1.9), since

$$T_t(x) - x = \bigl(\sinh(\tau + t) - \sinh \tau\bigr)\sqrt{1 + h^2}\, e,$$

and $\sinh(\tau + t) \geq \sinh \tau$ for $\tau + t \geq \tau$.

Suppose that $T : \mathbb{R} \to \text{Perm } X$ is an arbitrary translation group of X in the direction of e. Obviously, the group $\{T_t \mid t \in \mathbb{R}\}$ is a homomorphic image of the additive group of $\mathbb{R}$, and even an isomorphic image, since $T_s = T_0$ implies $T_s(x) = T_0(x)$ for all $x \in X$, i.e. $s = 0$, in view of (T2). Notice $T_0 = \text{id} \in \text{Perm } X$. The function $\varrho : H \times \mathbb{R} \to \mathbb{R}$,

$$\varrho(h, t) := [T_t(h) - h] \cdot e = T_t(h) \cdot e \tag{1.10}$$

is called the *kernel* of the translation group T. In the case (1.8), for instance, we get

$$\varrho(h, t) = \sinh t \cdot \sqrt{1 + h^2}. \tag{1.11}$$

Theorem 5. *The kernel $\varrho : H \times \mathbb{R} \to \mathbb{R}$ of a translation group T of X with axis e satisfies*

(i) *$\varrho(h, 0) = 0$ and $\varrho(h, t_1) < \varrho(h, t_2)$ for all $h \in H$ and all reals $t_1 < t_2$,*

(ii) *To $h \in H$ and $\xi \in \mathbb{R}$ there exists $t \in \mathbb{R}$ such that $\varrho(h, t) = \xi$.*

If, on the other hand, an arbitrary function $\varrho : H \times \mathbb{R} \to \mathbb{R}$ satisfies (i) and (ii), then

$$T_t\bigl(h + \varrho(h, \tau)\, e\bigr) = h + \varrho(h, \tau + t)\, e \tag{1.12}$$

defines a translation group of X in the direction of e; the kernel of this translation group is ϱ.

Proof. a) $\varrho(h, 0) = 0$ follows from (1.10) and $T_0 = \mathrm{id}$. Suppose $h \in H$ and $t_1 < t_2$ for $t_1, t_2 \in \mathbb{R}$. Hence, by (T3),

$$T_{t_2-t_1}\big(T_{t_1}(h)\big) - T_{t_1}(h) \in \mathbb{R}_{\geq 0}e,$$

i.e., by (T1), $T_{t_2}(h) - T_{t_1}(h) := \mu e$, $\mu \geq 0$. Since $t_2 \neq t_1$, we obtain $T_{t_2-t_1}(h) \neq h$, by (T2), i.e. $T_{t_2}(h) \neq T_{t_1}(h)$, i.e. $\mu \neq 0$, i.e. $\mu > 0$. Hence, by (1.10),

$$\begin{aligned}
\varrho(h, t_2) - \varrho(h, t_1) &= \big(T_{t_2}(h) - h\big)e - \big(T_{t_1}(h) - h\big)e \\
&= \big(T_{t_2}(h) - T_{t_1}(h)\big)e = \mu > 0.
\end{aligned}$$

If $h \in H$ and $\xi \in \mathbb{R}$, there exists, by (T2), $t \in \mathbb{R}$ with $T_t(h) = h + \xi e$. Hence, by (1.10),

$$\varrho(h, t) = \big(T_t(h) - h\big)e = \xi.$$

b) t in (ii) is uniquely determined: if also $\varrho(h, t')$ were equal to ξ with, say $t < t'$, then, by (i),

$$\xi = \varrho(h, t) < \varrho(h, t') = \xi.$$

If $x \in X$, the elements $h \in H$ and $\tau \in \mathbb{R}$ satisfying

$$x = h + \varrho(h, \tau)e$$

are uniquely determined because $h = \overline{x}$ and $\varrho(h, \tau) = x_0$ has exactly one solution τ. Hence T_t from (1.12) defines a mapping from X into X. Observe $T_{t+s} = T_t T_s$ from (1.12), and hence $T_t T_{-t} = T_0$. But, by (1.12), $T_0 = \mathrm{id}$. This implies that T_t is bijective.— In order to prove (T2), we consider $x, y \in \mathbb{R}$ with $y = x + \xi e$, $\xi \in \mathbb{R}$. Put $h := \overline{x}$, and determine, by (ii), the reals τ, t by means of

$$\varrho(h, \tau) = x_0 \text{ and } \varrho(h, \tau + t) = \xi + x_0.$$

Then

$$T_t(x) = T_t\big(h + \varrho(h, \tau)e\big) = h + \varrho(h, \tau + t)e = y.$$

Finally, we must prove (T3). Put again

$$x = h + \varrho(h, \tau)e.$$

For $t \geq 0$, we obtain

$$T_t(x) - x = \big(\varrho(h, \tau + t) - \varrho(h, \tau)\big)e \in \mathbb{R}_{\geq 0}\,e,$$

in view of (i).

The kernel of this translation group is, by (1.10), (1.12), $\varrho(h, 0) = 0$, i.e. $h = h + \varrho(h, 0)e$,

$$[T_t(h) - h]e = \varrho(h, t). \qquad \square$$

Remark. If T is an arbitrary translation group with kernel ϱ, then, of course, (1.12) holds true, since, by (T3), (1.10),

$$T_\tau(h) = h + \varrho(h, \tau)e, \ T_{\tau+t}(h) = h + \varrho(h, \tau + t)e,$$

and since $T_{\tau+t}(h) = T_t T_\tau(h)$.

1.8   Separable translation groups

The translation group T of X, in the direction of e, is called *separable* if, and only if, there exist functions

$$\varphi : \mathbb{R} \to \mathbb{R} \text{ and } \psi : H = e^{\perp} \to \mathbb{R}$$

such that $\varrho(h, t) = \varphi(t) \cdot \psi(h)$ holds true for all $(h, t) \in H \times \mathbb{R}$, where $\varrho(h, t)$ denotes the kernel of T.

If there existed $h_0 \in H$ with $\psi(h_0) = 0$, then $\varrho(h_0, 0) = \varrho(h_0, 1)$, contradicting (i). We put

$$\varphi_1(t) := \varphi(t) \cdot \psi(0) \text{ and } \psi_1(h) := \frac{\psi(h)}{\psi(0)}.$$

Again, $\varrho(h, t) = \varphi_1(t) \cdot \psi_1(h)$, and we will show $\psi_1(h) > 0$ for all $h \in H$, moreover, that φ_1 is an increasing bijection of $\mathbb{R}$ with $\varphi_1(0) = 0$. Since, by (i), (ii), $\varrho(0, t)$ is an increasing bijection of $\mathbb{R}$ with $\varrho(0, t) = \varphi_1(t) \psi_1(0) = \varphi_1(t)$, so must be $\varphi_1(t)$. If there existed $h_1 \in H$ with $\psi(h_1) < 0$, then, by (i),

$$0 = \varrho(h_1, 0) < \varrho_1(h_1, 1) = \varphi_1(1) \psi_1(h_1) < 0,$$

in view of $0 = \varphi_1(0) < \varphi_1(1)$, a contradiction.

So we may assume, without loss of generality, that the kernel of a separable translation group can be written in the form

$$\varrho(h, t) = \varphi(t) \cdot \psi(h)$$

with $\psi : H \to \mathbb{R}_{>0} := \mathbb{R}_{\geq 0} \backslash \{0\}$, $\psi(0) = 1$, and such that φ is an increasing bijection of $\mathbb{R}$ satisfying $\varphi(0) = 0$.

Of course, the translation groups with kernels t, $\sinh t \cdot \sqrt{1 + h^2}$, respectively, are separable. Separable is also the group with kernel

$$\left(\sinh t^{2n-1} \right) \cdot (1 + h^2)$$

for $n \in \{1, 2, 3, \ldots\}$, and non-separable the group, for instance, with kernel

$$\sinh \left(t^{2n-1} \cdot 2^{h^2} \right)$$

for $n \in \{1, 2, 3, \ldots\}$. Theorem 5 immediately shows that all these functions are indeed kernels of suitable translation groups.

The following theorem characterizes separability geometrically.

Theorem 6. *Suppose that T is a translation group of X in the direction of e. The group T is separable if, and only if,*

$$\frac{\|T_\alpha(h) - h\|}{\|T_\beta(h) - h\|} = \frac{\|T_\alpha(0)\|}{\|T_\beta(0)\|} \tag{1.13}$$

holds true for all $h \in H$ and $\alpha, \beta \in \mathbb{R}\backslash\{0\}$.

Proof. Since there is exactly one $t \in \mathbb{R}$ with $T_t(h) = h$, namely $t = 0$, we know $T_\beta(h) \neq h$ and $T_\beta(0) \neq 0$. Hence (1.13) is well-defined. From (i) (Theorem 5) and

$$T_t(h) = h + \varrho(h, t) e, \tag{1.14}$$

we obtain $\operatorname{sgn} \varrho(h, t) = \operatorname{sgn} t$ for $t \neq 0$. Assume now (1.13). Hence, by $\|T_t(h) - h\| = \operatorname{sgn} t \cdot \varrho(h, t)$,

$$\frac{\varrho(h, \alpha)}{\varrho(h, \beta)} = \frac{\varrho(0, \alpha)}{\varrho(0, \beta)}$$

for all $h \in H$ and $\alpha, \beta \in \mathbb{R}\backslash\{0\}$, and thus

$$\varrho(h, t) = \varrho(0, t) \cdot \frac{\varrho(h, 1)}{\varrho(0, 1)}$$

for $\alpha = t \neq 0$, $\beta = 1$. Of course, this formula also holds true for $t = 0$. Define

$$\varphi(t) := \varrho(0, t) \text{ and } \psi(h) := \frac{\varrho(h, 1)}{\varrho(0, 1)}.$$

Hence $\varrho(h, t) = \varphi(t) \cdot \psi(h)$, and thus T is separable. If, on the other hand, T is separable, we have to prove (1.13). As we already mentioned, we may assume, without loss of generality,

$$\varrho(h, t) = \varphi(t) \cdot \psi(h)$$

with $\psi : H \to \mathbb{R}_{>0}$, $\psi(0) = 1$. By (1.14),

$$\|T_t(h) - h\| = |\varphi(t)| \cdot \psi(h),$$

i.e. (1.13) holds true. $\qquad\square$

Remark. In view of (1.14), $\varrho(h, \alpha) \cdot \varrho(0, \beta) = \varrho(h, \beta) \cdot \varrho(0, \alpha)$ is equivalent with

$$T_\alpha(h) \cdot T_\beta(0) = T_\beta(h) \cdot T_\alpha(0),$$

by noticing $h \in e^\perp$. Hence (1.13) can be replaced by this latter equation which, of course, also holds true in the case $\alpha\beta = 0$. Moreover, by (1.14), formula (1.13) is equivalent with

$$T_\alpha(h_1) \cdot T_\beta(h_2) = T_\alpha(h_2) \cdot T_\beta(h_1)$$

(for all $\alpha, \beta \in \mathbb{R}$ and all $h_1, h_2 \in H$) as well, and also with

$$T_\alpha(h) \cdot T_1(0) = T_\alpha(0) \cdot T_1(h)$$

for all $\alpha \in \mathbb{R}$ and $h \in H$. To all these equations there correspond formulas like (1.13), for instance

$$\frac{\|T_\alpha(h_1) - h_1\|}{\|T_\beta(h_1) - h_1\|} = \frac{\|T_\alpha(h_2) - h_2\|}{\|T_\beta(h_2) - h_2\|}$$

for all real α, β with $\beta \neq 0$ and all $h_1, h_2 \in H$.

1.9   Geometry of a group of permutations

Let $S \neq \emptyset$ be a set and let G be a subgroup of Perm S. The structure (S, G) will be called a *geometry*. Suppose that $N \neq \emptyset$ is a set and

$$\varphi : G \times N \to N$$

an *action* of G on N, i.e. a mapping satisfying

$$\text{(i)} \quad \varphi(fg, l) = \varphi(f, \varphi(g, l)),$$

$$\text{(ii)} \quad \varphi(j, l) = l$$

for all $f, g \in G$ and $l \in N$ where j denotes the neutral element of G. We then call (N, φ) an *invariant notion* of the geometry (S, G). Instead of $\varphi(f, l)$ we often shall write $f(l)$. Hence (i), (ii) are given by

$$fg(l) := (fg)(l) = f(g(l))$$

and $j(l) = l$.

Let (N, φ) be an invariant notion of (S, G) and let W be a set. A function

$$h : N \to W$$

is called an *invariant* of the geometry (S, G) provided

$$h(f(l)) = h(l)$$

holds true for all $f \in G$ and all $l \in N$.

Geometries (S, G) and (S', G') are called *isomorphic* if, and only if, there exist bijections

$$\sigma : S \to S' \text{ and } \tau : G \to G'$$

such that the following equations hold true:

$$\tau(g_1 g_2) = \tau(g_1)\tau(g_2), \quad \sigma(g(s)) = \tau(g)(\sigma(s)) \tag{1.15}$$

for all $s \in S$ and $g_1, g_2, g \in G$. The mapping τ is hence an isomorphism between the groups G, G'. If (S, G), (S', G') are isomorphic, we shall write

$$(S, G) \cong (S', G').$$

The relation $\cong$ is reflexive, symmetric and transitive on every set of geometries (S, G). The isomorphism of geometries (S, G), (S', G') was already playing an important role in geometry by the 19th century. At that time geometers spoke of so-called *Übertragunsprinzipe* which means that two geometries, based on different terminologies, could turn out to coincide from a structural point of view, by just following a vocabulary which associates to the objects of one geometry the objects of the other geometry. We would like to present an example, connecting

the so-called Cayley–Klein-model of proper 1-dimensional hyperbolic geometry, (S, G),

with

the so-called Poincaré-model of proper 1-dimensional hyperbolic geometry, (S', G').

Define

$$S = \ ]-1, +1[:= \{r \in \mathbb{R} \mid -1 < r < +1\},$$
$$S' = \ ]0, \infty[:= \{r \in \mathbb{R} \mid r > 0\},$$
$$G = \{\varphi_p : S \to S \mid p \in S\}, \ \varphi_p(x) = \frac{x+p}{xp+1},$$
$$G' = \{\psi_q : S' \to S' \mid q \in S'\}, \ \psi_q(x) = qx,$$

and put

$$\sigma(s) = \frac{1+s}{1-s}, \ \tau(\varphi_p) = \psi_{\sigma(p)}$$

for $s \in S$ and $\varphi_p \in G$.

It is easy to verify that every φ_p, $p \in S$, is a bijection of S, that

$$\varphi_{p_1} \cdot \varphi_{p_2} = \varphi_{p_1 * p_2} \text{ with } p_1 * p_2 = \varphi_{p_1}(p_2)$$

holds true, that $\sigma : S \to S'$ is a bijection and $\tau : G \to G'$ an isomorphism satisfying

$$\sigma\left(\varphi_p(s)\right) = \tau\left(\varphi_p\right)\left(\sigma(s)\right)$$

for all $s \in S$ and $\varphi_{p_1}, \varphi_{p_2}, \varphi_p \in G$. We thus established isomorphism (or an Übertragungsprinzip) between the geometries (S, G) and (S', G'). (For more information about the *hyperbolic line* in connection with the present definitions see W. Benz [3], sections 2.1 to 2.5).

Isomorphic geometries (S, G) and (S', G') have, up to notation, the same invariant notions and the same invariants. Let (N, φ) be an invariant notion of (S, G), let N' be a set and

$$\nu : N \to N'$$

a bijection (for instance $N' = N$ and $\nu = \mathrm{id}$). We then define an invariant notion (N', φ') of (S', G'). Put $\varphi' : G' \times N' \to N'$ with

$$\varphi'\big(\tau\,(g),\,\nu\,(l)\big) = \nu\,\big(\varphi\,(g, l)\big)$$

for $g \in G$ and $l \in N$ where $\tau : G \to G'$ is a bijection satisfying (1.15). Obviously,

$$\varphi'\big(\tau\,(j), \nu\,(l)\big) = \nu\,\big(\varphi\,(j, l)\big) = \nu\,(l)$$

since $\varphi\,(j, l) = l$ (see (ii)). Moreover, we must prove

$$L := \varphi'\big(\tau\,(f)\,\tau\,(g),\,\nu\,(l)\big) = \varphi'\big(\tau\,(f),\,\varphi'[\tau\,(g),\,\nu\,(l)]\big) =: R,$$

i.e. (i) for $\varphi' : G' \times N' \to N'$. Notice

$$
\begin{aligned}
L &= \varphi'\big(\tau\,(fg),\,\nu\,(l)\big) = \nu\,\big(\varphi\,(fg, l)\big) = \nu\,\Big(\varphi\,\big(f, \varphi\,(g, l)\big)\Big) \\
&= \varphi'\big(\tau\,(f),\,\nu\,[\varphi\,(g, l)]\big) = \varphi'\big(\tau\,(f),\,\varphi'[\tau\,(g),\,\nu\,(l)]\big) = R.
\end{aligned}
$$

Now let

$$h : N \to W$$

be an invariant of (S, G) based on the invariant notion (N, φ) of (S, G). We then would like to define an invariant

$$h' : N' \to W$$

of (S', G'). (It also could be useful here to work with a set W' and a bijection $\mu : W \to W'$.) Put $h'(l') := h\,\big(\nu^{-1}(l')\big)$ for all $l' \in N'$, by observing that $\nu : N \to N'$ is a bijection. Then

$$
\begin{aligned}
h'\Big(\varphi'\big(\tau\,(g),\,\nu\,(l)\big)\Big) &= h'\Big(\nu\,\big(\varphi\,(g, l)\big)\Big) \\
&= h\,\big(\varphi\,(g, l)\big) = h\,(l) = h'\big(\nu\,(l)\big).
\end{aligned}
$$

h' is hence an invariant of (S', G'). If we rewrite the definition of φ', namely

$$\varphi'\big(\tau\,(g),\,\nu\,(l)\big) = \nu\,\big(\varphi\,(g, l)\big),$$

by using the abbreviations

$$\varphi\,(g, l) =: g\,(l)$$

and

$$\varphi'\big(\tau\,(g),\,\nu\,(l)\big) =: \tau\,(g)\big(\nu\,(l)\big),$$

we get

$$\tau\,(g)\big(\nu\,(l)\big) = \nu\,\big(g\,(l)\big)$$

for all $l \in N$ and $g \in G$. In the case that N is a set of subsets of S, or that it otherwise is based on S, the mapping ν might be taken equal to σ in view of (1.15), in order to construct the corresponding invariant notion of (N, φ) for (S', G') in terms of this latter geometry. Principally, however, the corresponding invariant notion of (N, φ) for (S', G') might be based on (N, φ') $(N' := N, \nu := \mathrm{id})$ with

$$\varphi'\big(\tau(g), l\big) := \varphi(g, l),$$

according to the proof above. (For more information in this connection compare W. Benz [3], chapter 1.)

In this context it might be interesting to look again at our previous example of two isomorphic geometries in connection with proper 1-dimensional hyperbolic geometry,

$$(S, G) \cong (S', G').$$

We are interested in a special invariant notion and in a special invariant. Define

$$N := S \times S,$$

and the action from $G \times N$ into N by

$$g(x, y) := \big(g(x), g(y)\big)$$

for all $g \in G$ and $x, y \in S$. Define, moreover,

$$W := \{r \in \mathbb{R} \mid r > 0\}$$

and $h : N \to W$ by

$$h(x, y) := \frac{(1 - x)(1 + y)}{(1 + x)(1 - y)}$$

for all $(x, y) \in N$. Obviously,

$$h(x, y) = h\big(g(x), g(y)\big)$$

for all $g \in G$ and $(x, y) \in N$, so that h is an invariant of (S, G). With respect to (S', G') define $N' := S' \times S'$ and $\nu : N \to N'$ by

$$\nu(x, y) := \big(\sigma(x), \sigma(y)\big) =: \sigma(x, y)$$

for all $(x, y) \in N$. Because of the general formula

$$h'\big(\nu(l)\big) := h(l),$$

we hence define in our present situation

$$h'\big(\sigma(x), \sigma(y)\big) := h(x, y),$$

i.e.

$$h'(v, w) = h\big(\sigma^{-1}(v), \sigma^{-1}(w)\big) = \frac{w}{v}$$

for all $v, w \in S'$. Clearly, $h'\big(\psi_q(v), \psi_q(w)\big) = h'(v, w)$ for all $q, v, w \in S'$.

1.10   Euclidean, hyperbolic geometry

Suppose that X is a real inner product space containing two linearly independent elements. Take a fixed element e of X with $e^2 = 1$, and a translation group T of X with axis e. Let $G\left(T, O\left(X\right)\right)$ be the group generated by T and $O\left(X\right)$, i.e. the subgroup of Perm X consisting of all finite products of elements of $T \cup O\left(X\right)$. We obtain a geometry

$$\left(X, G\left(T, O\left(X\right)\right)\right). \tag{1.16}$$

For T with $T_t(x) = x + te$, $t \in \mathbb{R}$, (1.16) is called *euclidean geometry* over the real inner product space X, and for T with kernel (1.11),

$$\varrho\left(h, t\right) = \sinh t \cdot \sqrt{1 + h^2},$$

(1.16) is called *hyperbolic geometry* over X.

Define $N := X \times X$ and $\varphi\left(f, l\right)$ for $f \in G\left(T, O\left(X\right)\right)$ and $l = (x, y) \in N$ by

$$\varphi\left(f, (x, y)\right) := \left(f\left(x\right), f\left(y\right)\right).$$

Hence (N, φ) is an invariant notion of (X, G). Define

$$d : N \to \mathbb{R}_{\geq 0} \tag{1.17}$$

by $d\left(x, y\right) = \|x - y\|$ in the case of euclidean geometry, and by

$$\cosh d\left(x, y\right) = \sqrt{1 + x^2}\, \sqrt{1 + y^2} - xy \tag{1.18}$$

for hyperbolic geometry, and we obtain important invariants of these geometries, namely their *distance functions*. Observe that the right-hand side of (1.18) is ≥ 1. This is trivial for $xy \leq 0$, and follows otherwise from $(xy)^2 \leq x^2 y^2$ and $(x - y)^2 \geq 0$, i.e. from $(1 + x^2)(1 + y^2) \geq (1 + xy)^2$. In the case of hyperbolic geometry and for

$$x =: h_1 + \varrho\left(h_1, \tau_1\right) e \text{ and } y =: h_2 + \varrho\left(h_2, \tau_2\right) e$$

with $h_1, h_2 \in H$ we get, by (1.11),

$$\sqrt{1 + x^2}\, \sqrt{1 + y^2} - xy = \cosh(\tau_1 - \tau_2) \cdot \sqrt{(1 + h_1^2)(1 + h_2^2)} - h_1 h_2,$$

i.e. for $x' := T_t(x) = h_1 + \varrho\left(h_1, \tau_1 + t\right) e$, $y' := T_t(y) = h_2 + \varrho\left(h_2, \tau_2 + t\right) e$ we get

$$\sqrt{1 + x'^2}\, \sqrt{1 + y'^2} - x'y'$$
$$= \cosh\left((\tau_1 + t) - (\tau_2 + t)\right) \cdot \sqrt{(1 + h_1^2)(1 + h_2^2)} - h_1 h_2.$$

Hence, by (1.18), (1.17), $d\left(x, y\right) = d\left(T_t(x), T_t(y)\right)$ for all $t \in \mathbb{R}$. That also $d\left(x, y\right) = d\left(\omega\left(x\right), \omega\left(y\right)\right)$ holds true for $\omega \in O\left(X\right)$, follows from $\omega\left(p\right)\omega\left(q\right) = pq$ for all $p, q \in X$.

Instead of $d\left(x, y\right)$ we mostly will write eucl $\left(x, y\right)$, hyp $\left(x, y\right)$, respectively, in the euclidean, hyperbolic case, and instead of the elements of X we also will speak of the *points* of X.

1.11   A common characterization

Let X be a real inner product space containing two linearly independent elements, and let e be a fixed element of X with $e^2 = 1$. The following result will be proved in this section (W. Benz [13]).

Theorem 7. *Let T be a separable translation group of X with axis e, and suppose that $d : X \times X \to \mathbb{R}_{\geq 0}$ is not identically 0 and satisfies*

(i) $d(x, y) = d\left(\omega(x), \omega(y)\right)$,

(ii) $d(x, y) = d\left(T_t(x), T_t(y)\right)$,

(iii) $d(\beta e, 0) = d(0, \beta e) = d(0, \alpha e) + d(\alpha e, \beta e)$

for all $x, y \in X$, $\omega \in O(X), t, \alpha, \beta \in \mathbb{R}$ with $0 \leq \alpha \leq \beta$. Then, up to isomorphism,

$$\Big(X, G\left(T, O(X)\right)\Big)$$

is the euclidean or the hyperbolic geometry over X. Moreover, there exist positive reals k, l and δ such that

$$\varrho(h, t) = lt, \quad d(x, y) = \frac{k}{l}\|x - y\|$$

hold true for all $x, y \in X$, $h \in e^{\perp}$, $t \in \mathbb{R}$, or

$$\varrho(h, t) = \frac{\sinh(lt)}{\sqrt{\delta}}\sqrt{1 + \delta h^2}, \quad d(x, y) = \frac{k}{l}\,\mathrm{hyp}\,(x\sqrt{\delta}, y\sqrt{\delta})$$

for all $x, y \in X$, $h \in e^{\perp}$, $t \in \mathbb{R}$.

Remark. Instead of (i) and (ii) it is possible to write

$$d\left(\omega(x), \omega(y)\right) = d\left(T_t(x), T_t(y)\right),$$

since (i) follows here from $t = 0$ and (ii) from $\omega = \mathrm{id}$.

Theorem 7 will be proved in several steps.

A. *If p is in X, there exists $\gamma \in O(X)$ with $\gamma(p) = \|p\| \cdot e$.*

Proof. This is trivial in the case $p = -\|p\|\,e$ by just applying $\gamma(x) = -x$. Otherwise put

$$b := p + \|p\|\,e \quad \text{and} \quad \|b\| \cdot a := b$$

and, moreover, $\gamma(x) := -x + 2(xa)\,a$. Now observe that γ is an involution, i.e. that it satisfies $\gamma^2 = \mathrm{id} \neq \gamma$. Hence γ is bijective and, obviously, it satisfies $xy = \gamma(x)\gamma(y)$ for all $x, y \in X$ by noticing that $\gamma(p) = \|p\|e$ follows from $2(pa)\,a = p + \|p\|e$. □

B. *Let d be a function as described in Theorem 7. Then there exists* $f : K \to \mathbb{R}_{\geq 0}$
with

$$K := \{(\xi_1, \xi_2, \xi_3) \in \mathbb{R}^3 \mid \xi_1, \xi_2 \in \mathbb{R}_{\geq 0} \text{ and } \xi_3^2 \leq \xi_1 \xi_2 \}$$

such that

$$d(x, y) = f(x^2, y^2, xy) \tag{1.19}$$

holds true for all $x, y \in X$.

Proof. Take $j \in X$ with $j^2 = 1$ and $ej = 0$. If (ξ_1, ξ_2, ξ_3) is in K and $\xi_1 = 0$, put $x_0 := 0$, $y_0 := e\sqrt{\xi_2}$ and

$$f(\xi_1, \xi_2, \xi_3) := d(x_0, y_0). \tag{1.20}$$

Observe here $\xi_3 = 0$, since $\xi_3^2 \leq \xi_1 \xi_2$. For $(\xi_1, \xi_2, \xi_3) \in K$ and $\xi_1 > 0$ define again f by (1.20), but now with

$$x_0 := e\sqrt{\xi_1} \text{ and } y_0\sqrt{\xi_1} := e\xi_3 + j\sqrt{\xi_1 \xi_2 - \xi_3^2}.$$

The function $f : K \to \mathbb{R}_{\geq 0}$ is hence determined for all elements of K, and we finally must prove (1.19). So let x, y be elements of X and put

$$\xi_1 := x^2, \; \xi_2 := y^2, \; \xi_3 := xy. \tag{1.21}$$

Because of the inequality of Cauchy–Schwarz, (ξ_1, ξ_2, ξ_3) must be in K. If we are able to prove that there exists $\omega \in O(X)$ with

$$\omega(x_0) = x \text{ and } \omega(y_0) = y, \tag{1.22}$$

where x_0, y_0 are the already defined elements with respect to (ξ_1, ξ_2, ξ_3), then

$$d(x, y) = d(\omega(x_0), \omega(y_0)) = d(x_0, y_0) = f(\xi_1, \xi_2, \xi_3) = f(x^2, y^2, xy),$$

and hence (1.19) holds true. In order to find $\omega \in O(X)$ with (1.22), observe, by (1.21),

$$x^2 = x_0^2, \; y^2 = y_0^2, \; xy = x_0 y_0. \tag{1.23}$$

If $x = 0$, take, by step A, $\gamma \in O(X)$ with $\gamma(y) = \|y\|e = \sqrt{\xi_2}\,e$. Hence

$$\gamma^{-1}(x_0) = \gamma^{-1}(0) = 0 = x \text{ and } \gamma^{-1}(y_0) = \gamma^{-1}(e\sqrt{\xi_2}) = y,$$

i.e. $\omega := \gamma^{-1}$ solves (1.22). So assume $x \neq 0$ and take $\gamma \in O(X)$ with $\gamma(x) = \|x\| \cdot e = e \cdot \sqrt{\xi_1} = x_0$. As soon as we have found $\tau \in O(X)$ with $\tau(x_0) = x_0$ and $\tau(y_0) = \gamma(y)$, we also have solved (1.22) for $x \neq 0$, namely by $\omega = \gamma^{-1}\tau$. Put $z := \gamma(y)$ and observe, by (1.21),

$$\xi_1 = x_0^2, \; \xi_2 = y^2 = (\gamma(y))^2 = z^2, \; \xi_3 = xy = \gamma(x)\gamma(y) = x_0 z. \tag{1.24}$$

In the case $z = y_0$, take $\tau = \mathrm{id}$. If $y_0 = 0$, we obtain, by (1.23), $y = 0$. Also here put $\tau = \mathrm{id}$. So we may assume

$$z \neq y_0 \neq 0.$$

Observe $X = s^\perp \oplus \mathbb{R}s$ for $s := z - y_0$. If $v \in X$, there exist uniquely determined $\alpha \in \mathbb{R}$ and $m \in s^\perp$ satisfying

$$v = m + \alpha s.$$

Define $\tau(v) := m - \alpha s$. Hence $\tau \in O(X)$. It remains to show $\tau(x_0) = x_0$ and $\tau(y_0) = z$. Observe $x_0 z = \xi_3$ from (1.24), and $x_0 y_0 = \xi_3$, by (1.23), (1.21). Hence $x_0 s = 0$, i.e. $x_0 \in s^\perp$, and thus $\tau(x_0) = x_0$. From $s = z - y_0$ we obtain

$$y_0 = \frac{z + y_0}{2} - \frac{s}{2}.$$

Since $(z + y_0) s = z^2 - y_0^2 = z^2 - y^2 = 0$, by (1.23), (1.24), we get

$$\tau(y_0) = \frac{z + y_0}{2} + \frac{s}{2} = z. \qquad \square$$

From now on we will work with the expression (1.19) for our distance function d. Since T is a separable translation group, we may assume

$$\varrho(h, t) = \varphi(t) \cdot \psi(h)$$

with $\psi : H \to \mathbb{R}_{>0}$, $\psi(0) = 1$, $H := e^\perp$, and such that φ is an increasing bijection of $\mathbb{R}$ with $\varphi(0) = 0$.

C. *There exists a real constant $k \geq 0$ with*

$$f\left(0, \varphi^2(\xi), 0\right) = k \cdot \xi$$

for all $\xi \geq 0$.

Proof. Given reals $0 \leq \alpha \leq \beta$, we get $0 = \varphi(0) \leq \varphi(\alpha) \leq \varphi(\beta)$. Hence (iii) of Theorem 7 yields, with $\varphi(\beta)$ instead of β and $\varphi(\alpha)$ instead of α,

$$d\left(0, \varphi(\beta) e\right) = d\left(0, \varphi(\alpha) e\right) + d\left(\varphi(\alpha) e, \varphi(\beta) e\right). \tag{1.25}$$

Since, by (1.12),

$$T_{-\alpha}\left(0 + \varphi(\beta) \psi(0) e\right) = 0 + \varphi(\beta - \alpha) \psi(0) e,$$

i.e. $T_{-\alpha}\left(\varphi(\beta) e\right) = \varphi(\beta - \alpha) e$, i.e. by (ii),

$$d\left(\varphi(\alpha) e, \varphi(\beta) e\right) = d\left(T_{-\alpha}(\varphi(\alpha) e), T_{-\alpha}(\varphi(\beta) e)\right) = d\left(0, \varphi(\beta - \alpha) e\right).$$

Hence, in view of (1.19), (1.25),

$$f\left(0,\varphi^2(\beta),0\right) = f\left(0,\varphi^2(\alpha),0\right) + f\left(0,\varphi^2(\beta-\alpha),0\right),$$

which implies, for $\xi,\eta \in \mathbb{R}_{\geq 0}$ with $\alpha := \xi$, $\beta := \xi+\eta$,

$$f\left(0,\varphi^2(\xi+\eta),0\right) = f\left(0,\varphi^2(\xi),0\right) + f\left(0,\varphi^2(\eta),0\right). \tag{1.26}$$

This proves C (see the following Remark). $\square$

Remark. Define $g(\xi) := f\left(0,\varphi^2(\xi),0\right)$ for $\xi \geq 0$ and observe $g : \mathbb{R}_{\geq 0} \to \mathbb{R}_{\geq 0}$, since $d : X \times X \to \mathbb{R}_{\geq 0}$, i.e. since $d\left(0,\varphi(\xi)e\right) \geq 0$. Because of (1.26), we get $g(\xi+\eta) = g(\xi) + g(\eta)$ for all $\xi,\eta \in \mathbb{R}_{\geq 0}$. Putting $\xi = \eta = 0$ we obtain $g(0) = 0$. Define $k := g(1)$. Hence $k \geq 0$. Equation (1.26) can be extended to

$$g(\xi_1 + \cdots + \xi_n) = \sum_{i=1}^{n} g(\xi_i)$$

for every positive integer n by induction. Hence

$$k = g(1) = g\left(\frac{1}{n} + \cdots + \frac{1}{n}\right) = n \cdot g\left(\frac{1}{n}\right),$$

i.e. $g\left(\frac{1}{n}\right) = \frac{k}{n}$ for every positive integer n. Thus

$$g\left(\frac{m}{n}\right) = g\left(\frac{1}{n} + \cdots + \frac{1}{n}\right) = mg\left(\frac{1}{n}\right) = \frac{m}{n}k$$

for $m \in \{1,2,3,\ldots\}$. This leads to $g(r) = kr$ for every rational number $r \geq 0$. Suppose that $0 \leq \xi \leq \eta$. Then $\eta - \xi \geq 0$ and hence $g(\eta - \xi) \geq 0$. This implies $g(\xi) \leq g(\eta)$ because

$$g(\xi) + g(\eta - \xi) = g(\eta).$$

Now let $\zeta > 0$ be a real number and let $r_1,r_2,r_3,\ldots$ and $s_1,s_2,s_3,\ldots$ be sequences of rational and non-negative numbers satisfying

$$\lim r_i = \zeta = \lim s_i$$

and $r_i \leq z \leq s_i$ for all $i = 1,2,3,\ldots$. Hence

$$kr_i = g(r_i) \leq g(\zeta) \leq g(s_i) = ks_i$$

and thus $g(\zeta) = \lim kr_i = k\zeta$. The equation $g(\xi+\eta) = g(\xi) + g(\eta)$ is called a Cauchy equation in the theory of functional equations. For the other Cauchy equations see J. Aczél [1].

D. a) *To the elements $x \neq y$ of X there exist $\omega_1, \omega_2 \in O(X)$ and $\lambda, t \in \mathbb{R}$ with $\lambda > 0$ and*

$$\omega_1 T_t \omega_2(x) = 0, \quad \omega_1 T_t \omega_2(y) = \lambda e.$$

b) *The constant k of statement C is positive.*

c) $d(x, y) = d(y, x)$ *for all $x, y \in X$.*

d) *If $x, y \in X$, then $d(x, y) = 0$ if, and only if, $x = y$.*

Proof. a) Because of A there exists $\omega_2 \in O(X)$ with $\omega_2(x) = \|x\| e$. Since $\|x\| e - 0 \in \mathbb{R}e$, (T2) implies the existence of $t \in \mathbb{R}$ with $T_t(\|x\| e) = 0$. Finally take $\omega_1 \in O(X)$ with $\omega_1(z) = \lambda e$, $\lambda := \|z\| \geq 0$, where $z := T_t \omega_2(y)$. Hence $g(x) = 0$ and $g(y) = \lambda e$ with $g := \omega_1 T_t \omega_2$. Since $x \neq y$, we obtain $\lambda > 0$. This proves a).

b) The distance function d is assumed to be not identically 0. There hence exist $p, q \in X$ with $d(p, q) > 0$. If $p = q$, take, in view of a), $\omega_1 T_t \omega_2 =: g$ with $g(p) = 0$. This implies, by (i), (ii),

$$0 < d(p, p) = d\big(g(p), g(p)\big) = d(0, 0).$$

But (iii) yields $d(0, 0) = 0$ for $\alpha = \beta = 0$. So $p = q$ is impossible. For $p \neq q$ take, in view of a), $g = \omega_1 T_t \omega_2$ with $g(p) = 0$, $g(q) = \lambda e$, $\lambda > 0$. Hence $0 < d(p, q) = d(0, \lambda e)$. If $\xi \in \mathbb{R}$ satisfies $\varphi(\xi) = \lambda > 0$, then $\xi > 0$, because $\xi \leq 0$ would imply $\varphi(\xi) \leq \varphi(0) = 0$. Hence

$$0 < d(0, \lambda e) = f(0, \lambda^2, 0) = f\big(0, \varphi^2(\xi), 0\big) = k\xi,$$

and thus $k > 0$.

c) For $x \neq y$ take $g = \omega_1 T_t \omega_2$, in view of a), with $g(x) = 0$, $g(y) = \lambda e$, $\lambda > 0$. Hence, by (iii),

$$d(x, y) = d(0, \lambda e) = d(\lambda e, 0) = d(y, x).$$

d) If $x \neq y$, we may work again with a mapping g getting

$$d(x, y) = d(0, \lambda e) > 0,$$

by $\lambda > 0$. On the other hand, by (iii) with $\alpha = \beta = 0$,

$$d(x, x) = d(0, 0) = 0. \qquad \square$$

E. a) $\varphi(-t) = -\varphi(t)$ *for all $t \in \mathbb{R}$.*

b) $\varphi\left(\dfrac{d(0, x)}{k}\right) = \|x\|$ *for all $x \in X$.*

Proof. a) Assume $t < 0$ and define $\varphi(\tau) := -\varphi(t)$. Hence $\tau > 0$. Observe, by D.c,

$$d\big(0, \varphi(-t) e\big) = d\Big(T_t(0), T_t\big(\varphi(-t) e\big)\Big) = d\big(\varphi(t) e, 0\big) = d\big(0, \varphi(t) e\big),$$

i.e. $f\left(0,\varphi^2(-t),0\right) = f\left(0,\varphi^2(t),0\right) = f\left(0,\varphi^2(\tau),0\right)$. Hence, from C we get $k \cdot (-t) = k\tau$, i.e. $\tau = -t$ in view of D.b. This proves E.a for $t < 0$. Thus $\varphi(-t) = -\varphi(t)$ for all reals t.

b) By C and D.b we obtain

$$\varphi\left(\frac{f\left(0,\varphi^2(\xi),0\right)}{k}\right) = \varphi(\xi)$$

for all $\xi \geq 0$, and hence $\varphi\left(\frac{1}{k} f\left(0,t^2,0\right)\right) = t$ for all $t \geq 0$. Applying this formula for $t := \|x\|$, $x \in X$, we obtain, by (1.19), statement E.b. $\qquad \square$

Let x, y be elements of X with $x \neq 0$. Observe $(xy)^2 \leq x^2 y^2$, i.e. $\lambda \geq 0$ where

$$\lambda x^2 := x^2 y^2 - (xy)^2.$$

Let $j \in H$ be given with $j^2 = 1$, and define $\eta \in \mathbb{R}$ by

$$\frac{xy}{\|x\|} = \varphi(\eta) \cdot \psi(\sqrt{\lambda}\, j). \tag{1.27}$$

Of course, η seems to depend on the chosen j. Put

$$\|x\| =: \varphi(\xi), \tag{1.28}$$

i.e. we obtain $\xi > 0$ and also $T_{-\xi}(\|x\|e) = 0$. Obviously, by (1.27), (1.28), (1.19), we obtain

$$d(x,y) = f\left(x^2, y^2, xy\right) = d\left(\varphi(\xi)\, e,\ \sqrt{\lambda}\, j + \varphi(\eta)\, \psi(\sqrt{\lambda}\, j)\, e\right),$$

i.e. transforming the elements of X of the right-hand side by $T_{-\xi}$,

$$d(x,y) = d\left(0, \sqrt{\lambda}\, j + \varphi(\eta - \xi)\, \psi(\sqrt{\lambda}\, j)\, e\right).$$

Hence, by E.b,

$$\varphi^2\left(\frac{d(x,y)}{k}\right) = \lambda + \varphi^2(\eta - \xi)\, \psi^2(\sqrt{\lambda}\, j). \tag{1.29}$$

If we repeat the same calculation for a $j' \in H$ satisfying $(j')^2 = 1$, we get instead of (1.29)

$$\varphi^2\left(\frac{d(x,y)}{k}\right) = \lambda + \varphi^2(\eta' - \xi)\, \psi^2(\sqrt{\lambda}\, j') \tag{1.30}$$

by observing

$$\frac{xy}{\|x\|} = \varphi(\eta') \cdot \psi(\sqrt{\lambda}\, j') \tag{1.31}$$

instead of (1.27). We apply (1.29) and (1.30) for $xy = 0$. Noticing that (1.27), (1.31) lead to $\eta = 0 = \eta'$, we obtain, in view of $\lambda = y^2$ and $\varphi^2(0 - \xi) = x^2$ from E.a,

$$\psi^2(\|y\|j) = \frac{1}{x^2}\left(\varphi^2\left(\frac{d(x,y)}{k}\right) - y^2\right) = \psi^2(\|y\|j').$$

So we get $\psi^2(\alpha j) = \psi^2(\alpha j')$ for all real $\alpha \geq 0$ and all $j, j' \in H$ with $j^2 = 1 = j'^2$. Since $\psi(h) > 0$ for $h \in H$ we obtain

$$\psi(h) = \psi(h') \tag{1.32}$$

for all $h, h' \in H$ with $h^2 = (h')^2$. Hence η in (1.27) does not depend on the chosen j.

F. a) *There exists a constant $\delta \geq 0$ such that for all $h \in H$,*

$$\psi(h) = \sqrt{1 + \delta h^2}.$$

b) *For all $x, y \in X$ with $x \neq 0$,*

$$\varphi^2\left(\frac{d(x,y)}{k}\right) = \lambda + \varphi^2(\eta - \xi) \cdot (1 + \delta\lambda)$$

holds true, where $\|x\| =: \varphi(\xi)$, $\lambda x^2 := x^2 y^2 - (xy)^2$ and

$$xy =: \varphi(\xi)\varphi(\eta)\sqrt{1 + \delta\lambda}.$$

Proof. Because of (1.32), formula (1.27) does not depend on the chosen $j \in H$ satisfying $j^2 = 1$, and thus η does not depend on this j. So we may define

$$\psi_0(\eta) := \psi(\sqrt{\eta}\,j)$$

for $\eta \geq 0$, where $j \in H$ is chosen arbitrarily with $j^2 = 1$. Hence, by (1.29),

$$\varphi^2\left(\frac{d(x,y)}{k}\right) = \lambda + \varphi^2(\eta - \xi)\,\psi_0^2(\lambda) \tag{1.33}$$

for all $x, y \in X$ with $x \neq 0$ and $\lambda x^2 = x^2 y^2 - (xy)^2$.

Take an arbitrary element $h \neq 0$ of H. We get $d(e, h) = d(h, e)$ by D.c, i.e. by (1.33),

$$h^2 + \varphi^2(\eta - \xi)\,\psi_0^2(h^2) = 1 + \varphi^2(\eta' - \xi')\,\psi_0^2(1) \tag{1.34}$$

with $1 = \varphi(\xi)$, $\|h\| = \varphi(\xi')$, $0 = \varphi(\eta)\,\psi_0^2(h^2)$, $0 = \varphi(\eta')\,\psi_0(1)$, i.e. $\eta = 0$ and $\eta' = 0$. Thus, by (1.34), E.a,

$$h^2 + \psi_0^2(h^2) = 1 + h^2\psi_0^2(1),$$

i.e. $\psi_0^2(h^2) = 1 + h^2\big(\psi_0^2(1) - 1\big)$. If $\psi_0^2(1)$ were < 1, then for sufficiently large h^2, $\psi_0^2(h^2)$ would become negative. So we get with $\psi_0^2(h^2) \geq 1$ for all $h \in H$,

$$\psi_0(h^2) = \sqrt{1 + \delta h^2}$$

with $\delta := \psi_0^2(1) - 1 \geq 0$, since $\psi_0(\eta) = \psi\left(\sqrt{\eta}\,j\right) \in \mathbb{R}_{>0}$ for $\eta \geq 0$. Hence

$$\psi(h) = \psi_0(h^2) = \sqrt{1 + \delta h^2}.$$

This proves F.a, and F.b follows from (1.33). $\square$

If $h \neq 0$ is in H and $t \in \mathbb{R}$, then D.c and (ii) imply

$$d(0, h) = d(h, 0) = d\big(T_t(h), T_t(0)\big).$$

Hence, by E.b, F.b,

$$
\begin{aligned}
h^2 \;&= \varphi^2\left(\tfrac{d(0,h)}{k}\right) = \varphi^2\left(\frac{d\big(h + \varphi(t)\,\psi(h)\,e,\varphi(t)\,e\big)}{k}\right)\\
&= \frac{h^2\varphi^2(t)}{\varphi^2(\xi)} + \varphi^2(\eta - \xi)\left(1 + \delta\,\frac{h^2\varphi^2(t)}{\varphi^2(\xi)}\right),
\end{aligned}
$$

where $\xi > 0$ and η are given by

$$
\begin{aligned}
\varphi^2(\xi) \quad &= \quad \big(h + \varphi(t)\,\psi(h)\,e\big)^2 = h^2 + \varphi^2(t)\,\psi^2(h),\\
\varphi^2(t)\,\psi(h) \quad &= \quad \varphi(\xi)\,\varphi(\eta)\sqrt{1 + \delta\,\frac{h^2\varphi^2(t)}{\varphi^2(\xi)}}.
\end{aligned}
$$

We thus get, by F.a,

$$h^2\big(\varphi^2(\xi) - \varphi^2(t)\big) = \varphi^2(\eta - \xi)\big(\varphi^2(\xi) + \delta h^2\varphi^2(t)\big), \tag{1.35}$$

where $\xi > 0$ and η satisfy

$$\varphi^2(\xi) \quad = \quad h^2 + \varphi^2(t)(1 + \delta h^2), \tag{1.36}$$

$$\varphi^2(t)\sqrt{1 + \delta h^2} \quad = \quad \varphi(\eta)\sqrt{\varphi^2(\xi) + \delta h^2\varphi^2(t)}. \tag{1.37}$$

Take a fixed $j \in H$ with $j^2 = 1$, and take an arbitrary real number $\mu > 0$. Put $h = \sqrt{\mu}\,j$. Then $0 \neq h \in H$. Defining $\xi > 0$ by

$$\varphi^2(\xi) = \mu + \varphi^2(t)(1 + \delta\mu) \tag{1.38}$$

(compare (1.36)), and η by (compare (1.37))

$$\varphi^2(t)\sqrt{1 + \delta\mu} = \varphi(\eta)\sqrt{\varphi^2(\xi) + \delta\mu\varphi^2(t)}, \tag{1.39}$$

we obtain from (1.35) and E.a,

$$\mu \cdot \left(\varphi^2(\xi) - \varphi^2(t)\right) = \varphi^2(\eta - \xi)\left(\varphi^2(\xi) + \delta\mu\varphi^2(t)\right). \tag{1.40}$$

G. *Given arbitrary real numbers $\xi > \eta \geq 0$, there exist real numbers $\mu > 0$ and t such that (1.38) and (1.39) hold true.*

Proof. $\xi > \eta \geq 0$ implies $\varphi(\xi) > \varphi(\eta) \geq 0$, i.e.

$$\varphi^2(\xi)\left(1 + \delta\varphi^2(\eta)\right) > \varphi^2(\eta)\left(1 + \delta\varphi^2(\xi)\right).$$

Therefore

$$\mu := \varphi^2(\xi) - \varphi(\xi)\,\varphi(\eta)\sqrt{\frac{1 + \delta\varphi^2(\xi)}{1 + \delta\varphi^2(\eta)}} > 0, \tag{1.41}$$

i.e. $\varphi^2(\xi) - \mu \geq 0$. There hence exists $t \in \mathbb{R}$ with

$$\varphi^2(t) := \frac{\varphi^2(\xi) - \mu}{1 + \delta\mu}. \tag{1.42}$$

Obviously, (1.42) implies (1.38). From (1.41) we obtain

$$\mu^2 - 2\varphi^2(\xi)\,\mu = \varphi^2(\xi)\frac{\varphi^2(\eta) - \varphi^2(\xi)}{1 + \delta\varphi^2(\eta)}, \tag{1.43}$$

i.e. (1.39), if we square both sides of (1.39) by observing $\varphi(\eta) \geq 0$, and replacing there $\varphi^2(t)$ by (1.42). □

Because of G, (1.40) holds true for arbitrarily given $\xi > \eta \geq 0$, if we define μ by (1.41), and $\varphi^2(t)$ by (1.42). Replacing these values in (1.40), we obtain with $\alpha := \varphi^2(\xi)$ and $\beta := \varphi^2(\eta)$,

$$\mu \cdot \left(\alpha - \frac{\alpha - \mu}{1 + \delta\mu}\right) = \varphi^2(\xi - \eta)\left(\alpha + \delta\mu \cdot \frac{\alpha - \mu}{1 + \delta\mu}\right),$$

i.e. $\mu^2(1 + \delta\alpha) = \varphi^2(\xi - \eta)(\alpha + \delta \cdot [2\alpha\mu - \mu^2])$, i.e., by (1.43),

$$\mu^2 \cdot (1 + \delta\beta) = \alpha\varphi^2(\xi - \eta),$$

i.e., by (1.41),

$$\left(\sqrt{\alpha(1 + \delta\beta)} - \sqrt{\beta(1 + \delta\alpha)}\right)^2 = \varphi^2(\xi - \eta). \tag{1.44}$$

$\varphi(\xi) > \varphi(\eta) \geq 0$ implies $\alpha > \beta$, i.e. $\alpha(1 + \delta\beta) > \beta(1 + \delta\alpha)$. Hence, by (1.44) and $\xi > \eta \geq 0$, the functional equation

$$\varphi(\xi - \eta) = \varphi(\xi)\sqrt{1 + \delta\varphi^2(\eta)} - \varphi(\eta)\sqrt{1 + \delta\varphi^2(\xi)} \tag{1.45}$$

holds true for all $\xi > \eta \geq 0$. Since $\varphi(0) = 0$, (1.45) also holds true for $\xi = \eta \geq 0$.

Since $\delta \geq 0$ (see F.a), we will distinguish two cases, namely $\delta = 0$ and $\delta > 0$. For $\delta = 0$, (1.45) yields

$$\varphi(\xi - \eta) = \varphi(\xi) - \varphi(\eta)$$

for all $\xi \geq \eta \geq 0$. Given arbitrarily $t, s \in \mathbb{R}_{\geq 0}$, put $\xi := t + s$ and $\eta := s$. Hence $\xi \geq \eta \geq 0$ and thus

$$\varphi(t + s) = \varphi(t) + \varphi(s).$$

Since $\varphi(r) \geq 0$ for $r \geq 0$, this implies (compare the Remark to C)

$$\varphi(t) = lt \tag{1.46}$$

for all $t \in \mathbb{R}_{\geq 0}$ with a constant $l > 0$, in view of $\varphi(1) > \varphi(0) = 0$. From F.a we get $\psi(h) = 1$ for all $h \in H$.

In the case $\delta > 0$ we will write $f(t) := \sqrt{\delta}\,\varphi(t)$ for $t \geq 0$. Hence, by (1.45),

$$f(\xi - \eta) = f(\xi)\sqrt{1 + f^2(\eta)} - f(\eta)\sqrt{1 + f^2(\xi)} \tag{1.47}$$

(see Aczél–Dhombres [1], Z. Daróczy [2], M. Kuczma [2])) for all $\xi \geq \eta \geq 0$. Since φ is an increasing bijection of $\mathbb{R}$, satisfying E.a, f must be an increasing bijection of $\mathbb{R}_{\geq 0}$. So define

$$f(\xi) =: \sinh g(\xi), \ \xi \geq 0,$$

and g must be an increasing bijection of $\mathbb{R}_{\geq 0}$ as well. (1.47) implies

$$\sinh g(\xi - \eta) = \sinh\big(g(\xi) - g(\eta)\big)$$

for all $\xi \geq \eta \geq 0$. Hence $g(\xi - \eta) = g(\xi) - g(\eta)$ and we obtain again

$$g(\xi) = l\xi$$

for all $\xi \geq 0$ with a constant $l > 0$. Thus

$$\varphi(t) = \frac{1}{\sqrt{\delta}} \sinh(lt) \tag{1.48}$$

for all $t \geq 0$. This implies, in view of E.a, that (1.48) holds true for all $t \in \mathbb{R}$ with a constant $l > 0$. From F.a we get $\psi(h) = \sqrt{1 + \delta h^2}$.

H. a) *In the case $\delta = 0$,*

$$d(x, y) = \frac{k}{l} \cdot \|x - y\|$$

holds true for all $x, y \in X$ and, moreover, $\varrho(h, t) = lt$.

b) *For $\delta > 0$ we get*

$$d(x, y) = \frac{k}{l} \operatorname{hyp}\left(x\sqrt{\delta},\, y\sqrt{\delta}\right)$$

for all $x, y \in X$ and

$$\cosh \operatorname{hyp}(p, q) = \sqrt{1 + p^2}\sqrt{1 + q^2} - pq,$$

$\operatorname{hyp}(p, q) \geq 0$, *for $p, q \in X$. Moreover,* $\varrho(h, t) = \frac{1}{\sqrt{\delta}} \sinh(lt) \cdot \sqrt{1 + \delta h^2}$.

Proof. H.a follows from E.b and F.b. Suppose now $\delta > 0$. From E.b and (1.48) we get

$$\delta x^2 = \sinh^2\left(l\frac{d(0, x)}{k}\right),$$

i.e. $1 + \delta x^2 = \cosh^2\left(l \cdot \frac{d(0,x)}{k}\right)$. Hence

$$\cosh\left(l \cdot \frac{d(0, x)}{k}\right) = \cosh \operatorname{hyp}(0, x\sqrt{\delta}),$$

and thus $d(0, x) = \frac{k}{l} \operatorname{hyp}(0, x\sqrt{\delta})$. From F.b and (1.48) we obtain for elements $x \neq 0$ and y of X with $\lambda x^2 := x^2 y^2 - (xy)^2$,

$$\frac{1}{\delta} \sinh^2\left(l \cdot \frac{d(x, y)}{k}\right) = \lambda + \varphi^2(\eta - \xi) \cdot (1 + \delta\lambda) \tag{1.49}$$

with $\|x\| \frac{1}{\sqrt{\delta}} \sinh(l\xi)$ and $xy = \|x\| \varphi(\eta)\sqrt{1 + \delta\lambda}$. Hence

$$\sqrt{\delta}\, \varphi(\eta - \xi) = \sinh(l\eta - l\xi) = \sinh(l\eta) \cdot \cosh(l\xi) - \cosh(l\eta)\sinh(l\xi)$$

$$= \frac{\sqrt{\delta}\, xy}{\|x\|\sqrt{1 + \delta\lambda}} \cdot \sqrt{1 + \delta x^2} - \sqrt{1 + \frac{\delta(xy)^2}{x^2(1 + \delta\lambda)}} \cdot \sqrt{\delta}\,\|x\|.$$

Observing $\lambda x^2 = x^2 y^2 - (xy)^2$ this implies

$$\lambda + \varphi^2(\eta - \xi)(1 + \delta\lambda) = \frac{1}{\delta}\left[\left(\sqrt{1 + \delta x^2}\sqrt{1 + \delta y^2} - \delta xy\right)^2 - 1\right]$$

$$= \frac{1}{\delta}\left[\left(\cosh \operatorname{hyp}(x\sqrt{\delta}, y\sqrt{\delta})\right)^2 - 1\right].$$

Thus, by (1.49), we obtain H.b also for $x \neq 0$. □

A surjective mapping $f : X \to X$ is called a *motion* of $\left(X, G\left(T, O\left(X\right)\right)\right)$ provided

$$d(x, y) = d\left(f(x), f(y)\right)$$

holds true for all $x, y \in X$.

I. a) $G\left(T, O\left(X\right)\right) = \{\alpha T_t \beta \mid \alpha, \beta \in O\left(X\right), t \in \mathbb{R}\}$.

b) *In all cases $\delta \geq 0$, $G\left(T, O\left(X\right)\right)$ is the group of motions of (X, G) where $G = G\left(T, O\left(X\right)\right)$.*

Proof. a) If γ is an element of $G\left(T, O\left(X\right)\right)$, take, by A, $\omega \in O\left(X\right)$ with $\omega\left(p\right) = \|p\|\, e$ where $p := \gamma\left(0\right)$. In view of (T2), there exists $s \in \mathbb{R}$ with

$$T_s\left(\omega\left(p\right)\right) = T_s(\|p\|\, e) = 0.$$

Hence $\lambda\left(0\right) = 0$ for $\lambda := T_s \omega \gamma$.

Case $\delta = 0$. By H.a, (i), (ii), we obtain

$$\frac{k}{l}\, \|x - y\| = d\left(x, y\right) = d\left(\lambda\left(x\right), \lambda\left(y\right)\right) = \frac{k}{l}\, \|\lambda\left(x\right) - \lambda\left(y\right)\|.$$

Hence, by Proposition 3, λ must be orthogonal. Since λ is bijective, we get $\lambda \in O\left(X\right)$ and $\gamma = \omega^{-1} T_{-s} \lambda$.

Case $\delta > 0$. By H.b, (i), (ii), we obtain with $d\left(x, y\right) = d\left(\lambda\left(x\right), \lambda\left(y\right)\right)$,

$$\sqrt{1 + \delta x^2}\, \sqrt{1 + \delta y^2} - \delta xy = \sqrt{1 + \delta \lambda^2(x)}\, \sqrt{1 + \delta \lambda^2(y)} - \delta \lambda\left(x\right) \lambda\left(y\right).$$

For $y = 0$ we get $x^2 = [\lambda\left(x\right)]^2$ with $\lambda\left(0\right) - 0$. This and the previous equation then imply $xy = \lambda\left(x\right) \lambda\left(y\right)$ for all $x, y \in X$. Hence

$$
\begin{aligned}
(x - y)^2 &= x^2 - 2xy + y^2 \\
&= \lambda^2(x) - 2\lambda\left(x\right) \lambda\left(y\right) + \lambda^2(y) = \left(\lambda\left(x\right) - \lambda\left(y\right)\right)^2,
\end{aligned}
$$

and thus, as in Case $\delta = 0$, $\lambda \in O\left(X\right)$.

b) In view of (i), (ii), obviously, $G\left(T, O\left(X\right)\right)$ consists only of motions. Suppose that γ is an arbitrary motion. It is now possible to follow, mutatis mutandis, the proof of I.a. So having $\lambda := T_s \omega \gamma$ with $\lambda\left(0\right) = 0$ as in a), we obtain for $\delta = 0$ by H.a and the fact that λ is a motion,

$$\|x - y\| = \|\lambda\left(x\right) - \lambda\left(y\right)\|,$$

and for $\delta > 0$, $xy = \lambda\left(x\right) \lambda\left(y\right)$ for all $x, y \in X$, by H.b and by applying that λ is a motion. $\qquad\square$

J. a) *Let T be the translation group with axis e and kernel $\varrho\left(h, t\right) = t$ and, moreover, T' the group with axis e and kernel $\varrho\left(h, t\right) = lt$, $l > 0$. Then*

$$\left(X, G\left(T, O\left(X\right)\right)\right) \cong \left(X, G\left(T', O\left(X\right)\right)\right).$$

b) *Also here the underlying axis is supposed to be e. Let now T, T' be the group with kernel*

$$\sinh t \cdot \sqrt{1 + h^2}, \quad \frac{1}{\sqrt{\delta}} \sinh(lt) \cdot \sqrt{1 + \delta h^2},$$

respectively, with $\delta > 0$ and $l > 0$. Then

$$\left(X, G\left(T, O\left(X\right)\right)\right) \cong \left(X, G\left(T', O\left(X\right)\right)\right).$$

Proof. a) The distance functions of the two geometries are, by H.a,

$$\text{eucl}\,(x,y) = \|x - y\| \text{ and } d\,(x,y) = \frac{k}{l}\,\|x - y\|.$$

Define $\sigma\,(x) := \frac{l}{k} \cdot x$ for $x \in X$. Hence

$$d\left(\sigma\,(x),\,\sigma\,(y)\right) = \text{eucl}\,(x,y) \tag{1.50}$$

holds true for all $x, y \in X$. In view of I, the groups G, G' of motions of our two geometries are $G\left(T, O\,(X)\right)$, $G\left(T', O\,(X)\right)$, respectively. If $g \in G$, then $\tau\,(g) := \sigma g \sigma^{-1}$ is in G' since (1.50) and

$$\text{eucl}\,(x,y) = \text{eucl}\left(g\,(x),\,g\,(y)\right)$$

imply, by $z' := \sigma\,(z)$ for $z \in X$,

$$\begin{aligned}
d\left(\tau\,(x'),\,\tau\,(y')\right) &= d\left(\sigma g\,(x),\,\sigma g\,(y)\right) = \text{eucl}\left(g\,(x),\,g\,(y)\right) \\
&= \text{eucl}\,(x,y) = d\,(x',y')
\end{aligned}$$

for all $x', y' \in X$. Also $\gamma \in G'$ implies $\sigma^{-1}\gamma\sigma \in G$. Hence $\tau : G \to G'$ is an isomorphism satisfying

$$\sigma\left(g\,(x)\right) = \tau\,(g)\,\sigma\,(x)$$

for all $x \in X$ and $g \in G$.

b) The proof of b) is, mutatis mutandis, the same as that one of a). The distance functions of the two geometries are, by H.b,

$$\text{hyp}\,(x,y) \text{ and } d\,(x,y) = \frac{k}{l}\,\text{hyp}\,(x\sqrt{\delta},\,y\sqrt{\delta}).$$

Define $\sigma\,(x) = \frac{x}{\sqrt{\delta}}$ for $x \in X$ and let G, G' be the groups of motions of the two geometries in question. Observe again I. If $g \in G$, then $\tau\,(g) = \sigma g \sigma^{-1}$ is in G' since

$$d\left(\sigma\,(x)\,\sigma\,(y)\right) = \frac{k}{l}\,\text{hyp}\,(x,y)$$

and

$$\text{hyp}\,(x,y) = \text{hyp}\left(g\,(x),\,g\,(y)\right)$$

imply, by $z' := \sigma\,(z)$ for $z \in X$,

$$\begin{aligned}
d\left(\tau\,(x'),\,\tau\,(y')\right) &= d\left(\sigma g\,(x),\,\sigma g\,(y)\right) = \frac{k}{l}\,\text{hyp}\left(g\,(x),\,g\,(y)\right) \\
&= \frac{k}{l}\,\text{hyp}\,(x,y) = d\,(x',y')
\end{aligned}$$

for all $x', y' \in X$. Also $\gamma \in G'$ implies $\sigma^{-1}\gamma\sigma \in G$. Hence $\tau : G \to G'$ is an isomorphism satisfying

$$\sigma\left(g\,(x)\right) = \tau\,(g)\,\sigma\,(x)$$

for all $x \in X$ and $g \in G$. $\qquad\square$

This finishes the proof of Theorem 7.

1.12   Other directions, a counterexample

Proposition 8. *Let T be a translation group of X with axis e, $e^2 = 1$, and kernel $\varrho(h, t)$ for all $t \in \mathbb{R}$ and $h \in e^\perp$. If $\omega \in O(X)$, then*

$$\{\omega T_t \omega^{-1} \mid t \in \mathbb{R}\}$$

is a translation group with axis $\omega(e)$ and kernel

$$\varrho(h', t) = \varrho\left(\omega^{-1}(h'), t\right)$$

for all $t \in \mathbb{R}$ and $h' \in [\omega(e)]^\perp = \omega(e^\perp)$.

Proof. $t \to \omega T_t \omega^{-1}$, $t \in \mathbb{R}$, defines a translation group of X with axis $\omega(e)$. This is shown as soon as (T1), (T2), (T3) are verified for $\omega(e)$ instead of e. Of course,

$$\omega T_{t+s} \omega^{-1} = \omega T_t \omega^{-1} \cdot \omega T_s \omega^{-1},$$

i.e. (T1), holds true. If $x - y \in \mathbb{R}_{\geq 0} \omega(e)$, we get

$$x = h' + \gamma_1 \omega(e), \ y = h' + \gamma_2 \omega(e)$$

with $\gamma_1, \gamma_2 \in \mathbb{R}$ and $h' \in [\omega(e)]^\perp = \omega(e^\perp)$. Define τ by $\gamma_1 = \varrho(h, \tau)$ where $h := \omega^{-1}(h')$, and hence $h \in e^\perp$. The equation

$$y = \omega T_t \omega^{-1}(x)$$

implies $h' + \gamma_2 \omega(e) = \omega T_t \big(h + \varrho(h, \tau) e\big) = h' + \varrho(h, \tau + t) \omega(e)$. Since t is uniquely determined by $\gamma_2 = \varrho(h, \tau + t)$, (T2) holds true. Moreover, by $t \geq 0$ and with the notations before,

$$\omega T_t \omega^{-1}\big(h' + \varrho(h, \tau) \omega(e)\big) - \big(h' + \varrho(h, \tau) \omega(e)\big)$$
$$= \ [\varrho(h, \tau + t) - \varrho(h, \tau)] \omega(e) \in \mathbb{R}_{\geq 0} \omega(e),$$

in view of (i) of Theorem 5. This proves (T3). The kernel of $\{\omega T_t \omega^{-1} \mid t \in \mathbb{R}\}$ is given by

$$\varrho(h', t) \ = \ [\omega T_t \omega^{-1}(h')] \cdot \omega(e)$$
$$= \ \omega\big(h + \varrho(h, t) e\big) \cdot \omega(e) = \varrho(h, t),$$

in view of (1.10), for all $h' \in [\omega(e)]^\perp$ and $t \in \mathbb{R}$. $\square$

Proposition 9. *Let T^1, T^2 be translation groups of X such that e_i with $e_i^2 = 1$ is the axis and*

$$\varrho(h_i, t) = \sinh t \cdot \sqrt{1 + h_i^2} \text{ for } t \in \mathbb{R}, \ h_i \in e_i^\perp,$$

the kernel of T^i, $i = 1, 2$. If $\omega \in O(X)$ satisfies $\omega(e_1) = e_2$ (such an ω exists because of step A of the proof of Theorem 7), then

$$T_t^2 = \omega T_t^1 \omega^{-1}$$

for all $t \in \mathbb{R}$.

Proof. We know from Proposition 8 that $\{\omega T_t^1 \omega^{-1} \mid t \in \mathbb{R}\}$ is a translation group in the direction of $\omega(e_1) = e_2$ with kernel

$$\varrho(h', t) = \sinh t \cdot \sqrt{1 + [\omega^{-1}(h')]^2} = \sinh t \cdot \sqrt{1 + (h')^2},$$

in view of $h' \cdot h' = \omega^{-1}(h') \cdot \omega^{-1}(h')$, for $t \in \mathbb{R}$ and $h' \in e_2^{\perp}$. Since T^2 and $\{\omega T_t^1 \omega^{-1} \mid t \in \mathbb{R}\}$ have the same axis and the same kernel, they must coincide (compare (1.12) and the Remark to Theorem 5). $\qquad\square$

The arbitrary motion $\alpha T_t \beta$ (see step I of the proof of Theorem 7) can be written as

$$\alpha T_t \alpha^{-1} \cdot \gamma = \gamma \cdot \beta^{-1} T_t \beta$$

with $\gamma := \alpha\beta \in O(X)$, where $\alpha T_t \alpha^{-1}$, $\beta^{-1} T_t \beta$ are translations in the direction of $\alpha(e)$, $\beta^{-1}(e)$, respectively.

Let (X, G) be a geometry (1.16) as defined at the beginning of section 10. The group G is generated by $O(X)$ and a translation group T with axis $e \in X$, $e^2 = 1$. The stabilizer of G in $a \in X$ consists of all $g \in G$ satisfying $g(a) = a$.

Proposition 10. *Suppose that $G = O(X) \cdot T \cdot O(X)$. The stabilizer of G in 0 is then $O(X)$, and that one in $a \in X$ is isomorphic to $O(X)$.*

Proof. 1. Assume $g(0) = 0$ for $g \in G$. Since g is of the form $\alpha\tau\beta$ with $\tau \in T$ and $\alpha, \beta \in O(X)$, we get $\tau\beta(0) = \alpha^{-1}(0)$, i.e. $\tau(0) = 0$, i.e. $\tau = T_0 = $ id. Hence $g = \alpha\beta \in O(X)$.

2. If $a \in X$, take $\omega \in O(X)$ with $\omega(a) = \|a\| \cdot e$ (see step A of the proof of Theorem 7), and, by (T 2), T_t with $\tau(a) = 0$, $\tau := T_t\omega$. The stabilizer of G in a is then given by $\tau^{-1} O(X) \tau$, the τ-conjugate of the stabilizer of G in 0. $\qquad\square$

Because of I of the proof of Theorem 7, Proposition 10 applies to euclidean as well as to hyperbolic geometry.

We now will present an example of a geometry (1.16) where T is even separable, such that there exists $g \in G$ satisfying $g(0) = 0$ and $g \notin O(X)$. Take the $\mathbb{R}^2$ (see section 3) and, by Theorem 5, the translation group T with axis $e := (1, 0)$ and kernel

$$\varrho((0, x_2), t) := t \cdot (1 + x_2^2)$$

with $x_2, t \in \mathbb{R}$, i.e. $T_t(x_1, x_2) = (x_1 + t[1 + x_2^2], x_2)$. Define

$$g := T_{-\frac{1}{\sqrt{2}}} \cdot R\left(-\frac{\pi}{2}\right) \cdot T_{-\frac{\sqrt{2}}{3}} \cdot R\left(\frac{\pi}{4}\right) \cdot T_1,$$

where $R(\alpha)$ designates the rotation (in the positive sense) about 0 with angle α, i.e.

$$R\left(\tfrac{\pi}{4}\right)(x_1, x_2) = \tfrac{1}{\sqrt{2}}(x_1 - x_2, x_1 + x_2),$$
$$R\left(-\tfrac{\pi}{2}\right)(x_1, x_2) = (x_2, -x_1).$$

Hence $g(0) = 0$ and $g(1,0) = \left(\frac{1}{\sqrt{2}}, 0\right)$. Since

$$\text{eucl}\left(0, (1,0)\right) \neq \text{eucl}\left(g(0), g(1,0)\right),$$

g cannot be in $O(X)$.

Proposition 11. *Given again a geometry (1.16), (X, G). If the stabilizer of G in 0 is $O(X)$, then to every $g \in G$ there exist $\alpha, \beta \in O(X)$ and $\tau \in T$ with $g = \alpha\tau\beta$.*

Proof. Put $a := g(0)$ and take $\alpha \in O(X)$ and $\tau \in T$ with $\alpha(\|a\|e) = a$ and $\tau(0) = \|a\|e$. Hence $\tau^{-1}\alpha^{-1}g(0) = 0$ and thus $\beta := \tau^{-1}\alpha^{-1}g \in O(X)$. $\square$

Chapter 2

Euclidean and Hyperbolic Geometry

X designates again an arbitrary real inner product space containing two linearly independent elements. As throughout the whole book, we do not exclude the case that there exists an infinite and linearly independent subset of X.

A natural and satisfactory definition of hyperbolic geometry over X was already given by Theorem 7 of chapter 1. If T is a separable translation group of X, and d an appropriate distance function of X invariant under T and $O(X)$, then there are, up to isomorphism, exactly two geometries

$$\left(X, G\left(T, O\left(X\right)\right)\right).$$

These geometries are called euclidean, hyperbolic geometry over X. Their distance functions are eucl (x, y), hyp (x, y), respectively.

2.1 Metric spaces

A set $S \neq \emptyset$ together with a mapping $d : S \times S \to \mathbb{R}$ is called a *metric space* (S, d) provided

(i) $d(x, y) = 0$ if, and only if, $x = y$,

(ii) $d(x, y) = d(y, x)$,

(iii) $d(x, y) \leq d(x, z) + d(z, y)$

hold true for all $x, y, z \in S$.

Observe $d(x, y) \geq 0$ for all $x, y \in S$, since (i), (ii), (iii) imply

$$0 = d(x, x) \leq d(x, y) + d(y, x) = 2d(x, y).$$

(i) is called the *axiom of coincidence*, (ii) the *symmetry axiom* and (iii) the *triangle inequality*.

Proposition 1. *$(X,$ eucl), $(X,$ hyp) are metric spaces, called the euclidean, hyperbolic metric space, respectively, over X.*

Proof. Axioms (i), (ii) hold true for both structures $(X,$ eucl), $(X,$ hyp), because of D.c and D.d of step D of the proof of Theorem 7. The triangle inequality of section 1.1 implies

$$\|(x - z) + (z - y)\| \le \|x - z\| + \|z - y\|$$

for $x, y, z \in X$, i.e. eucl $(x, y) \le$ eucl $(x, z)+$ eucl (z, y). It remains to prove (iii) for $(X,$ hyp). We may assume $x \ne z$. Because of D.a and the invariance of hyp under T and $O(X)$, it is sufficient to show (iii) for $z = 0$ and $x = \lambda e$ with $\lambda > 0$, i.e. to prove

$$L := \text{hyp}(\lambda c, y) \le \text{hyp}(\lambda e, 0) + \text{hyp}(0, y) =: R$$

or, equivalently, $\cosh L \le \cosh R$. Obviously, this latter inequality can be written as

$$\sqrt{1 + \lambda^2}\,\sqrt{1 + y^2} - (\lambda e)\, y \le \sqrt{1 + \lambda^2}\,\sqrt{1 + y^2} + \sqrt{\lambda^2}\,\sqrt{y^2}.$$

So observe finally $-(\lambda e)\, y \le |(\lambda e)\, y| \le \sqrt{(\lambda e)^2}\,\sqrt{y^2}$ from the inequality of Cauchy–Schwarz.   □

2.2   The lines of L.M. Blumenthal

Let (S, d) be a metric space and $x : \mathbb{R} \to S$ a function satisfying

$$d\left(x\left(\xi\right),\, x\left(\eta\right)\right) = |\xi - \eta| \tag{2.1}$$

for all real ξ, η. Then $\{x(\xi) \mid \xi \in \mathbb{R}\}$ is called a (Blumenthal) *line* of (S, d) (L.M. Blumenthal, K. Menger [1], p. 238). Observe that $x : \mathbb{R} \to S$ must be injective in view of the axiom of coincidence and (2.1).

Lemma 2. *If $\|x + y\| = \|x\| + \|y\|$ holds true for $x, y \in X$, then x, y are linearly dependent.*

Proof. Squaring both sides, we obtain $xy = \|x\|\,\|y\|$. Now apply Lemma 1 of chapter 1.   □

We would like to determine all solutions $x : \mathbb{R} \to X$ of the functional equation (2.1) in the case of $(X$ eucl). Let x be a solution. If $\alpha < \beta < \gamma$ are reals, then, by (2.1),

$$\|x(\gamma) - x(\alpha)\| = \gamma - \alpha, \quad \|x(\gamma) - x(\beta)\| = \gamma - \beta, \quad \|x(\beta) - x(\alpha)\| = \beta - \alpha,$$

i.e., by Lemma 2, $x(\gamma) - x(\beta)$, $x(\beta) - x(\alpha)$ must be linearly dependent. Put

$$p := x(0), \quad q := x(1) - x(0).$$

Hence, by (2.1), $\|q\| = 1$. If $0 < 1 < \xi$, we obtain

$$x(\xi) - x(1) = \varrho \cdot \big(x(1) - x(0)\big) = \varrho q$$

for a suitable $\varrho \in \mathbb{R}$. Thus $\xi - 1 = \|x(\xi) - x(1)\| = \|\varrho q\| = |\varrho|$. Moreover,

$$\xi - 0 = \|x(\xi) - p\| = \|x(1) + \varrho q - p\| = |1 + \varrho|.$$

Hence $\varrho = \xi - 1$ and thus $x(\xi) = x(1) + \varrho q = p + \xi q$ for $\xi > 1$, a formula which holds also true for $\xi = 1$, $\xi = 0$, but also in the cases $0 < \xi < 1$, $\xi < 0 < 1$ as similar arguments show. That, on the other hand,

$$x(\xi) := p + \xi q, \quad \|q\| = 1,$$

solves (2.1), is obvious. Hence

$$\{(1 - \lambda)a + \lambda b \mid \lambda \in \mathbb{R}\}$$

with $a, b \in \mathbb{R}$ and $a \neq b$ are the *euclidean* lines of (X, eucl) by writing $p := a$, $q \cdot \|b - a\| := b - a$, $\xi := \lambda \cdot \|b - a\|$.

Theorem 3. *The (hyperbolic) lines of (X, hyp) are given by all sets*

$$\{p \cosh \xi + q \sinh \xi \mid \xi \in \mathbb{R}\},$$

where p, q are elements of X with $pq = 0$ and $q^2 = 1$.

Proof. Let p, q be elements of X satisfying $pq = 0$ and $q^2 = 1$. Define $x : \mathbb{R} \to X$ by

$$x(\xi) = p \cosh \xi + q \sinh \xi \tag{2.2}$$

and observe $\mathrm{hyp}\,\big(x(\xi), x(\eta)\big) = |\xi - \eta|$ for all $\xi, \eta \in \mathbb{R}$. Hence (2.2) is the equation of a line of (X, hyp). Suppose now that $x : \mathbb{R} \to X$ solves (2.1) in the case of (X, hyp). Since x is injective, choose a real ξ_0 with $x(\xi_0) \neq 0$ and put

$$e := \frac{x(\xi_0)}{\sinh t_0}, \quad \sinh t_0 := \|x(\xi_0)\|.$$

Define the translation group

$$T_t(h + \sinh \tau \cdot \sqrt{1 + h^2}\, e) = h + \sinh(\tau + t) \cdot \sqrt{1 + h^2}\, e$$

for all $h \in e^{\perp}$ and $\tau, t \in \mathbb{R}$. Since

$$\mathrm{hyp}\,\big(T_t(y), T_t(z)\big) = \mathrm{hyp}\,(y, z) \tag{2.3}$$

holds true for all $y, z \in X$,

$$\xi \rightarrow \overline{x}\left(\xi\right) := T_{-t_0}\left(x\left(\xi + \xi_0\right)\right)$$

must be a solution of (2.1) as well: by (2.3),

$$\mathrm{hyp}\left(\overline{x}\left(\xi\right), \overline{x}\left(\eta\right)\right) = \mathrm{hyp}\left(x\left(\xi + \xi_0\right), x\left(\eta + \xi_0\right)\right) = |\xi - \eta|. \tag{2.4}$$

Notice $T_{t_0}(0) = x\left(\xi_0\right)$, i.e. $T_{-t_0}\left(x\left(\xi_0\right)\right) = 0$, i.e. $\overline{x}\left(0\right) = 0$. By (2.4),

$$\cosh(\xi - \eta) = \sqrt{1 + \overline{x}^2(\xi)}\sqrt{1 + \overline{x}^2(\eta)} - \overline{x}\left(\xi\right)\overline{x}\left(\eta\right).$$

For $\eta = 0$ we get $\cosh \xi = \sqrt{1 + \overline{x}^2(\xi)}$. Thus $\overline{x}^2(\xi) = \sinh^2 \xi$ and

$$\overline{x}\left(\xi\right)\overline{x}\left(\eta\right) = \cosh \xi \cosh \eta - \cosh\left(\xi - \eta\right) = \sinh \xi \sinh \eta, \tag{2.5}$$

i.e. $[\overline{x}\left(\xi\right)\overline{x}\left(\eta\right)]^2 = \sinh^2 \xi \sinh^2 \eta = \overline{x}^2(\xi)\,\overline{x}^2(\eta)$ for all real ξ, η. Hence, by Lemma 1, chapter 1, $\overline{x}\left(\xi\right), \overline{x}\left(\eta\right)$ must be linearly dependent. Since $\overline{x}$ is injective and $\overline{x}\left(0\right) = 0$, we obtain $\overline{x}\left(1\right) \neq 0$. Put $a \cdot \|\overline{x}\left(1\right)\| := \overline{x}\left(1\right)$. Thus

$$\overline{x}\left(\xi\right) = \varphi\left(\xi\right) \cdot a \tag{2.6}$$

with a suitable function $\varphi : \mathbb{R} \rightarrow \mathbb{R}$ satisfying $\varphi\left(0\right) = 0$, in view of the fact that $\overline{x}\left(\xi\right), \overline{x}\left(1\right)$ are linearly dependent. (2.5), (2.6) imply

$$\varphi\left(\xi\right)\varphi\left(1\right) = \sinh \xi \sinh 1$$

for all real ξ, so especially $\varphi^2(1) = \sinh^2 1$, i.e.

$$\overline{x}\left(\xi\right) \quad = \sinh \xi \cdot \tfrac{\sinh 1}{\varphi\left(1\right)}\, a$$
$$= p \cdot \cosh \xi + q \cdot \sinh \xi$$

with $p := 0$, $\varphi\left(1\right) q := a \sinh 1$, i.e. $pq = 0$ and $q^2 = 1$. Hence $\overline{x}\left(\xi\right)$ is of type (2.2), and we finally must show that

$$x\left(\xi\right) = T_{t_0}\left(\overline{x}\left(\xi - \xi_0\right)\right) = T_{t_0}\left(q \sinh(\xi - \xi_0)\right)$$

is of type (2.2) as well. This turns out to be a consequence of the following Lemma 4.   □

Lemma 4. *Let T be the translation group*

$$T_t(h + \sinh \tau \cdot \sqrt{1 + h^2}\, e) = h + \sinh(\tau + t) \cdot \sqrt{1 + h^2}\, e$$

with axis e, $e^2 = 1$, for all $h \in e^\perp$ and $\tau, t \in \mathbb{R}$. If $q \neq 0$ is in X and s in $\mathbb{R}$, there exist $a, b \in X$ with $ab = 0$, $b^2 = 1$ and

$$\{a \cosh \eta + b \sinh \eta \mid \eta \in \mathbb{R}\} = \{T_s(\mu q) \mid \mu \in \mathbb{R}\}.$$

Proof. There is nothing to prove for $q \in \mathbb{R}e$ or $s = 0$. So assume $s \neq 0$, and that q, e are linearly independent. Hence $q \neq (qe)\, e$. Because of

$$\{T_s(\mu q) \mid \mu \in \mathbb{R}\} = \{T_s(\mu \cdot \beta q) \mid \mu \in \mathbb{R}\}$$

for a fixed real $\beta \neq 0$, we may assume $\|q - (qe)\, e\| = 1$, without loss of generality. Put

$$S := \sinh s, \ C := \cosh s, \ j := q - (qe)\, e, \ \alpha := qe,$$

and observe $S \neq 0$, $C > 1$, $j^2 = 1$, $je = 0$, $q = \alpha e + j$, $q^2 = 1 + \alpha^2$. Since (1.8), (1.9) represent the same T_t, we obtain

$$T_s(\mu q) = \mu q + [\mu \alpha\, (C - 1) + \sqrt{1 + \mu^2 q^2}\, S]\, e =: x_1 e + x_2 j$$

with $x_1(\mu) = \mu \alpha C + \sqrt{1 + \mu^2(1 + \alpha^2)}\, S$ and $x_2(\mu) = \mu$. We hence get

$$x_1^2 - 2\alpha C x_1 x_2 + (\alpha^2 - S^2)\, x_2^2 = S^2$$

with the branch $\operatorname{sgn}\, (x_1 - \mu \alpha C) = \operatorname{sgn}\, S$, and also

$$\frac{y_1^2}{k} - y_2^2 = 1, \ \operatorname{sgn}\, y_1 = \operatorname{sgn}\, S,$$

$q^2 k := S^2 > 0$, by applying the orthogonal mapping ω of the subspace Σ of X, spanned by e, j, namely

$$\delta y_1 \ = \ x_1 C - x_2 \alpha,$$
$$\delta y_2 \ = \ x_1 \alpha + x_2 C,$$

$\delta := \sqrt{\alpha^2 + C^2}$. In order to find the interesting branch $\operatorname{sgn}\, y_1 = \operatorname{sgn}\, S$ of the hyperbola

$$\left\{ y_1 e + y_2 j \in \Sigma \ \middle| \ \frac{y_1^2}{k} - y_2^2 = 1 \right\},$$

observe $x_1(\mu) - \mu \alpha C = \sqrt{1 + \mu^2(1 + \alpha^2)}\, S$, $x_2(\mu) = \mu$, and hence

$$\delta y_1 \ = x_1 C - x_2 \alpha = C\, (x_1 - x_2 \alpha C) + x_2 \alpha S^2$$
$$= \left(C\sqrt{1 + \mu^2(1 + \alpha^2)} + x_2 \alpha S \right) S,$$

i.e. $\operatorname{sgn}\, y_1 = \operatorname{sgn}\, S$, if the coefficient of S is positive. But

$$0 < C^2(1 + x_2^2) + x_2^2 \alpha^2 (C^2 - S^2) = C^2(1 + x_2^2) + x_2^2 \alpha^2,$$

i.e. $x_2^2 \alpha^2 S^2 < C^2 \left(1 + x_2^2(1 + \alpha^2) \right)$, i.e.

$$-x_2 \alpha S \leq |x_2 \alpha S| < C\sqrt{1 + x_2^2(1 + \alpha^2)}.$$

Obviously, $l := \{T_s(\mu q) \mid \mu \in \mathbb{R}\} \subset \Sigma$. Hence

$$\omega(l) = \{\operatorname{sgn} S \cdot \sqrt{k}\, e \cosh\eta + j \sinh\eta \mid \eta \in \mathbb{R}\},$$

i.e.

$$l = \{\operatorname{sgn} S \cdot \sqrt{k}\, \omega^{-1}(e) \cosh\eta + \omega^{-1}(j) \sinh\eta \mid \eta \in \mathbb{R}\}$$

with

$$
\begin{aligned}
\delta\omega^{-1}(e) &= Ce - \alpha j,\\
\delta\omega^{-1}(j) &= \alpha e + Cj.
\end{aligned}
$$

So the line l is given by

$$\{a \cosh\eta + b \sinh\eta \mid \eta \in \mathbb{R}\}$$

with $a := \operatorname{sgn} S \cdot \sqrt{k} \cdot \omega^{-1}(e)$, $b := \omega^{-1}(j)$. Notice $ab = 0$, in view of $\omega^{-1}(e)\omega^{-1}(j) = ej$, and $b^2 = 1$. $\quad\square$

That images of lines under motions are lines follows immediately from the definition of lines. In fact! If $l = \{x(\xi) \mid \xi \in \mathbb{R}\}$ is a line and $f : X \to X$ a motion, then, by (2.1),

$$d\Big(f(x(\xi)),\, f(x(\eta))\Big) = d\big(x(\xi),\, x(\eta)\big) = |\xi - \eta|$$

for all $\xi, \eta \in \mathbb{R}$. This holds true in euclidean as well as in hyperbolic geometry. In both geometries also holds true the

Proposition 5. *If $a \neq b$ are elements of X, there is exactly one line l through a, b, i.e. with $l \ni a, b$.*

Proof. From D.a (section 1.3) we know that there exists a motion f such that $f(a) = 0$ and $f(b) = \lambda e$, $\lambda > 0$, e a fixed element of X with $e^2 = 1$. In the euclidean case there is exactly one line

$$\{(1 - \alpha)\, p + \alpha q \mid \alpha \in \mathbb{R}\},$$

$p \neq q$, through $0, \lambda e$, namely $\{\beta e \mid \beta \in \mathbb{R}\}$. There hence is exactly one line, namely $f^{-1}(\mathbb{R}e)$ through a, b. In the hyperbolic case there is also exactly one line

$$\{v \cosh\xi + w \sinh\xi \mid \xi \in \mathbb{R}\},\ vw = 0,\ w^2 = 1,$$

through $0, \lambda e$, namely $\mathbb{R}e$. This implies that $f^{-1}(\mathbb{R}e)$ is the uniquely determined line through a, b. $\quad\square$

2.3   The lines of Karl Menger

Let (S, d) be a metric space. If $a \neq b$ are elements of S, then

$$[a, b] := \{x \in S \mid d(a, x) + d(x, b) = d(a, b)\}$$

is called the *interval* (the Menger interval) $[a, b]$ (Menger [1], [2]). Observe $a, b \in [a, b] = [b, a]$. Moreover,

$$l(a, b) := \{z \in S \backslash \{b\} \mid a \in [z, b]\} \cup [a, b] \cup \{z \in S \backslash \{a\} \mid b \in [a, z]\}$$

is called a (Menger) *line* of (S, d).

In the euclidean case (X, eucl), the interval $[a, b]$ consists of all $x \in X$ with

$$\|(a - x) + (x - b)\| = \|a - b\| = \|a - x\| + \|x - b\|. \tag{2.7}$$

Hence, by Lemma 2, the elements $a - x$ and $x - b$ are linearly dependent. If $x \notin \{a, b\}$, then $x - b = \lambda(a - x)$ with a suitable real $\lambda \notin \{0, -1\}$, i.e.

$$x = \frac{\lambda}{1 + \lambda} a + \frac{1}{1 + \lambda} b = a + \frac{b - a}{1 + \lambda}.$$

For $\lambda > 0$ equation (2.7) holds true, but not for $\lambda \in]-1, 0[$ or $\lambda < -1$. Hence

$$[a, b] = \{a + \mu(b - a) \mid 0 \leq \mu \leq 1\},$$

and $l(a, b) = \{a + \mu(b - a) \mid \mu \in \mathbb{R}\}$. In the case (X, eucl) the Menger lines are thus exactly the previous lines. The same holds true for (X, hyp) as will be proved in Theorem 6.

If $a \neq b$ are elements of X and if

$$\{p \cosh \xi + q \sinh \xi \mid \xi \in \mathbb{R}\}, \tag{2.8}$$

$pq = 0$, $q^2 = 1$, is the hyperbolic line through a, b, then

$$a = p \cosh \alpha + q \sinh \alpha,$$
$$b = p \cosh \beta + q \sinh \beta$$

with uniquely determined reals α, β. If $\beta < \alpha$ we will replace ξ in (2.8) by $\xi' = -\xi$ and q by $q' = -q$. So without loss of generality we may assume $\alpha < \beta$.

Theorem 6. *Let* $x(\xi) = p \cosh \xi + q \sinh \xi$ *be the equation of the line through* $a \neq b$ *with* $a = x(\alpha)$, $b = x(\beta)$, $\alpha < \beta$. *Then*

$$[a, b] = \{x(\xi) \mid \alpha \leq \xi \leq \beta\} \tag{2.9}$$

and $l(a, b) = \{x(\xi) \mid \xi \in \mathbb{R}\}$.

Proof. The right-hand side of (2.9) is a subset of $[a, b]$. This follows from $\alpha \leq \xi \leq \beta$ and

$$\text{hyp}\left(x\left(\alpha\right), x\left(\beta\right)\right) = |\alpha - \beta| = \beta - \alpha,$$
$$\text{hyp}\left(x\left(\alpha\right), x\left(\xi\right)\right) = \xi - \alpha,$$
$$\text{hyp}\left(x\left(\xi\right), x\left(\beta\right)\right) = \beta - \xi.$$

Let now z be an element of X with $z \in [a, b]$, i.e. with

$$\beta - \alpha = \text{hyp}\left(x\left(\alpha\right) x\left(\beta\right)\right) = \text{hyp}\left(x\left(\alpha\right), z\right) + \text{hyp}\left(z, x\left(\beta\right)\right).$$

Define $\xi := \alpha + \text{hyp}\left(x\left(\alpha\right), z\right)$. Obviously, $\xi - \alpha \geq 0$, and

$$\beta - \xi = \text{hyp}\left(z, x\left(\beta\right)\right) \geq 0,$$

i.e. $\alpha \leq \xi \leq \beta$. Hence $x\left(\xi\right)$ is an element of the right-hand side of (2.9). Observe

$$\text{hyp}\left(x\left(\alpha\right), z\right) = \xi - \alpha = \text{hyp}\left(x\left(\alpha\right), x\left(\xi\right)\right), \tag{2.10}$$

$$\text{hyp}\left(z, x\left(\beta\right)\right) = \beta - \xi = \text{hyp}\left(x\left(\xi\right), x\left(\beta\right)\right). \tag{2.11}$$

We take a motion f with $f(a) = 0$ and $f(b) = \lambda e$, $\lambda > 0$. Since $x\left(\xi\right)$ is on the line through a, b and

$$\text{hyp}\left(a, b\right) = \text{hyp}\left(a, x\left(\xi\right)\right) + \text{hyp}\left(x\left(\xi\right), b\right)$$

holds true, we obtain that $f\left(x\left(\xi\right)\right)$ is on the line $\mathbb{R}e$ through 0 and λe, and that

$$\text{hyp}\left(0, \lambda e\right) = \text{hyp}\left(0, f\left(x\left(\xi\right)\right)\right) + \text{hyp}\left(f\left(x\left(\xi\right)\right), \lambda e\right),$$

i.e. that $f(a) = e \sinh \eta_1$, $f\left(x\left(\xi\right)\right) = e \sinh \eta_2$, $f(b) = e \sinh \eta_3$ with $\eta_3 = |\eta_2| + |\eta_3 - \eta_2|$ and $\lambda = \sinh \eta_3$. Hence $0 = \eta_1 \leq \eta_2 \leq \eta_3$ and $f\left(x\left(\xi\right)\right) =: \mu e$ with $0 \leq \mu \leq \lambda$. If we take the images of $x\left(\alpha\right), z, \ldots$ in (2.10), (2.11), we get from these equations with $\overline{z} := f(z)$,

$$\sqrt{1 + \overline{z}^2} = \sqrt{1 + \mu^2},$$
$$\sqrt{1 + \overline{z}^2}\sqrt{1 + \lambda^2} - \overline{z}\lambda e = \sqrt{1 + \mu^2}\sqrt{1 + \lambda^2} - \mu\lambda,$$

i.e. $\overline{z}^2 = \mu^2$ and $\overline{z}e = \mu$. Thus $(\overline{z}e)^2 = \overline{z}^2 e^2$, i.e. $\overline{z} \in \mathbb{R}e$, by Lemma 1, chapter 1, i.e. $\overline{z} = \mu e$, by $\overline{z}e = \mu$. Hence $f(z) = \overline{z} = f\left(x\left(\xi\right)\right)$, i.e. $z = x\left(\xi\right) \in [a, b]$.

We finally must show that the Menger lines of (X, hyp) are the hyperbolic lines. If $l\left(a, b\right)$ is a Menger line, designate by g the hyperbolic line through a, b. If $z \in X\backslash\{b\}$ with $a \in [z, b]$, then the hyperbolic line through z, b must contain a since, by (2.9), intervals are subsets of hyperbolic lines. Hence, by Proposition 5, $z \in g$. Moreover, $z \in X\backslash\{a\}$ with $b \in [a, z]$ belongs also to g, i.e. $l\left(a, b\right) \subseteq g$. If $x\left(\xi\right) \in g$, we distinguish three cases $\xi < \alpha$, $\alpha \leq \xi \leq \beta$, $\beta < \xi$ with $a = x\left(\alpha\right)$, $b = x\left(\beta\right)$, $\alpha < \beta$. In the first case we get

$$x\left(\xi\right) \in X\backslash\{x\left(\beta\right)\} \text{ with } x\left(\alpha\right) \in [x\left(\xi\right), x\left(\beta\right)],$$

in the last $x\left(\xi\right) \in X\backslash\{x\left(\alpha\right)\}$ with $x\left(\beta\right) \in [x\left(\alpha\right), x\left(\xi\right)]$. $\qquad\square$

2.4 Another definition of lines

We proposed the following definition of a line, W. Benz [1, 6]. Suppose that (S, d) is a metric space and that $c \in S$ and $\varrho \geq 0$ is in $\mathbb{R}$. Then

$$B(c, \varrho) := \{x \in S \mid d(c, x) = \varrho\}$$

is defined to be the *ball* with *center* c and *radius* ϱ. Obviously, $B(c, 0) = \{c\}$. If a, b are distinct elements of S, we will call

$$g(a, b) := \{x \in S \mid B(a, d(a, x)) \cap B(b, d(b, x)) = \{x\}\}$$

a *g-line*. Notice $a, b \in g(a, b) = g(b, a)$.

Let S contain exactly three distinct elements a, b, c and define

$$d(a, b) = 3, \ d(a, c) = 4, \ d(b, c) = 5$$

and $d(x, x) = 0$, $d(x, y) = d(y, x)$ for all $x, y \in S$. Hence (S, d) is a metric space. Of course, (S, d) does not contain a line in the sense of L.M. Blumenthal. The Menger line $l(a, b)$ is given by

$$l(a, b) = \{a, b\},$$

and the *g*-line $g(a, b)$ by $\{a, b, c\}$.

Define $\Sigma = (X, \text{eucl})$ and $\Sigma' = (X, d)$ with

$$d(x, y) = \frac{\|x - y\|}{1 + \|x - y\|}$$

for all $x, y \in X$. The *g*-lines of the metric spaces Σ, Σ' coincide. Every Menger line of Σ' contains exactly two distinct elements. There do not exist lines of Σ' in the sense of L.M. Blumenthal, because

$$\frac{\|x(\xi) - x(\eta)\|}{1 + \|x(\xi) - x(\eta)\|} = |\xi - \eta|, \text{for all } \xi, \eta \in \mathbb{R},$$

cannot be true for $\xi = 1$ and $\eta = 0$.

Theorem 7. *Let Σ be one of the metric spaces (X, eucl), (X, hyp). Then $l(a, b) = g(a, b)$ for all $a \neq b$ of X, where $l(a, b)$ designates the Menger line through a, b.*

Proof. If $g(a, b)$, $a \neq b$, is a *g*-line, then $x \in X$ is in $g(a, b)$ if, and only if,

$$\forall_{z \in X} \ [d(a, z) = d(a, x)] \text{ and } [d(b, z) = d(b, x)] \text{ imply } z = x. \tag{2.12}$$

As a consequence we get

$$f(g(a, b)) = g(f(a), f(b)), \ a \neq b,$$

for every g-line g and motion f. In order to prove $l(a,b) = g(a,b)$, it is hence sufficient to prove $l(0, \lambda e) = g(0, \lambda e)$ for $\lambda > 0$, i.e. $g(0, \lambda e) = \{\mu e \mid \mu \in \mathbb{R}\}$.

a) Euclidean case. (2.12) has for $a = 0$ and $b = \lambda e$ the form

$$\forall_{z \in X} \ z^2 = x^2 \text{ and } ez = ex \text{ imply } z = x. \tag{2.13}$$

$x = \mu e$ belongs to $g(0, \lambda e)$, because $z^2 = \mu^2$ and $ez = \mu$ imply $(ez)^2 = e^2 z^2$, i.e., by Lemma 1, chapter 1, $z \in \mathbb{R}e$, i.e. $z = \mu e$, by $ez = \mu$. If $x \in g(0, \lambda e)$ put

$$z := -(x - (xe)\,e) + (xe)\,e, \tag{2.14}$$

and observe $z^2 = x^2$, $ez = ex$, i.e., by (2.13), $z = x$. Hence, by (2.14), $x = (xe)\,e \in \mathbb{R}e$.

b) Hyperbolic case. (2.12) has for $a = 0$ and $b = \lambda e$ also the form (2.13). So also here we get $g(0, \lambda e) = \mathbb{R}e$. □

2.5   Balls, hyperplanes, subspaces

Proposition 8. *Suppose that* $B(c, \varrho)$, $B(c', \varrho')$ *are balls of* (X, eucl) *satisfying* $\varrho > 0$ *and*

$$B(c, \varrho) \subseteq B(c', \varrho').$$

Then $c = c'$ *and* $\varrho = \varrho'$.

Proof. $c + \frac{\varrho x}{\|x\|} \in B(c, \varrho)$ implies

$$\frac{(c - c')\,x}{\|x\|} = \frac{1}{2\varrho} \left(\varrho'^2 - \varrho^2 - (c - c')^2 \right)$$

for all elements $x \neq 0$ of X. If $c - c'$ were $\neq 0$, the left-hand side of this equation would be 0 for $0 \neq x \perp (c - c')$ and $\neq 0$ for $x = c - c'$ which is impossible, since the right-hand side of the equation does not depend on x. (Notice that $a \perp b$ stands for $ab = 0$.) Hence $c - c' = 0$, and thus

$$0 = \varrho'^2 - \varrho^2 - (c - c')^2 = \varrho'^2 - \varrho^2. \qquad \square$$

Proposition 9. *Let* $B(c, \varrho)$, $\varrho > 0$, *be a ball of* (X, hyp). *Then*

$$B(c, \varrho) = \{x \in X \mid \|x - a\| + \|x - b\| = 2\alpha\}$$

with $a := ce^{-\varrho}$, $b := ce^{\varrho}$ *and* $\alpha := \sinh \varrho \cdot \sqrt{1 + c^2}$, *where* e^t *denotes the exponential function* $\exp(t)$ *for* $t \in \mathbb{R}$.

Proof. Put $S := \sinh \varrho$, $C := \cosh \varrho$ and $p := x - cC$. Observe $C + S = e^{\varrho}$ and $C - S = e^{-\varrho}$.

a) Assume $\|x - a\| = 2\alpha - \|x - b\|$ for a given $x \in X$. Squaring this equation yields

$$S(1 + c^2) - cp = \sqrt{1 + c^2}\,\|x - b\|.$$

Observing $x - b = p - Sc$ and squaring again, we get $(cp)^2 = (p^2 - S^2)(1 + c^2)$. This implies

$$|cx + C| = \sqrt{1 + c^2}\,\sqrt{1 + x^2}, \tag{2.15}$$

since $cx + C = cp + C(1 + c^2)$. If $-cx - C$ were equal to $\sqrt{1 + c^2}\,\sqrt{1 + x^2}$, then

$$1 \le \cosh\ \mathrm{hyp}\,(c, -x) = \sqrt{1 + c^2}\,\sqrt{1 + x^2} + cx = -C$$

would follow, a contradiction. Hence, by (2.15),

$$\cosh\ \mathrm{hyp}\,(c, x) = C = \cosh\varrho,$$

i.e. $x \in B(c, \varrho)$.

b) Assume vice versa $C = \sqrt{1 + c^2}\,\sqrt{1 + x^2} - cx$, i.e. $x \in B(\varrho, c)$, for a given $x \in X$. A similar calculation as in step a), but now in the other direction, leads to

$$\sqrt{(p + cS)^2}\,\sqrt{(p - cS)^2} = |S^2(2 + c^2) - p^2|. \tag{2.16}$$

If $S^2(2 + c^2) \ge p^2$, then $\|x - a\| + \|x - b\| = 2\alpha$ follows from (2.16). So observe, by the inequality of Cauchy–Schwarz,

$$(cx)^2 \le c^2 x^2 + S^2,$$

i.e., by $(cx + C)^2 = (1 + c^2)(1 + x^2)$,

$$x^2 - 2(cx)\,C + c^2 = (cx)^2 + S^2 - c^2 x^2 \le 2S^2,$$

i.e. $S^2(2 + c^2) \ge p^2$. $\qquad\square$

Suppose $a, b \in X$ and let γ be a positive real number. Then

$$\{x \in X \mid \|x - a\| + \|x - b\| = \gamma\}$$

is called a *hyperellipsoid* in euclidean geometry, i.e. in (X, eucl). Let now $B(c, \varrho)$, $\varrho > 0$, be a hyperbolic ball. If $c = 0$, then, in view of Proposition 9, it is also a euclidean ball with center 0 and radius $\sinh\varrho$. In the case $c \ne 0$, the hyperbolic ball $B(c, \varrho)$ is a euclidean hyperellipsoid such that its *foci* $ce^{-\varrho}$, ce^{ϱ} are in

$$\mathbb{R}_{>0}c = \{\lambda c \mid 0 < \lambda \in \mathbb{R}\}.$$

Observe $\tau_0 > 1$ for $ce^{\varrho} =: \tau_0(ce^{-\varrho})$.

Lemma 10. *Let $a \neq 0$ be an element of X and $\tau > 1$ be a real number. Then*

$$\{x \in X \mid \|x - a\| + \|x - \tau a\| = 2\alpha\} \tag{2.17}$$

is the hyperbolic ball $B\left(a\sqrt{\tau}, \ln \sqrt{\tau}\right)$ if

$$2\alpha = (\tau - 1)\sqrt{\frac{1}{\tau} + a^2}.$$

Proof. Since $\{x \in X \mid \|x - ce^{-\varrho}\| + \|x - ce^{\varrho}\| = 2\sinh \varrho \cdot \sqrt{1 + c^2}\}$ is $B\left(c, \varrho\right)$, we get with $c := a\sqrt{\tau}, \varrho := \ln \sqrt{\tau}$, obviously,

$$a = ce^{-\varrho}, \ \tau a = ce^{\varrho}, \ 2\alpha = (e^{\varrho} - e^{-\varrho})\sqrt{1 + c^2} = (\tau - 1)\sqrt{\frac{1}{\tau} + a^2}. \qquad \square$$

Proposition 11. *Suppose that $B\left(c, \varrho\right)$, $B\left(c', \varrho'\right)$ are hyperbolic balls satisfying $\varrho > 0$ and*

$$B\left(c, \varrho\right) \subseteq B\left(c', \varrho'\right). \tag{2.18}$$

Then $c = c'$ and $\varrho = \varrho'$.

Proof. Assume that there exist balls $B\left(c, \varrho\right)$, $B\left(c', \varrho'\right)$ with (2.18), $c \neq c'$ and $\varrho > 0$. If $j \in X$ is given with $j^2 = 1$, there exists, by D.a, a motion μ such that $\mu\left(c\right) = 0$, $\mu\left(c'\right) = \lambda j$, $\lambda > 0$. Hence $B\left(0, \varrho\right) \subseteq B\left(\lambda j, \varrho'\right)$, i.e.

$$\text{hyp}\,(0, x) = \varrho \ \text{ implies } \ \text{hyp}\,(\lambda j, x) = \varrho'$$

for all $x \in X$, i.e.

$$\sqrt{1 + x^2} = \cosh \varrho \ \text{ implies } \ \sqrt{1 + \lambda^2}\sqrt{1 + x^2} - \lambda j x = \cosh \varrho'.$$

Applying this implication twice, namely for $x = j \sinh \varrho$ and for $x = i \sinh \varrho$ with $i \in X$, $i^2 = 1$, $ij = 0$ we obtain

$$\sqrt{1 + \lambda^2}\cosh \varrho - \lambda \sinh \varrho = \cosh \varrho' = \sqrt{1 + \lambda^2}\cosh \varrho,$$

a contradiction, since $\lambda > 0$ and $\varrho > 0$. Thus $c = c'$. Take now $j \in X$ with $j^2 = 1$ and $jc = 0$, and observe for $x := \sinh \varrho \cdot j + \cosh \varrho \cdot c$,

$$\text{hyp}\,(c, x) = \varrho,$$

i.e., by (2.18), hyp $(c, x) = \varrho'$. Hence $\varrho = \varrho'$. $\qquad \square$

If $a \neq 0$ is in X and $\alpha \in \mathbb{R}$, then we will call

$$H\left(a, \alpha\right) := \{x \in X \mid ax = \alpha\}$$

a *euclidean hyperplane* of X.

If $e \in X$ satisfies $e^2 = 1$, if $t \in \mathbb{R}$ and $\omega_1, \omega_2 \in O(X)$, then

$$\omega_1 T_t \omega_2(e^{\perp}) = \{\omega_1 T_t \omega_2(x) \mid x \in e^{\perp}\}$$

will be called a *hyperbolic hyperplane*, where $\{T_t \mid t \in \mathbb{R}\}$ is based on the axis e and the kernel $\sinh \varrho \cdot \sqrt{1 + h^2}$. Of course, mutatis mutandis, also the euclidean hyperplanes can be described this way.

In Proposition 17 parametric representations of hyperbolic hyperplanes will be given.

Proposition 12. *If $H(a, \alpha)$ and $H(b, \beta)$ are euclidean hyperplanes with $H(a, \alpha) \subseteq H(b, \beta)$, then $H(a, \alpha) = H(b, \beta)$ and there exists a real $\lambda \neq 0$ with $b = \lambda a$ and $\beta = \lambda \alpha$.*

Proof. If a, b are linearly dependent, then there exists a real $\lambda \neq 0$ with $b = \lambda a$ since a, b are both unequal to 0. Put $x_0 a^2 := \alpha a$. Hence

$$x_0 \in H(a, \alpha) \subseteq H(b, \beta),$$

i.e. $\beta = b x_0 = \lambda a \cdot x_0 = \lambda \alpha$, and thus $H(a, \alpha) = H(b, \beta)$. If a, b were linearly independent, then

$$q := x_0 + b - \frac{ab}{a^2} a \in H(a, \alpha) \subseteq H(b, \beta),$$

i.e. $b x_0 = \beta = bq$, i.e.

$$\left(b - \frac{ab}{a^2} a\right)^2 = b^2 - \frac{(ab)^2}{a^2} = b(q - x_0) = 0,$$

i.e. $b - \frac{ab}{a^2} a = 0$ would hold true. $\qquad\square$

If $a \neq 0$ is in X and $a^2 = 1$, then the hyperplanes of (X, hyp) can also be defined by

$$\alpha T_t \beta(a^{\perp}) \text{ with } \alpha, \beta \in O(X) \text{ and } t \in \mathbb{R}:$$

take $\omega \in O(X)$ with $a = \omega(e)$ and observe

$$\alpha T_t \beta\left([\omega(e)]^{\perp}\right) = \alpha T_t \beta\left(\omega(e^{\perp})\right) = \alpha T_t \beta \omega(e^{\perp}).$$

Obviously, $\omega(H(a, \alpha)) = H(\omega(a), \alpha)$ for $\omega \in O(X)$, where $H(a, \alpha)$ is a euclidean hyperplane. The image of $H(a, \alpha)$ under $y = x + t$, $t \in X$, is $H(a, at + \alpha)$. Of course, if μ is a hyperbolic motion, then $\mu[\omega_1 T_t \omega_2(e^{\perp})]$ is again a hyperbolic hyperplane since $\mu \cdot \omega_1 T_t \omega_2$ is also a motion (see I of the proof of Theorem 7 of chapter 1).

A *subspace* of (X, eucl) (or (X, hyp)) is a set $\Gamma \subseteq X$ such that for all $a \neq b$ in Γ the euclidean (hyperbolic) line through a, b is a subset of Γ. Of course, $\emptyset$ and X are subspaces, also every point of X, but lines as well. Since every euclidean (hyperbolic) line is contained in a one- or a two-dimensional subspace of the vector space X, the following proposition must hold true.

Proposition 13. *All euclidean (hyperbolic) subspaces are given by the subspaces of the vector space X and their images under motions.*

A *spherical subspace* of (X, eucl) or (X, hyp) is a set

$$\Gamma \cap B(c, \varrho),$$

where $\Gamma \ni c$ is a subspace and $B(c, \varrho)$ a ball of (X, eucl), (X, hyp), respectively. Without loss of generality we may assume $c = 0$. Hence the following proposition holds true.

Proposition 14. *All spherical subspaces of X are given by the spherical subspaces $\Gamma \cap B(c, \varrho)$ with $c = 0 \in \Gamma$ and their images under motions.*

A subspace V of the vector space X is called *maximal* if, and only if, $V \neq X$ and, moreover, every subspace $W \supseteq V$ of X is equal to X or V. If $0 \neq a \in X$, then $a^\perp$ is a maximal subspace of the vector space X: observe

1) $x, y \in a^\perp$ implies $x + y \in a^\perp$ and $\lambda x \in a^\perp$ for every $\lambda \in \mathbb{R}$,

2) if $W \supseteq a^\perp$ is a subspace of X and $x \in W \backslash a^\perp$, then $xa \neq 0$ and $-x + \frac{xa}{a^2} a \in a^\perp \subseteq W$, i.e. $\frac{xa}{a^2} a = x + \left(-x + \frac{xa}{a^2} a \right) \in W$, i.e. $a \in W$, i.e. $X = a^\perp \oplus \mathbb{R}a \subseteq W$, i.e. $X = W$.

Maximal subspaces of X and their images under euclidean (hyperbolic) motions will be called euclidean (hyperbolic) *quasi-hyperplanes*. Since $a^\perp$ with $0 \neq a \in X$ is maximal, hyperplanes are quasi-hyperplanes. But there are quasi-hyperplanes which are not hyperplanes.

2.6   A special quasi-hyperplane

Let X be the set of all power series with real coefficients and radius of convergence greater than 1,

$$A(\xi) = a_0 + a_1 \xi + a_2 \xi^2 + \cdots,$$

which will be of interest for us in the interval $[0, 1]$. Define

$$
\begin{aligned}
\lambda A(\xi) &= \lambda a_0 + \lambda a_1 \xi + \lambda a_2 \xi^2 + \cdots, \\
A(\xi) + B(\xi) &= (a_0 + b_0) + (a_1 + b_1)\xi + (a_2 + b_2)\xi^2 + \cdots
\end{aligned}
$$

and $AB = \int_0^1 A(\xi) B(\xi) \, d\xi$. Observe that the following set of elements of X, namely

$$e^\xi, 1, \xi, \xi^2, \xi^3, \ldots,$$

$e^\xi := \exp(\xi)$, is linearly independent: if

$$k e^\xi + k_0 \cdot 1 + k_1 \cdot \xi + \cdots + k_n \cdot \xi^n = 0 \tag{2.19}$$

for all $\xi \in [0,1]$ where $k, k_0, \ldots \in \mathbb{R}$, then differentiating (2.19) $(n+1)$-times yields $ke^\xi = 0$, i.e. $k = 0$, and differentiating it n-times, $k_n = 0$, and so on, $k_{n-1} = \cdots = k_0 = 0$. Let B be a basis of X which contains the functions $e^\xi, 1, \xi, \xi^2, \ldots$. Let V be the subspace of X generated by B' which is defined by B without the function e^ξ. Hence V is maximal. Since, of course, $0 \in V$, V must be a euclidean subspace of X. We would like to show that there is no $a \neq 0$ in X such that

$$V = H(a, 0), \tag{2.20}$$

i.e. that V is a quasi-hyperplane which is not a hyperplane. Assume that (2.20) holds true for an element $a \neq 0$ in X. Put

$$a(\xi) = a_0 + a_1 \xi + \cdots$$

and notice

$$0 < \int_0^1 a(\xi) a(\xi) d(\xi) = \sum_{i=0}^\infty \int_0^1 a_i a(\xi) \xi^i d\xi = 0,$$

since the functions ξ^i, $i = 0, 1, \ldots$, belong to B' and hence to V.

2.7 Orthogonality, equidistant surfaces

Let l_1, l_2 be lines through $s \in X$. We will say that l_1 is *orthogonal* to l_2 and write $l_1 \perp l_2$ if, and only if, there exist

$$p_1 \in l_1 \backslash \{s\}, \; p_2 \in l_2 \backslash \{s\}$$

such that (see (2.21) for the euclidean and (2.22) for the hyperbolic case)

$$\|p_1 - p_2\|^2 = \|p_1 - s\|^2 + \|s - p_2\|^2, \tag{2.21}$$

$$\cosh \hyp (p_1, p_2) = \cosh \hyp (p_1, s) \, \cosh \hyp (s, p_2). \tag{2.22}$$

Since $(p_1 - p_2)^2 = \big((p_1 - s) + (s - p_2)\big)^2$, we also may write $(p_1 - s)(s - p_2) = 0$ instead of (2.21). Formula (2.22) is the so-called theorem of Pythagoras of hyperbolic geometry (see, for instance, W. Benz [4], p. 153) for the triangle $p_1 s p_2$. If $s = 0$ in (2.22), then this formula reduces to $p_1 p_2 = 0$, i.e. that in 0 euclidean and hyperbolic orthogonality coincide. Observe that $l_1 \perp l_2$ implies $l_2 \perp l_1$. Moreover, there is no line l orthogonal to itself, $l \not\perp l$: if

$$l = \{p + \xi q \mid \xi \in \mathbb{R}\}, \; q^2 = 1, \tag{2.23}$$

in the euclidean case or

$$l = \{x(\xi) = p \cosh \xi + q \sinh \xi \mid \xi \in \mathbb{R}\}, \; pq = 0, \; q^2 = 1, \tag{2.24}$$

in the hyperbolic case, $l \perp l$ would imply

$$(\pi_1 - \sigma)(\sigma - \pi_2) = 0 \text{ for } s = p + \sigma q,\ p_i = p + \pi_i q \neq s\ (i = 1, 2),$$

a contradiction, or for $s = x(\sigma)$, $p_i = x(\pi_i) \neq s\ (i = 1, 2)$, by (2.1),

$$\cosh(\pi_1 - \pi_2) = \cosh(\pi_1 - \sigma) \cosh(\sigma - \pi_2).$$

Put $\alpha := \pi_1 - \sigma$ and $\beta := \sigma - \pi_2$, observe

$$\cosh(\alpha + \beta) = \cosh\alpha \cosh\beta + \sinh\alpha \sinh\beta,$$

i.e. $\sinh(\pi_1 - \sigma) \sinh(\sigma - \pi_2) = 0$, which is also a contradiction.

Since $l \not\perp l$ holds true for every line l, we obtain, by Proposition 5, that $l_1 \perp l_2$ implies $\#(l_1 \cap l_2) = 1$, i.e. that l_1, l_2 have one single point in common.

If l_1, l_2 are lines with $l_1 \perp l_2$ and μ is a motion, then $\mu(l_1) \perp \mu(l_2)$. This follows from (2.21), (2.22) since distances are invariant under motions.

Let l_1, l_2 be lines through s with $l_1 \perp l_2$. If $a_i \in l_i \backslash \{s\}$, $i = 1, 2$, are arbitrary points, then

$$\|a_1 - a_2\|^2 = \|a_1 - s\|^2 + \|s - a_2\|^2 \tag{2.25}$$

holds true in the euclidean case, and

$$\cosh \mathrm{hyp}\,(a_1, a_2) = \cosh \mathrm{hyp}\,(a_1, s) \cosh \mathrm{hyp}\,(s, p_2) \tag{2.26}$$

in the hyperbolic case.

Equation (2.25) follows from $q_1 q_2 = 0$ (see (2.23)). In order to prove (2.26) we may assume $s = 0$ by applying a suitable motion. As we already know, $l_1 \perp l_2$ is in this case equivalent with $a_1 a_2 = 0$. But (2.26) is given for $s = 0$ by

$$\sqrt{1 + a_1^2}\,\sqrt{1 + a_2^2} - a_1 a_2 = \sqrt{1 + a_1^2}\,\sqrt{1 + a_2^2}.$$

Proposition 15. *Let l be a line and $a \notin l$ a point. Then there exists exactly one line g through a with $g \perp l$.*

Proof. Hyperbolic case. Without loss of generality we may assume $a = 0$. Then l is of the form (2.24) with $p \neq 0$. If l_1 is the line through 0 and p, it is trivial to verify $l_1 \perp l$. So assume that there is another line l_2 through 0 and $x(\alpha) \neq p = x(0)$, i.e. $\alpha \neq 0$, with $l_2 \perp l$. This implies

$$\cosh \mathrm{hyp}\,(0, p) = \cosh \mathrm{hyp}\,\big(0, x(\alpha)\big) \cosh \mathrm{hyp}\,\big(x(0), x(\alpha)\big),$$

i.e. $\sqrt{1 + p^2} = \cosh\alpha \cdot \sqrt{1 + p^2} \cdot \cosh\alpha$, i.e. $\alpha = 0$, i.e. $x(\alpha) = p$, a contradiction.

Also in the euclidean case we may assume $a = 0$ and that l is of the form (2.23) with $l \not\ni 0$, i.e. that p, q are linearly independent. Obviously, $l \perp \mathbb{R}w$ with $w := p - (pq)\,q$. Moreover, $\mathbb{R}\,(p + \xi_0 q) \perp l$ implies $(p + \xi_0 q)\,q = 0$. $\qquad\square$

If Γ is a subspace of (X, d), where d stands for eucl or hyp, and l a line with $l \cap \Gamma = \{s\}$, then l is called *orthogonal* to Γ, or Γ to l, provided $l \perp g$ holds true for all lines $g \subseteq \Gamma$ passing through s.

Proposition 16. *Let p be a point and H a hyperplane. Then there exists exactly one line $l \ni p$ with $l \perp H$.*

Proof. Case $p \in H$.

Without loss of generality we may assume $p = 0$. Hence in both cases (X, eucl), (X, hyp), H is a euclidean hyperplane $a^\perp$ with $0 \neq a \in X$. The line $\mathbb{R}a$ is orthogonal to every line $g \ni 0$ with $g \subseteq H$: if $g = \mathbb{R}b$, then $g \subseteq a^\perp$ implies $ab = 0$, i.e. $g \perp \mathbb{R}a$. If $\mathbb{R}c$ is orthogonal to every $\mathbb{R}b$ with $ab = 0$, then $b \in a^\perp$ implies $b \in c^\perp$, i.e. $H(a, 0) \subseteq H(c, 0)$, i.e. $\mathbb{R}c = \mathbb{R}a$, in view of Proposition 12.

Case $p \notin H$.

Since a point of H can be transformed into 0 by a motion, we may assume without loss of generality $H = H(a, 0)$ in both cases, i.e. in the euclidean as well in the hyperbolic case. Let

$$p + \mathbb{R}q := \{p + \lambda q \mid \lambda \in \mathbb{R}\}$$

be a euclidean line l orthogonal to H. Hence $l \perp r + \mathbb{R}b$ for all $b \in a^\perp$ where $r \in l \cap H$, i.e. $a^\perp \subseteq q^\perp$, i.e. $l = p + \mathbb{R}a$, by applying Proposition 12. On the other hand, $p + \mathbb{R}a \perp H(a, 0)$. The point of intersection is $r = p - \frac{pa}{a^2} a$. It remains to consider $p \notin H$ in the hyperbolic case. Put $H = H(a, 0)$, $a^2 = 1$. If $p - (pa)a \neq 0$, we define

$$j := \frac{p - (pa)a}{\|p - (pa)a\|}.$$

Take $\omega \in O(X)$ with $\omega(e) = j$, where e is the axis of our underlying translation group, and $t \in \mathbb{R}$ with $\omega T_t \omega^{-1}(p) = (pa)a$, in view of (T2) for j. Because of

$$\omega T_t \omega^{-1}(x) = x + [(xj)(\cosh t - 1) + \sqrt{1 + x^2} \sinh t] j$$

for $x \in X$, we obtain $\omega T_t \omega^{-1}(H) = H$ on account of $j \in a^\perp$. There hence exists a motion

$$\mu := \begin{cases} \omega T_t \omega^{-1} & \text{for } p \neq (pa)a \\ \mathrm{id} & \text{for } p = (pa)a \end{cases}$$

with $\mu(H) = H$ and $\mu(p) \in \mathbb{R}a \backslash H$, by $p \notin H$, i.e. $pa \neq 0$. So we assume, without loss of generality, $H = a^\perp$ and $p = \lambda a$, $\lambda \neq 0$. There hence is a hyperbolic line, namely $l = \mathbb{R}a$ with $p \in l \perp H$. Assume now that there is another hyperbolic line $g \ni p$ with $l \neq g \perp H$. Hence $0 \notin g$ because all hyperbolic lines through 0 are of the form $\mathbb{R}b$. Put $g \cap H =: \{r\}$. Hence

$$\cosh \mathrm{hyp}\,(0, p) = \cosh \mathrm{hyp}\,(r, 0) \cdot \cosh \mathrm{hyp}\,(r, p),$$

i.e. $\sqrt{1 + p^2} = \sqrt{1 + r^2}\,(\sqrt{1 + r^2}\,\sqrt{1 + p^2} - rp)$. But $p \in \mathbb{R}a$, $r \in H$ implies $pr = 0$. Thus $1 + r^2 = 1$, i.e. $r = 0$, a contradiction. $\qquad\square$

The distance $d\,(p, H)$ between a point p and a hyperplane H is defined by $d\,(p, r)$, where r is the point of intersection of H and the line $l \ni p$ orthogonal to H. This applies for (X, eucl) as well as for (X, hyp).

Let $\varrho > 0$ be a real number and H be a hyperplane. An interesting set of points is then given by

$$D_\varrho(H) = \{x \in X \mid d\,(x, H) = \varrho\},$$

a so-called *equidistant surface* or *hypercycle* of H. We will look to these sets in the case $0 \in H$. In the euclidean case we get with $a \in X$, $a^2 = 1$,

$$D_\varrho\big(H\,(a, 0)\big) = H\,(a, \varrho) \cup H\,(a, -\varrho),$$

i.e. we obtain the union of two euclidean hyperplanes parallel to $a^\perp$, since the euclidean hyperplanes H_1, H_2 are called *parallel*, $H_1 \parallel H_2$, provided $H_1 = H_2$ or $H_1 \cap H_2 = \emptyset$ hold true. Of course, $H\,(a, \alpha) \parallel H\,(b, \beta)$ is satisfied if, and only if, $\mathbb{R}a = \mathbb{R}b$. In hyperbolic geometry we obtain for $\varrho > 0$, $H = H\,(a, 0)$, $a^2 = 1$, as we will show,

$$D_\varrho\big(H\,(a, 0)\big) = H\,(a, \sinh \varrho) \cup H\,(a, -\sinh \varrho).$$

As a matter of fact, this is again the union of two euclidean hyperplanes, and not, say, of two hyperbolic hyperplanes. The point

$$p \in \{a \sinh \varrho, \, -a \sinh \varrho\}$$

has distance ϱ from H. Take $\omega_j \in O\,(X)$ with $\omega_j(e) = j$ for a given $j \in X$ with $j^2 = 1$ and $aj = 0$, i.e. $j \in H$. Now

$$\mu\,(0) = j \sinh t, \;\; \mu\,(p) = p + j \cosh \varrho \sinh t$$

where $\mu := \omega_j T_t \omega_j^{-1}$, $t \in \mathbb{R}$, holds true, and the line through $\mu\,(0)$, $\mu\,(p)$ must be orthogonal to H, in view of $\mathbb{R}p \perp H$. Since $\mu\,(0)$ runs over H by varying j and t, $\mu\,(p)$ runs over $D_\varrho(H)$ on account of

$$\mathrm{hyp}\,\big(\mu\,(0), \mu\,(p)\big) = \mathrm{hyp}\,(0, p) = \varrho :$$

through $h \in H$ there is exactly one hyperbolic line l orthogonal to H, and on l there are exactly two points of distance ϱ from H. Hence

$$D_\varrho(H) = (a \sinh \varrho + H) \cup (-a \sinh \varrho + H)$$

where $p + H := \{p + h \mid h \in H\}$.

2.8 A parametric representation

Proposition 17. *If H is a hyperbolic hyperplane, there exist $p \in X$ with $p^2 = 1$ and $\gamma \in \mathbb{R}_{\geq 0}$ with $H = \Pi\,(p, \gamma)$, where*

$$\Pi\,(p, \gamma) := \{\gamma p \cosh \xi + y \sinh \xi \mid \xi \in \mathbb{R}, \, y \in p^\perp \text{ with } y^2 = 1\}. \tag{2.27}$$

On the other hand, every $\Pi(p, \gamma)$ *is a hyperbolic hyperplane provided* $\gamma \in \mathbb{R}_{\geq 0}$ *and* $p \in X$ *satisfies* $p^2 = 1$. *Moreover,*

$$\Pi(p, 0) = p^{\perp} = \{x - (xp)\, p \mid x \in X\}, \tag{2.28}$$

and for $\gamma > 0$,

$$\Pi(p, \gamma) = \{\gamma p \cosh \xi + \frac{x - (xp)\, p}{\sqrt{x^2 - (xp)^2}} \sinh \xi \mid \xi \in \mathbb{R},\ x \in X \backslash \mathbb{R}p\}. \tag{2.29}$$

Proof. 1. Let $\gamma \geq 0$ and $p \in X$, $p^2 = 1$, be given. Take $\omega \in O(X)$ with $\omega(e) = p$ and $t \in \mathbb{R}$ with $\sinh t = \gamma$. Then $\omega T_t \omega^{-1}(p^{\perp})$ must be a hyperbolic hyperplane. Observe

$$\omega T_t \omega^{-1}(p^{\perp}) = \{h + \sinh t \cdot \sqrt{1 + h^2}\, p \mid h \in p^{\perp}\}.$$

With $h = y \sinh \xi$, $y \in p^{\perp}$, $y^2 = 1$, we obtain

$$\omega T_t \omega^{-1}(p^{\perp}) = \{\gamma p \cosh \xi + y \sinh \xi \mid \xi \in \mathbb{R},\ y \in p^{\perp} \text{ with } y^2 = 1\},$$

i.e. $\omega T_t \omega^{-1}(p^{\perp}) = \Pi(p, \gamma)$, i.e. $\Pi(p, \gamma)$ is a hyperbolic hyperplane.

2. *Given a hyperbolic line l and a point $r \in l$. Then there exists exactly one hyperbolic hyperplane $H \ni r$ with $l \perp H$.* In order to prove this statement, take $b \in l \backslash \{r\}$ and a motion μ with $\mu(r) = 0$, $\mu(b) =: a$. Since $r \neq b$ we get $0 \neq a$. There is exactly one hyperbolic hyperplane through 0 which is orthogonal to the line through a and 0, namely $a^{\perp}$. There hence is exactly one hyperbolic hyperplane through r, namely $\mu^{-1}(a^{\perp})$, which is orthogonal to l.

3. Let now H be an arbitrary hyperbolic hyperplane. Because of Proposition 16 there exists exactly one hyperbolic line l through 0 which is orthogonal to H. Let r be the point of intersection of l and H. Let r be the point of intersection of l and H. Because of step 2 we know that H is uniquely determined as the hyperbolic hyperplane through r which is orthogonal to l. But we already know a hyperbolic hyperplane of this kind, namely $a^{\perp}$ for $r = 0$ and $l = \mathbb{R}a$, and $\Pi(p, \gamma)$ for $r \neq 0$, $p := \frac{r}{\|r\|}$, $\gamma := \|r\|$. In fact, $r = \gamma p \in \Pi$ for $\xi = 0$, and $g \perp l$ for all hyperbolic lines g through r and $s := \gamma p \cosh \xi + y \sinh \xi$ with $\xi \neq 0$, $y \in p^{\perp}$, $y^2 = 1$ on account of

$$\cosh \mathrm{hyp}(0, s) = \cosh \mathrm{hyp}(0, r) \cosh \mathrm{hyp}(r, s).$$

Hence $H = \Pi(p, \gamma)$.

4. Since $[x - (xp)\, p]p = 0$, we get $x - (xp)\, p \in p^{\perp}$ for all $x \in X$. If $y \in p^{\perp}$ we obtain $yp = 0$, i.e. $y = y - (yp)\, p$. This proves (2.28). In order to get (2.29) we must show

$$\left\{ \frac{x - (xp)\, p}{\sqrt{x^2 - (xp)^2}} \,\middle|\, x \in X \backslash \mathbb{R}p \right\} = \{y \in X \mid y \in p^{\perp} \text{ and } y^2 = 1\}.$$

Because of Lemma 1, chapter 1, we have $x^2 = (xp)^2$ if, and only if, $x \in \mathbb{R}p$, in view of $p^2 = 1$. Obviously,

$$y = \frac{x - (xp)\, p}{\sqrt{x^2 - (xp)^2}}$$

satisfies $y \in p^\perp$ and $y^2 = 1$. Given, finally, $y \in X$ with $y \in p^\perp$ and $y^2 = 1$, we obtain

$$y = \frac{y - (yp)\, p}{\sqrt{y^2 - (yp)^2}}$$

with $y \notin \mathbb{R}p$.   □

From (2.27) we obtain for $\omega \in O(X)$,

$$\omega\left(\Pi\left(p,\gamma\right)\right) = \{\gamma\omega\left(p\right)\cosh\xi + z\sinh\xi \mid \xi \in \mathbb{R},\, z \in \left[\omega\left(p\right)\right]^\perp \text{with } z^2 = 1\},$$

i.e.

$$\omega\left(\Pi\left(p,\gamma\right)\right) = \Pi\left(\omega\left(p\right),\gamma\right).$$

In Theorem 26 we will prove that

$$T_t\left(\Pi\left(p,\gamma\right)\right) = \Pi\left(\varepsilon p',|\gamma'|\right)$$

holds true for all $t,\gamma \in \mathbb{R}$ with $\gamma \geq 0$, and all $p \in X$ with $p^2 = 1$, where

$$\gamma' \quad := \quad \gamma\cosh t + (pe)\sqrt{1+\gamma^2}\sinh t,$$

$$\varepsilon \quad := \quad \operatorname{sgn} \gamma' \text{ for } \gamma' \neq 0$$

and $p' \cdot \|A\| := A := p + \left[\dfrac{\gamma}{\sqrt{1+\gamma^2}}\sinh t + (pe)(\cosh t - 1)\right]e$, by observing $A \neq 0$. In the case $\gamma' = 0$ the value of $\varepsilon \neq 0$ plays no role, since $\Pi\left(\varepsilon p',0\right) = (p')^\perp$. In Proposition 27 we will show that $\Pi\left(p,\gamma\right) \subseteq \Pi\left(q,\delta\right)$ and $\gamma > 0$ imply $p = q$ and $\gamma = \delta$.

Remark. A parametric representation of euclidean hyperplanes will be given in section 2, chapter 3.

2.9   Ends, parallelity, measures of angles

The notion of an *end* as introduced by David Hilbert (1862–1943) concerns hyperbolic geometry. If $w \in X\backslash\{0\}$, then we will call

$$\mathbb{R}_{\geq 0}w := \{\lambda w \mid \lambda \in \mathbb{R} \text{ and } \lambda \geq 0\}$$

an *end* of X. Two ends $\mathbb{R}_{\geq 0}w_1$, $\mathbb{R}_{\geq 0}w_2$ are equal if, and only if, there exists $\lambda > 0$ with $w_2 = \lambda w_1$. To every hyperbolic line l there will be associated two ends, the so-called ends of l. For

$$l_{p,q} = l = \{p\cosh\xi + q\sinh\xi \mid \xi \in \mathbb{R}\} \tag{2.30}$$

with $p, q \in X$ and $pq = 0$, $q^2 = 1$, the two ends of l are

$$\mathbb{R}_{\geq 0}(p + q), \ \mathbb{R}_{\geq 0}(p - q). \tag{2.31}$$

Note that p is the only element y in l with $\|y\| = \min_{x \in l}\|x\|$. This implies for $p, p', q, q' \in X$ with $pq = 0 = p'q'$ and $q^2 = 1 = q'^2$ that the lines $l_{p,q}$ and $l_{p',q'}$ coincide if and only if $p = p'$ and $q' \in \{q, -q\}$. If $l = \mathbb{R}q$, then $l = (\mathbb{R}_{\geq 0}q) \cup (\mathbb{R}_{\geq 0}(-q))$. In the case $p \neq 0$,

$$\mathbb{R}_{\geq 0}(p + q) \cup \mathbb{R}_{\geq 0}(-p - q),$$
$$\mathbb{R}_{\geq 0}(p - q) \cup \mathbb{R}_{\geq 0}(-p + q)$$

are the two asymptotes of the hyperbola of which (2.30) is a branch. Obviously, (2.31) are the limiting positions of $\mathbb{R}_{\geq 0}(p \cosh \xi + q \sinh \xi)$ for $\xi \to +\infty$, $\xi \to -\infty$, respectively:

$$\mathbb{R}_{\geq 0}\left(p + q \frac{\sinh \xi}{\cosh \xi}\right) \to \begin{array}{l} \mathbb{R}_{\geq 0}(p + q) \text{ for } \xi \to +\infty \\ \mathbb{R}_{\geq 0}(p - q) \text{ for } \xi \to -\infty \end{array}.$$

Proposition 18. *Let $E_i = \mathbb{R}_{\geq 0}w_i$, $i = 1, 2$, be distinct ends. Then there is exactly one hyperbolic line, of which E_1, E_2 are the ends.*

Proof. If $E_2 = -E_1$, i.e. if $\mathbb{R}_{\geq 0}w_2 = \mathbb{R}_{\geq 0}(-w_1)$, then $\mathbb{R}w_1$ is the uniquely determined line with the ends E_1, E_2. In the case that w_1, w_2 are linearly independent, we must solve

$$2\lambda_1 w_1 = p + q, \ 2\lambda_2 w_2 = p - q$$

in $\lambda_1, \lambda_2, p, q$ with $\lambda_1 > 0$, $\lambda_2 > 0$, $pq = 0$, $q^2 = 1$. This implies, by assuming $w_1^2 = 1 = w_2^2$, without loss of generality,

$$p = \lambda_1 w_1 + \lambda_2 w_2, \qquad q = \lambda_1 w_1 - \lambda_2 w_2,$$
$$2\lambda_1^2(1 - w_1 w_2) = 1, \quad \lambda_1 = \lambda_2,$$

with a uniquely determined solution

$$\left\{ \frac{(w_1 + w_2)\cosh \xi + (w_1 - w_2)\sinh \xi}{\sqrt{2(1 - w_1 w_2)}} \ \Big| \ \xi \in \mathbb{R} \right\},$$

in view of $w_1 w_2 \leq |w_1 w_2| < \|w_1\|\,\|w_2\| = 1$ since w_1, w_2 are linearly independent. $\square$

Let E be an end of X and μ be a hyperbolic motion. We would like to define the end $\mu(E)$. If $E = \mathbb{R}_{\geq 0}a$, $a^2 = 1$, put $\omega(E) := \mathbb{R}_{\geq 0}\omega(a)$ for $\omega \in O(X)$. Suppose $t \in \mathbb{R}$ and that T_t is a translation of (X, hyp) with axis e. Then

$$T_t(\{\lambda a \mid \lambda \geq 0\}) = \{\lambda a + [\lambda(ae)(\cosh t - 1) + \sqrt{1 + \lambda^2}\sinh t]e \mid \lambda \geq 0\}.$$

We are now interested in the question whether

$$\mathbb{R}_{\geq 0}\big(T_t(\lambda a)\big) \tag{2.32}$$

tends to a limiting position for $0 < \lambda \to +\infty$. Instead of (2.32) we may write

$$\mathbb{R}_{\geq 0}\left(a + \left[(ae)(\cosh t - 1) + \sqrt{\frac{1}{\lambda^2} + 1}\,\sinh t\right]e\right),$$

and we obtain as limiting position

$$\mathbb{R}_{\geq 0}\big(a + [(ae)(\cosh t - 1) + \sinh t]\,e\big) \tag{2.33}$$

which we define as the end $T_t(E) = T_t(\mathbb{R}_{\geq 0}a)$. In the case $0 > \lambda \to -\infty$ we observe

$$\tfrac{1}{\lambda}\big(\lambda a + [\lambda\,(ae)(\cosh t - 1) + \sqrt{1 + \lambda^2}\,\sinh t]\,e\big)$$
$$= \quad a + \left[(ae)(\cosh t - 1) - \sqrt{\tfrac{1}{\lambda^2} + 1}\,\sinh t\right]e$$
$$\to \quad a + [(ae)(\cosh t - 1) - \sinh t]\,e,$$

a result which corresponds to (2.33), replacing there a by $-a$, i.e. substituting $\mathbb{R}_{\geq 0}\big(T_t[\lambda \cdot (-a)]\big)$, $0 < \lambda \to +\infty$, for $\mathbb{R}_{\geq 0}\big(T_t(\lambda a)\big)$, $0 > \lambda \to -\infty$.

Proposition 19. *If E is an end of the line l and μ a motion, then $\mu\,(E)$ is an end of $\mu\,(l)$.*

Proof. Let $x\,(\xi) = p\cosh\xi + q\sinh\xi$ be the equation of l, and let E be given, say, by $\mathbb{R}_{\geq 0}(p + q)$ thus considering the case $\xi \to +\infty$. If $\mu \in O\,(X)$, we obtain $\mathbb{R}_{\geq 0}\big(\mu\,(p) + \mu\,(q)\big)$ as end of $\mu\,(l)$ for $\xi \to +\infty$, i.e. we get the end

$$\mathbb{R}_{\geq 0}\big(\mu\,(p + q)\big) = \mu\,(E).$$

Suppose now that $\mu = T_t$. We already know, by (2.33), with $\frac{p+q}{\sqrt{p^2+q^2}}$, i.e. $\frac{p+q}{\sqrt{1+p^2}}$ instead of a,

$$T_t(E) = \mathbb{R}_{\geq 0}\left(\frac{p + q}{\sqrt{1 + p^2}} + \left[\frac{(p+q)\,e}{\sqrt{1 + p^2}}\,(\cosh t - 1) + \sinh t\right]e\right).$$

Moreover, $\mathbb{R}_{\geq 0}\big(T_t(p\cosh\xi + q\sinh\xi)\big)$ is given by

$$\mathbb{R}_{\geq 0}\big(p + q\tanh\xi + [(p + q\tanh\xi)\,e\,(\cosh t - 1) + \sqrt{1 + p^2}\,\sinh t]\,e\big),$$

which tends to

$$\mathbb{R}_{\geq 0}\big(p + q + [(p + q)\,e\,(\cosh t - 1) + \sqrt{1 + p^2}\,\sinh t]\,e\big)$$

for $\xi \to +\infty$, i.e. which tends to $T_t(E)$. Hence $\mu\,(E)$ is an end of $\mu\,(l)$. $\square$

Two euclidean lines

$$l_i := \{p_i + \xi q_i \mid \xi \in \mathbb{R}\}$$

are called *parallel*, $l_1 \parallel l_2$, provided $\mathbb{R}q_1 = \mathbb{R}q_2$. Parallelity is an equivalence relation on the set of euclidean lines of X. If $l = p + \mathbb{R}q$ is a euclidean line and r a point, there exists exactly one euclidean line, namely $g = r + \mathbb{R}q$ through r, parallel to l.

Two hyperbolic lines of X are called *parallel* provided they have at least one end in common. If l_1, l_2 are hyperbolic lines, of course, $l_1 \parallel l_1$ holds true and also that $l_1 \parallel l_2$ implies $l_2 \parallel l_1$. However, parallelity need not be transitive. In order to verify this statement take elements a, b of X with $a^2 = 1 = b^2$ and $ab = 0$. Define

$$l_1 \;=\; \{a \cosh \xi + b \sinh \xi \mid \xi \in \mathbb{R}\},$$
$$l_2 \;=\; \{-a \cosh \xi + b \sinh \xi \mid \xi \in \mathbb{R}\},$$

and $l = \mathbb{R}(a+b)$. We obtain $l_1 \parallel l$, because these lines have $\mathbb{R}_{\geq 0}(a+b)$ in common, moreover, $l \parallel l_2$ since $\mathbb{R}_{\geq 0}(-a - b)$ is an end of both lines. But $l_1 \parallel l_2$ does not hold true: the ends of l_1 are $\mathbb{R}_{\geq 0}(a + b)$, $\mathbb{R}_{\geq 0}(a - b)$, and those of l_2 are

$$\mathbb{R}_{\geq 0}(-a + b), \; \mathbb{R}_{\geq 0}(-a - b).$$

If p is a point and $E := \mathbb{R}_{\geq 0}\, a$ an end, there is exactly one hyperbolic line through p having E as an end. In order to prove this statement take a motion μ with $\mu(p) = 0$. Of course, there is exactly one line through 0 having $\mu(E) =: \mathbb{R}_{\geq 0}b$ as an end, namely $\mathbb{R}b$. Hence, by Proposition 19, there is exactly one line, namely $\mu^{-1}(\mathbb{R}b)$, through p with E as an end.

If l is a line and $p \notin l$ a point, there are exactly two lines $l_1 \neq l_2$ through p which are parallel to l: take the two distinct ends E_1, E_2 associated with l, and then the lines l_1, l_2 through p with E_1, E_2, respectively, as an end.

Let $l = \{x(\xi) = p \cosh \xi + q \sinh \xi \mid \xi \in \mathbb{R}\}$ be a hyperbolic line and $a = x(\alpha)$ be a point of l. The two sets

$$\{x(\xi) \mid \xi \geq \alpha\}, \; \{x(\xi) \mid \xi \leq \alpha\} \qquad (2.34)$$

are called (hyperbolic) *rays* with *starting point* $x(\alpha)$. If $l = \{x(\xi) = p + \xi q \mid \xi \in \mathbb{R}\}$ is a euclidean line and $x(\alpha) = p + \alpha q$ a point a of l, then (2.34) are said to be (euclidean) *rays* with *starting point* $x(\alpha)$. Images $\mu(R)$ of rays R under motions μ are rays, and if a is the starting point of R, then $\mu(a)$ is the starting point of $\mu(R)$.

It is clear how to associate each of the ends of a hyperbolic line l to the two rays $R_1, R_2 \subset l$ of l with the same starting point. In this connection we will speak of *the end of a ray* or of a *ray through an end*.

Let R_1, R_2 be rays with the same starting point v such that $R_1 \cup R_2$ is not a line. The triple (R_1, R_2, v) consisting of the (unordered) pair R_1, R_2 and the point

v will be called an *angle*. If $p_i \in R_i$, $i = 1, 2$, is the point with

$$d(v, p_i) = 1, \; i = 1, 2,$$

then the measure $\angle(R_1, R_2, v)$ of the angle (R_1, R_2, v) is defined by $\angle(R_1, R_2, v) \in [0, \pi]$ and

$$1 - \cos \angle(R_1, R_2, v) = \begin{cases} \dfrac{1}{2} \left[\text{eucl } (p_1, p_2)\right]^2 \\ 2\dfrac{\cosh \text{hyp } (p_1, p_2) - 1}{\cosh 2 - 1} \end{cases} \tag{2.35}$$

for the euclidean, hyperbolic case, respectively. (For an axiomatic definition of measures of angles in 2-dimensional euclidean or hyperbolic geometry see, for instance, the book [4] of the author.)

If R_1, R_2 are rays both with starting point v and μ a motion, then

$$\angle(R_1, R_2, v) = \angle\big(\mu(R_1), \, \mu(R_2), \, \mu(v)\big).$$

This is clear since distances are preserved under motions.

Let a, b, v be elements of X with $a \neq 0 \neq b$ and R_1, R_2 the rays

$$v + \mathbb{R}_{\geq 0} a, \; v + \mathbb{R}_{\geq 0} b,$$

respectively. Define $p_1 = v + \frac{1}{\|a\|} a$, $p_2 = v + \frac{1}{\|b\|} b$, $\gamma = \angle(R_1, R_2, v)$. Hence

$$\begin{aligned} ab \;\; &= \|a\| \cdot \|b\| \cdot (p_1 - v)(p_2 - v) \\ &= \tfrac{1}{2} \|a\| \cdot \|b\| \cdot \big((p_1 - v)^2 + (p_2 - v)^2 - [(p_1 - v) - (p_2 - v)]^2\big) \\ &= \|a\| \cdot \|b\| \cdot \big(1 - \tfrac{1}{2} [p_1 - p_2]^2\big), \end{aligned}$$

i.e. $ab = \|a\| \cdot \|b\| \cdot \cos \gamma$, in view of (2.35). As a consequence we get the so-called cosine theorem:

$$\begin{aligned} \left[\text{eucl } (v + a, \, v + b)\right]^2 \;\; &= [(v + a) - (v + b)]^2 = (a - b)^2 \\ &= a^2 + b^2 - 2\|a\| \cdot \|b\| \cdot \cos \gamma, \end{aligned}$$

i.e., by $A = \text{eucl } (v, v + a)$, $B = \text{eucl } (v, v + b)$,

$$\left[\text{eucl } (v + a, \, v + b)\right]^2 = A^2 + B^2 - 2AB \cos \angle(R_1, R_2, v).$$

Similarly, we would like to consider the case of hyperbolic geometry.

Let a, b, v be elements of X with $a \neq v \neq b$. If l_1 is the hyperbolic line through v, a, and l_2 the one through v, b, if R_1, R_2 are the (hyperbolic) rays with starting point v and $a \in R_1$, $b \in R_2$, then the cosine theorem of hyperbolic geometry holds true:

$$\cosh C = \cosh A \cdot \cosh B - \sinh A \cdot \sinh B \cdot \cos \gamma$$

where $C = \text{hyp } (a, b)$, $A = \text{hyp } (v, a)$, $B = \text{hyp } (v, b)$, $\gamma = \sphericalangle(R_1, R_2, v)$.

For the proof of this statement we may assume $v = 0$ without loss of generality, since distances and measures of angles are preserved under motions. So put

$$l_i := \{x_i(\xi) = q_i \sinh \xi \mid \xi \in \mathbb{R}\}, \; i = 1, 2,$$

with $q_i^2 = 1$, $i = 1, 2$, and with a sign for q_i such that $\xi \geq 0$ describes R_i for $i = 1, 2$. Hence $x_1(0) = v = x_2(0)$ and

$$p_i = x_i(1), \; i = 1, 2, \; a =: x_1(\alpha), \; b =: x_2(\beta),$$

with $\alpha > 0$, $\beta > 0$, and thus $\|a\| = \sinh \alpha$, $\|b\| = \sinh \beta$,

$$\cosh C = \sqrt{1 + a^2} \, \sqrt{1 + b^2} - ab,$$

$\cosh A = \sqrt{1 + a^2}$, $\cosh B = \sqrt{1 + b^2}$, $\cosh \text{hyp } (p_1, p_2) = (\cosh 1)^2 - q_1 q_2 (\sinh 1)^2$. Moreover, $\sinh A = \|a\|$, $\sinh B = \|b\|$,

$$ab = x_1(\alpha) \, x_2(\beta) = q_1 q_2 \sinh \alpha \sinh \beta,$$

and, by (2.35), $\cosh 2 = 1 + 2 \sinh^2 1$,

$$
\begin{aligned}
\cos \gamma &= 1 - 2 \, \frac{\cosh \text{hyp } (p_1, p_2) - 1}{\cosh 2 - 1} \\
&= 1 - 2 \, \frac{(1 - q_1 q_2)[\sinh 1]^2}{\cosh 2 - 1} = q_1 q_2.
\end{aligned}
$$

Hence $\sqrt{1 + a^2} \, \sqrt{1 + b^2} - ab = \cosh A \cdot \cosh B - \sinh A \cdot \sinh B \cdot \cos \gamma$, since $\sinh A = \sinh \alpha$ and $\sinh B = \sinh \beta$, q.e.d.

Remark. Measures of angles $(R_1, R_2, 0)$ coincide in euclidean and hyperbolic geometry because of the previous formulas $\cos \gamma = q_1 q_2$ and $q_1 q_2 = \|q_1\| \cdot \|q_2\| \cos \gamma$. Notice, moreover, that the cosine theorem in both geometries leads for $\gamma = \frac{\pi}{2}$ to (2.21), (2.22), respectively.

2.10 Angles of parallelism, horocycles

Proposition 20. *Let $k \neq l$ be parallel hyperbolic lines with E as common end, $p \in l \backslash k$ a point, $a \ni p$ the line orthogonal to k, and r the point of intersection of k and a. If $R_1 \subset a$ is the ray through r with starting point p, and $R_2 \subset l$ the ray through E, also with starting point p, then*

$$\tan \frac{1}{2} \left(\sphericalangle(R_1, R_2, p) \right) = e^{-\text{hyp } (p, r)}.$$

Proof. Without loss of generality we may assume $p = 0$,

$$k = \{r \cosh \xi + q \sinh \xi \mid \xi \in \mathbb{R}\}, \ rq = 0, \ q^2 = 1,$$

$a = \mathbb{R}r$, $R_1 = \mathbb{R}_{\geq 0}r$, $l = \mathbb{R}(r + q)$, $R_2 = \mathbb{R}_{\geq 0}(r + q)$. Put $\gamma := \sphericalangle(R_1, R_2, p)$. Observe $r \neq 0 = p$, since $r \in k \not\ni p$. From (2.35) we obtain

$$1 - \cos \gamma = \frac{\cosh \operatorname{hyp}(p_1, p_2) - 1}{\sinh^2 1} = 1 - \frac{\|r\|}{\sqrt{1 + r^2}},$$

in view of $rq = 0$, $q^2 = 1$, $p_1 = \frac{r}{\|r\|} \sinh 1$, $p_2 = \frac{r+q}{\sqrt{1+r^2}} \sinh 1$. We hence get

$$\cos \gamma = \frac{\|r\|}{\sqrt{1 + r^2}},$$

i.e. $\gamma \in \left]0, \frac{\pi}{2}\right[$ because of $0 < \frac{\|r\|}{\sqrt{1+r^2}} < 1$. From

$$\cosh \operatorname{hyp}(p, r) = \sqrt{1 + r^2}$$

we obtain

$$e^{-\operatorname{hyp}(p,r)} = \sqrt{1 + r^2} - \|r\|, \ e^{\operatorname{hyp}(p,r)} = \sqrt{1 + r^2} + \|r\|,$$

i.e.

$$\tan \frac{1}{2}\gamma = \sqrt{\frac{1 - \cos \gamma}{1 + \cos \gamma}} = \sqrt{\frac{\sqrt{1 + r^2} - \|r\|}{\sqrt{1 + r^2} + \|r\|}} = e^{-\operatorname{hyp}(p,r)}. \qquad \square$$

Proposition 21. *Let l be a hyperbolic line and $R \subset l$ a ray with starting point v. There exists a paraboloid as limiting position for the balls $B\left(c, \operatorname{hyp}(c, v)\right)$ with $c \in R$ and $\operatorname{hyp}(c, v) \to \infty$. This limiting position is called a horocycle.*

Proof. If $l = \{x(\xi) := p \cosh \xi + q \sinh \xi \mid \xi \in \mathbb{R}\}$, $pq = 0$, $q^2 = 1$, and $v = x(\alpha)$, we may assume $R = \{x(\xi) \mid \xi \geq \alpha\}$, without loss of generality. Put $c =: x(\alpha + \varrho)$, $\varrho > 0$. Then

$$B_\varrho := B\left(c, \ \operatorname{hyp}(c, v)\right) = B\left(x(\alpha + \varrho), \varrho\right) = \{x \in X \mid \operatorname{hyp}\left(x(\alpha + \varrho), x\right) = \varrho\},$$

i.e. $B_\varrho = \{x \in X \mid \cosh(\alpha + \varrho)\sqrt{1 + p^2}\sqrt{1 + x^2} - x(\alpha + \varrho)x = \cosh \varrho\}$ holds true. This implies

$$\sqrt{1 + x^2}\sqrt{1 + p^2} - x\left(p + q \tanh(\alpha + \varrho)\right) = \frac{\cosh \varrho}{\cosh(\alpha + \varrho)},$$

i.e. $\sqrt{1 + x^2}\sqrt{1 + p^2} - x(p + q) = e^{-\alpha}$ for $\varrho \to +\infty$. Hence the limiting position for B_ϱ, $\varrho \to \infty$, is

$$B_\infty := \{x \in X \mid \sqrt{1 + x^2} - xm = \tau\} \qquad (2.36)$$

with $m := \frac{p+q}{\sqrt{1+p^2}}$, i.e. $m^2 = 1$, and $\tau := \frac{e^{-\alpha}}{\sqrt{1+p^2}} > 0$.

In view of $X = m^\perp \oplus \mathbb{R}m$, we will write $x =: \bar{x} + x_0 m$ with $\bar{x} \in m^\perp$ and $x_0 \in \mathbb{R}$. Thus

$$B_\infty = \{x \in X \mid \bar{x}^2 - 2\tau x_0 + 1 = \tau^2\}$$

by observing $x_0 + \tau \geq 0$ for an element x of B_∞: assuming $x_0 + \tau < 0$ would lead to

$$\tau^2 = \bar{x}^2 - 2\tau x_0 + 1 > \bar{x}^2 + 2\tau^2 + 1,$$

i.e. to $\bar{x}^2 + \tau^2 + 1 < 0$. The surface S of X,

$$S := \{\xi w + \eta m \mid \xi, \eta \in \mathbb{R}, \ w \in m^\perp, \ w^2 = 1, \ \xi^2 = 2\tau\eta + \tau^2 + 1\},$$

is called a *paraboloid*, and $B_\infty = S$ holds true. $\qquad\square$

If H^1, H^2 are horocycles, there exists a hyperbolic motion μ with $\mu(H^1) = H^2$. If H^i, $i = 1, 2$, is based on the ray R_i with starting point v_i, we take points $p_i \in R_i$, $i = 1, 2$, satisfying hyp $(v_i, p_i) = 1$. Moreover, we take a motion μ with

$$\mu(v_1) = v_2, \ \mu(p_1) = p_2.$$

Hence $\mu(R_1) = R_2$ and

$$\mu\Big(B\big(c_1, \text{hyp}\,(c_1, v_1)\big)\Big) = B\Big(\mu(c_1), \ \text{hyp}\,\big(\mu(c_1), v_2\big)\Big)$$

for all $c_1 \in R_1$ with hyp $(c_1, v_1) \to \infty$, i.e. for all $c_2 := \mu(c_1) \in R_2$ with hyp $(c_2, v_2) \to \infty$. Thus $\mu(H^1)$ and H^2 coincide.

2.11 Geometrical subspaces

If $S \neq \emptyset$ is a set of hyperplanes of (X, d), i.e. of (X, eucl) or (X, hyp), the intersection

$$\Sigma = \bigcap_{H \in S} H$$

will be called a *geometrical subspace* of (X, d). In this case we often will write $\Sigma \in \Gamma(X, d)$. We also define $X \in \Gamma(X, d)$. Let $a \neq 0$ be an element of X. Because of

$$H(a, 0) \cap H(a, 1) = \emptyset,$$

we obtain $\emptyset \in \Gamma(X, \text{eucl})$. Similarly,

$$H(a, 0) \cap \Pi(a, 1) = \emptyset,$$

i.e. $\emptyset \in \Gamma(X, \text{hyp})$. If $\Sigma \notin \{\emptyset, X\}$ is in $\Gamma(X, d)$, let μ be a motion with $\mu(p) = 0$ for a fixed element p of Σ. Hence $\mu(\Sigma)$ is an intersection

$$\mu(\Sigma) = \bigcap_{a \in S} a^\perp$$

with $0 \notin S \subseteq X$. Observing $0^\perp = X$, we obtain

Proposition 22. *All geometrical subspaces of (X, d) are given by $\emptyset$, moreover by*

$$\bigcap_{a \in S} a^{\perp}, \quad \emptyset \neq S \subseteq X,$$

and their images under motions.

We would like to show $D := \bigcap_{a \in X} a^{\perp} = \{0\}$. In fact! If $p \neq 0$ were in D, then $p \in a^{\perp}$ for all $a \in X$ would imply $p \in p^{\perp}$, i.e. $p^2 = 0$, i.e. $p = 0$, a contradiction. Hence, by Proposition 22, every set consisting of one single point is in $\Gamma(X, d)$.

Proposition 23. *Let V, $\dim V \geq 1$, be a finite-dimensional subspace $\neq X$ of the vector space X. Then the images of V under motions are in $\Gamma(X, d)$. So especially the lines of (X, d) are geometrical subspaces.*

Proof. Let $I(V)$ be the intersection of all hyperplanes containing the finite-dimensional subspace V, $\dim V \geq 1$, of the vector space X. Of course, we assume $n := \dim V < \dim X$ in the case that X is finite-dimensional. Hence $V \subseteq I(V)$. As a matter of fact, even $V = I(V)$ holds true. So assume there would exist

$$r \in I(V) \backslash V. \tag{2.37}$$

Let $b_1, \ldots, b_n$ be a basis of V satisfying

$$b_i b_j = \begin{cases} 0 \text{ for } i \neq j \\ 1 \text{ for } i = j \end{cases},$$

$i, j \in \{1, \ldots, n\}$. Notice $V = \{\sum_{i=1}^n \xi_i b_i \mid \xi_i \in \mathbb{R}\} \ni 0$ and put

$$z := \sum_{i=1}^n (r b_i) \, b_i.$$

Obviously, $(r - z) b_i = 0$, $i = 1, \ldots, n$, and $z \neq r$ since $z \in V$ and $r \notin V$, by (2.37). Hence $V \subseteq H(r - z, 0)$, i.e. $r \in H(r - z, 0)$, by (2.37), and thus $(r - z) r = 0$. We obtain, by $z \in V$, i.e. by $z \in H(r - z, 0)$,

$$(r - z)^2 = (r - z) \, r - (r - z) \, z = 0,$$

i.e. $r = z \in V$, a contradiction. Hence $V = I(V)$. Thus V must be a geometrical subspace of (X, d). Now apply Proposition 22. $\qquad \square$

The geometrical subspaces as described in Proposition 23 are given in the case (X, hyp) as follows. Let $p \in X$, $\gamma \in \mathbb{R}$ satisfy $p^2 = 1$ and $\gamma \geq 0$. Suppose that W, $n := \dim W \geq 1$, is a finite-dimensional subspace of the vector space $p^{\perp}$. Then

$$\{\gamma p \cosh \xi + y \sinh \xi \mid \xi \in \mathbb{R}, \ y \in W \text{ with } y^2 = 1\}$$

will be called an n-dimensional (geometrical) subspace of (X, hyp).

Not every subspace of (X, d) for $d = \text{eucl}$ or $d = \text{hyp}$ needs to be a geometrical subspace. Assume that $Q \ni 0$ is a quasi-hyperplane which is not a hyperplane. If

$$Q \subseteq H(a, 0), \tag{2.38}$$

$a \neq 0$, holds true, then $Q = H(a, 0)$ or $H(a, 0) = X$, since Q is a maximal subspace of X. Hence (2.38) is impossible and, as a consequence, Q cannot be a geometrical subspace of (X, d).

Other interesting geometrical subspaces occur in the case that X is not finite-dimensional, in the form

$$\bigcap_{i=1}^{n} a_i^{\perp},$$

where n is a positive integer and where $a_1, \ldots, a_n \in X$ are linearly independent, satisfying $a_i^2 = 1$ for $i = 1, \ldots, n$. It will be easy to prove:

$$\bigcap_{i=1}^{n} a_i^{\perp} = \left\{ x - \sum_{i=1}^{n} \alpha_i a_i \mid x \in X \text{ and } M \cdot \begin{pmatrix} \alpha_1 \\ \vdots \\ \alpha_n \end{pmatrix} = \begin{pmatrix} xa_1 \\ \vdots \\ xa_n \end{pmatrix} \right\},$$

where M is given by the regular matrix

$$M = \begin{pmatrix} a_1^2 & a_1 a_2 & \cdots & a_1 a_n \\ a_2 a_1 & a_2^2 & \cdots & a_2 a_n \\ \vdots & & & \\ a_n a_1 & a_n a_2 & \cdots & a_n^2 \end{pmatrix}$$

(det M is called Gram's determinant). In fact, take an element $x \in X$ with the described $\alpha_1, \ldots, \alpha_n$. Hence

$$\begin{pmatrix} (x - \Sigma \alpha_i a_i) & a_1 \\ \vdots & \\ (x - \Sigma \alpha_i a_i) & a_n \end{pmatrix} = \begin{pmatrix} xa_1 \\ \vdots \\ xa_n \end{pmatrix} - M \begin{pmatrix} \alpha_1 \\ \vdots \\ \alpha_n \end{pmatrix} = 0,$$

i.e. $(x - \Sigma \alpha_i a_i) a_j = 0$ for $j = 1, \ldots, n$, i.e. $x - \Sigma \alpha_i a_i \in \bigcap_{j=1}^{n} a_j^{\perp}$.

If, on the other hand, $x \in \bigcap_{i=1}^{n} a_i^{\perp}$ holds true, $xa_j = 0$ is satisfied for $j = 1, \ldots, n$. From

$$M \begin{pmatrix} \alpha_1 \\ \vdots \\ \alpha_n \end{pmatrix} = \begin{pmatrix} xa_1 \\ \vdots \\ xa_n \end{pmatrix} = 0,$$

we then obtain $\alpha_1 = \cdots = \alpha_n = 0$. Hence x has the required form

$$x - \sum_{i=1}^{n} \alpha_i a_i.$$

2.12   The Cayley–Klein model

The Weierstrass map $w : X \to X$,

$$w(x) := \frac{x}{\sqrt{1 + x^2}}, \tag{2.39}$$

is a bijection between X and $P := \{x \in X \mid x^2 < 1\}$. In view of

$$[w(x)]^2 = \frac{x^2}{1 + x^2} < 1,$$

we obtain $w(x) \in P$ for $x \in X$. Moreover,

$$w^{-1}(x) = \frac{x}{\sqrt{1 - x^2}} \in X$$

is the uniquely determined $y \in X$ satisfying $w(y) = x$ for $x \in P$. Defining

$$g(x, y) := \mathrm{hyp}\left(w^{-1}(x),\, w^{-1}(y)\right)$$

for $x, y \in P$, we get

$$\cosh g(x, y) = \frac{1 - xy}{\sqrt{1 - x^2}\,\sqrt{1 - y^2}}. \tag{2.40}$$

If $x \neq y$ are elements of P, then $(x - y)^2 > 0$, $1 - x^2 > 0$ and hence

$$D := [x(x - y)]^2 + (1 - x^2)(x - y)^2 > 0.$$

Put $\{a, b\} := \{x + \xi(y - x) \mid \xi \in \mathbb{R}\} \cap \{z \in X \mid z^2 = 1\}$, i.e. put

$$\{a, b\} := \{x + \xi_1(y - x),\; x + \xi_2(y - x)\}$$

with $\{(x - y)^2 \xi_1,\, (x - y)^2 \xi_2\} = \{x(x - y) \pm \sqrt{D}\}$. We now would like to determine

$$|\ln\{a, b; x, y\}|,$$

where $\ln \xi$ for $0 < \xi \in \mathbb{R}$ is defined by the real number η satisfying $\exp(\eta) = \xi$, and where

$$\{z_1, z_2; z_3, z_4\} := \frac{\lambda_1 - \lambda_3}{\lambda_1 - \lambda_4} : \frac{\lambda_2 - \lambda_3}{\lambda_2 - \lambda_4}$$

designates the *cross ratio* of the ordered quadruple z_1, z_2, z_3, z_4 of four distinct points

$$z_i = p + \lambda_i q,\; i = 1, 2, 3, 4,$$

on the line $p + \mathbb{R}q$, $q \neq 0$, which does not depend on the representation of the line

$$p + \mathbb{R}q = p' + \mathbb{R}q'.$$

Writing the points a, b, x, y of $\{a, b; x, y\}$ in the form

$$x + \xi\,(y - x),$$

we obtain $0 < \{a, b; x, y\} \in \{L, L^{-1}\}$ with

$$L := \frac{\xi_1}{\xi_1 - 1} \cdot \frac{\xi_2 - 1}{\xi_2}.$$

Observe here $\xi_1 < 0$ and $1 < \xi_2$ or $\xi_2 < 0$ and $1 < \xi_1$. The exact value L or L^{-1} of $\{a, b; x, y\}$ depends on how we associate a, b to ξ_1, ξ_2. Put

$$(x - y)^2 \xi_i = x\,(x - y) + \varepsilon_i \sqrt{D},$$

$i = 1, 2$, with $\varepsilon_1 = -\varepsilon_2 = 1$. Then we get

$$\frac{1}{2}\ln L = \ln \frac{1 - xy + \sqrt{D}}{\sqrt{1 - x^2}\,\sqrt{1 - y^2}} =: R$$

by observing $xy \le |xy| \le \sqrt{x^2}\,\sqrt{y^2} < 1$. Because of $|\ln L^{-1}| = |\ln L|$, we obtain

$$\frac{1}{2}|\ln \{a, b; x, y\}| = |R|, \qquad (2.41)$$

independent of how we associate a, b to ξ_1, ξ_2. Now, by (2.40),

$$\cosh |R| = \frac{e^R + e^{-R}}{2} = \frac{1 - xy}{\sqrt{1 - x^2}\,\sqrt{1 - y^2}} = \cosh g\,(x, y),$$

since $1 - xy + \sqrt{D} = e^R \sqrt{1 - x^2}\,\sqrt{1 - y^2}$. Hence, by (2.41),

$$g\,(x, y) = \frac{1}{2}|\ln \{a, b; x, y\}|. \qquad (2.42)$$

It is certainly more convenient to work with the expression (2.40) than with (2.42), since there the elements a, b must be determined before $g\,(x, y)$ can be calculated.

We now would like to look to different notions like translation and hyperplane as they appear in the Cayley–Klein model.

If ω is a surjective orthogonal mapping, i.e. a bijective orthogonal mapping, we obtain

$$w\,(x) = \frac{x}{\sqrt{1 + x^2}}, \; w\,(\omega\,(x)) = \frac{\omega\,(x)}{\sqrt{1 + [\omega\,(x)]^2}} = \omega\,(w\,(x)),$$

since $[\omega\,(x)]^2 = x^2$. Hence, if $x \in X$ goes over in $w\,(x)$, then $w\,(x)$ in $w\omega\,(x) = \omega w\,(x)$. Thus ω remains an orthogonal mapping, however, restricted on P. If $x \in X$ goes over in $T_t(x)$, then $w\,(x)$ in

$$w\,(T_t(x)) = wT_t w^{-1}\,(w\,(x)).$$

Thus the translation T_t, say, based on $e \in X$, $e^2 = 1$, as axis, corresponds to

$$z \to w T_t w^{-1}(z) =: T'_t(z)$$

for all $z \in P$. This implies for $z \in P$,

$$T'_t(z) = \frac{z + [(ze)(\cosh t - 1) + \sinh t]\, e}{\cosh t + (ze) \sinh t}, \tag{2.43}$$

by observing $\cosh t + (ze) \sinh t > 0$, which holds, since

$$-(ze) \tanh t \geq 1$$

would contradict

$$|-(ze) \tanh t| \leq \sqrt{z^2}\, \sqrt{e^2} \cdot 1 < 1.$$

Notice

$$T'_t T'_s = w T_t w^{-1} \cdot w T_s w^{-1} = w T_{t+s} w^{-1} = T'_{t+s},$$

and also $T'_0 = \mathrm{id}$ on P, and $T'_t T'_{-t} = T'_0$. If

$$\{p \cosh \xi + q \sinh \xi \mid \xi \in \mathbb{R}\},\ pq = 0,\ q^2 = 1, \tag{2.44}$$

is a line, we obtain its image in P as the set

$$\left\{ \frac{p}{\sqrt{1+p^2}} + \frac{q}{\sqrt{1+p^2}} \tanh \xi \mid \xi \in \mathbb{R} \right\}. \tag{2.45}$$

This is the segment of the euclidean ball $B(0,1)$ connecting its points

$$\frac{p}{\sqrt{1+p^2}} - \frac{q}{\sqrt{1+p^2}} \quad \text{and} \quad \frac{p}{\sqrt{1+p^2}} + \frac{q}{\sqrt{1+p^2}}. \tag{2.46}$$

If $u \neq v$ are points on $B(0,1)$, i.e. if they satisfy $u^2 = 1 = v^2$, then $\left(\frac{v+u}{2}\right)^2 < 1$, and (2.44) with

$$p = \frac{v+u}{2k},\ q = \frac{v-u}{2k},\ k = \sqrt{1 - \left(\frac{v+u}{2}\right)^2} = \sqrt{\frac{1-uv}{2}} \tag{2.47}$$

is the inverse image of the segment $\{u + \lambda(v-u) \mid 0 < \lambda < 1\}$.

Obviously, the two ends of $\{p \cosh \xi + q \sinh \xi \mid \xi \in \mathbb{R}\}$ can be described by the points (2.46) of $B(0,1)$,

$$\frac{p \pm q}{\sqrt{1+p^2}}.$$

Proposition 24. *The image of* $\Pi(p,\gamma)$ *(see (2.27) under the mapping* $w : X \to P$ *is given by*

$$\Pi'(p,\gamma) := \{x \in P \mid px = \frac{\gamma}{\sqrt{1+\gamma^2}}\}.$$

Proof. Because of

$$z := w\,(\gamma p\cosh\xi + y\sinh\xi) = \frac{\gamma p}{\sqrt{1+\gamma^2}} + \frac{y}{\sqrt{1+\gamma^2}}\tanh\xi$$

for $y \in p^\perp$, $y^2 = 1$, we get

$$pz = \frac{\gamma}{\sqrt{1+\gamma^2}}, \tag{2.48}$$

i.e. $z \in \Pi'(p,\gamma)$. Let now $z \in P$ be given satisfying (2.48). Hence, by $p^2 = 1$,

$$p\left(z - \frac{\gamma p}{\sqrt{1+\gamma^2}}\right) = 0,$$

i.e. $Y := z - \frac{\gamma p}{\sqrt{1+\gamma^2}} \in p^\perp$. For $Y = 0$ we get $w^{-1}(z) = \gamma p \in \Pi(p,\gamma)$. Suppose now $Y \neq 0$. We obtain, by $z \in P$ and $Y \perp p$,

$$0 < 1 - z^2 = 1 - \left(Y + \frac{\gamma p}{\sqrt{1+\gamma^2}}\right)^2 = \frac{1 - (1+\gamma^2)\,Y^2}{1+\gamma^2},$$

i.e. $0 < 1 - (1+\gamma^2)\,Y^2 < 1$. There hence exists $\xi > 0$ with

$$\cosh\xi = \frac{1}{\sqrt{1 - (1+\gamma^2)\,Y^2}}.$$

This implies $\tanh\xi = \sqrt{1+\gamma^2} \cdot \|Y\|$. Put $y = \frac{1}{\|Y\|} Y$ and observe $y \in p^\perp$, $y^2 = 1$ and

$$z = \frac{\gamma p}{\sqrt{1+\gamma^2}} + \|Y\| \cdot y = \frac{\gamma p}{\sqrt{1+\gamma^2}} + \frac{y}{\sqrt{1+\gamma^2}}\tanh\xi,$$

i.e. $w^{-1}(z) = \gamma p\cosh\xi + y\sinh\xi \in \Pi(p,\gamma)$. $\qquad\square$

Remark. Notice $\gamma p\cosh\xi + y\sinh\xi = \gamma p\cosh(-\xi) + (-y)\sinh(-\xi)$ and that $y \in p^\perp$, $y^2 = 1$ implies $(-y) \in p^\perp$, $(-y)^2 = 1$, so that, for instance, ξ could be chosen always non-negative. But in this case not all $y \in p^\perp$ with $y^2 = 1$ occur in the representation of $\Pi(p,\gamma)$.

Let $H(p,\alpha)$ be an arbitrary hyperplane of (X, eucl) with $p^2 = 1$. We will assume $\alpha \geq 0$, because otherwise we could work with $H(-p,-\alpha)$. If there is at least one point a in

$$H(p,\alpha) \cap B(0,1), \tag{2.49}$$

then $0 \leq \alpha \leq 1$: this follows from $pa = \alpha$ and $a^2 = 1$ by means of $\alpha \geq 0$ and

$$\alpha = pa \leq |pa| \leq \sqrt{p^2}\,\sqrt{a^2} = 1.$$

The intersection (2.49) contains exactly one point if, and only if, $\alpha = 1$. For $\alpha = 1$ we only have p in this intersection, since $px = 1$, $x^2 = 1$ imply

$$1 = px = |px| = \sqrt{p^2}\,\sqrt{x^2},$$

i.e. $x \in \{p, -p\}$, i.e. $x = p$ because of $p(-p) = -1$. If $0 \leq \alpha < 1$, take $r \in X$ with $pr = 0$ and $r^2 = 1$. Then

$$p\left(\alpha p \pm \sqrt{1 - \alpha^2}\,r\right) = \alpha, \quad \left(\alpha p \pm \sqrt{1 - \alpha^2}\,r\right)^2 = 1$$

lead to distinct points in (2.49).

Proposition 25. *If $p \in X$ satisfies $p^2 = 1$ and if $0 \leq \alpha < 1$ holds true for $\alpha \in \mathbb{R}$, then $\Pi(p, \gamma)$ with*

$$\gamma = \frac{\alpha}{\sqrt{1 - \alpha^2}}$$

is the image of $\{x \in P \mid px = \alpha\}$ under $w^{-1} : P \to X$.

Proof. The proof follows from Proposition 24, in view of $\dfrac{\gamma}{\sqrt{1+\gamma^2}} = \alpha$. $\qquad\square$

2.13 Hyperplanes under translations

Theorem 26. *Let $t, \gamma \in \mathbb{R}$ be given with $\gamma \geq 0$ and $p \in X$ with $p^2 = 1$. Suppose that*

$$T_t(x) = x + \left((xe)(\cosh t - 1) + \sqrt{1 + x^2}\,\sinh t\right)e,$$

$x \in X$, is a hyperbolic translation based on the axis e, $e^2 = 1$. Define

$$\gamma' \quad := \quad \gamma \cosh t + (pe)\sqrt{1 + \gamma^2}\,\sinh t,$$

$$\varepsilon \quad := \quad \operatorname{sgn}\gamma' \text{ for } \gamma' \neq 0,\ \varepsilon \in \mathbb{R}\backslash\{0\} \text{ for } \gamma' = 0,$$

and $p' \cdot \|A\| := A := p + \left[\dfrac{\gamma}{\sqrt{1+\gamma^2}}\sinh t + (pe)(\cosh t - 1)\right]e$ by observing $A \neq 0$. Then

$$T_t\big(\Pi(p, \gamma)\big) = \Pi\big(\varepsilon p', |\gamma'|\big) \tag{2.50}$$

holds true.

Proof. Notice $A^2 = \frac{1}{1+\gamma^2} + \left(\dfrac{\gamma}{\sqrt{1+\gamma^2}}\cosh t + (pe)\sinh t\right)^2 > 0$, i.e. $A \neq 0$ and, moreover,

$$\sqrt{1 + \gamma'^2} = \|A\| \cdot \sqrt{1 + \gamma^2}. \tag{2.51}$$

Instead of (2.50) we prove

$$w\left[\Pi\left(\varepsilon p',\,|\gamma'|\right)\right]=wT_tw^{-1}\left[\left\{x\in P\mid px=\frac{\gamma}{\sqrt{1+\gamma^2}}\right\}\right],\qquad(2.52)$$

since, by Proposition 25,

$$w^{-1}\left(\left\{x\in P\mid px=\frac{\gamma}{\sqrt{1+\gamma^2}}\right\}\right)=\Pi\left(p,\gamma\right).$$

From Proposition 24 we obtain

$$w\left[\Pi\left(\varepsilon p',\,|\gamma'|\right)\right]=\left\{x\in P\mid p'x=\frac{\gamma'}{\sqrt{1+\gamma'^2}}\right\},\qquad(2.53)$$

since $\varepsilon p'x=\frac{|\gamma'|}{\sqrt{1+\gamma'^2}}$ can be rewritten as $p'x=\frac{\gamma'}{\sqrt{1+\gamma'^2}}$. So in order to prove (2.50), we show, with (2.43),

$$\left\{x\in P\mid p'x=\frac{\gamma'}{\sqrt{1+\gamma'^2}}\right\}=T'_t\left[\left\{z\in P\mid pz=\frac{\gamma}{\sqrt{1+\gamma^2}}\right\}\right].\qquad(2.54)$$

Applying the decomposition $X=p^\perp\oplus\mathbb{R}p$, we will write

$$z=\bar{z}+z_0p,\ \bar{z}\in p^\perp,\ z_0\in\mathbb{R},$$

for $z\in X$. Hence

$$\left\{z\in P\mid pz=\frac{\gamma}{\sqrt{1+\gamma^2}}\right\}=\left\{\bar{z}+\frac{\gamma}{\sqrt{1+\gamma^2}}\,p\mid\bar{z}\in p^\perp,\ \|\bar{z}\|<\frac{1}{\sqrt{1+\gamma^2}}\right\}.$$

T'_t is a bijection of P. With $\alpha:=\frac{\gamma}{\sqrt{1+\gamma^2}}$, we obtain

$$T'_t\left[\left\{\bar{z}+\alpha p\mid\bar{z}\in p^\perp,\ \|\bar{z}\|<\sqrt{1-\alpha^2}\right\}\right],$$

by (2.43), as the set of all points

$$u\left(z\right):=\frac{\bar{z}+\alpha p+\left[(\bar{z}e+\alpha pe)(\cosh t-1)+\sinh t\right]e}{\cosh t+(\bar{z}e+\alpha pe)\sinh t}\qquad(2.55)$$

with $\|\bar{z}\|^2<1-\alpha^2$. In view of (2.54), we will show

$$p'\cdot u\left(z\right)=\frac{\gamma'}{\sqrt{1+\gamma'^2}},$$

i.e. by (2.51),

$$A \cdot u(z) = \frac{\gamma' \cdot \|A\|}{\sqrt{1 + \gamma'^2}} = \frac{\gamma'}{\sqrt{1 + \gamma^2}}.$$

Calling the nominator, denominator of the right-hand side of (2.55) $N(z)$, $D(z)$, respectively, the equation

$$A \cdot N(z) = D(z) \cdot (\alpha \cosh t + (pe) \sinh t)$$

must be verified, which can be accomplished easily. Observe, finally, that T_t maps hyperbolic hyperplanes onto such hyperplanes, and that consequently T'_t maps images (under w) of hyperbolic hyperplanes of X onto images of such hyperplanes. $\qquad \square$

Proposition 27. *Let $p, q \in X$ and $\gamma, \delta \in \mathbb{R}$ be given with $p^2 = 1 = q^2$, $\gamma \geq 0$ and $\delta \geq 0$. If*

$$\Pi(p, \gamma) \subseteq \Pi(q, \delta) \tag{2.56}$$

and $\gamma > 0$ hold true, then $p = q$ and $\gamma = \delta$. If (2.56) and $\gamma = 0$ hold true, then $p = \pm q$ and $\delta = 0$.

Proof. Instead of (2.56) we will consider $w\left(\Pi(p, \gamma)\right) \subseteq w\left(\Pi(q, \delta)\right)$, i.e.

$$L := \left\{ x \in P \mid px = \frac{\gamma}{\sqrt{1 + \gamma^2}} \right\} \subseteq \left\{ x \in P \mid qx = \frac{\delta}{\sqrt{1 + \delta^2}} \right\} =: R.$$

If $v \neq 0$ is in $p^{\perp}$, we obtain, by $\frac{1/2}{1 + \gamma^2} + \frac{\gamma^2}{1 + \gamma^2} < 1$,

$$\pm \frac{v}{\|v\| \sqrt{2(1 + \gamma^2)}} + \frac{\gamma p}{\sqrt{1 + \gamma^2}} \in L.$$

These two points x_1, x_2 must hence be elements of R, i.e. $q \cdot (x_1 - x_2) = 0$, i.e. $v \in q^{\perp}$. Thus $p^{\perp} \subseteq q^{\perp}$, i.e. $H(p, 0) \subseteq H(q, 0)$, i.e., by Proposition 12 we get $p = q$ or $p = -q$. Now

$$\frac{\gamma p}{\sqrt{1 + \gamma^2}} \in L \subseteq R \text{ implies } \frac{\gamma pq}{\sqrt{1 + \gamma^2}} = \frac{\delta}{\sqrt{1 + \delta^2}}.$$

Hence, if $\gamma > 0$, we obtain $p = q$, i.e. $\gamma = \delta$, and if $\gamma = 0$, $p = \pm q$ and $\delta = 0$ is the consequence. $\qquad \square$

2.14 Lines under translations

Let $\{p \cosh \xi + q \sinh \xi \mid \xi \in \mathbb{R}\}$ be a hyperbolic line l with elements $p, q \in X$ satisfying $pq = 0$, $q^2 = 1$. For $w \in O(X)$ we obtain

$$w(l) = \{w(p) \cosh \xi + w(q) \sinh \xi \mid \xi \in \mathbb{R}\} \tag{2.57}$$

with $0 = pq = \omega(p)\omega(q)$, $1 = qq = \omega(q)\omega(q)$. Suppose that $e \in X$, $t \in \mathbb{R}$ are given with $e^2 = 1$. We are then interested in the image $T_t(l)$ of l under the hyperbolic translation T_t with axis e.

Theorem 28. *Define*

$$
\begin{aligned}
p' &:= p + [pe(\cosh t - 1) + \sqrt{1+p^2}\,\sinh t]\,e, \\
q' &:= q + (qe)(\cosh t - 1)\,e, \\
A &:= qe\sinh t, \\
B &:= pe\sinh t + \sqrt{1+p^2}\,\cosh t.
\end{aligned}
$$

Then $|A| < B$. Define $\alpha \in \mathbb{R}$ by $B \cdot \tanh\alpha := -A$. Then

$$T_t(l) = \{p^* \cosh\eta + q^* \sinh\eta \mid \eta \in \mathbb{R}\} \tag{2.58}$$

with

$$
\begin{aligned}
p^* &:= p'\cosh\alpha + q'\sinh\alpha, \\
q^* &:= p'\sinh\alpha + q'\cosh\alpha
\end{aligned}
$$

and $p^*q^* = 0$, $(q^*)^2 = 1$.

Proof. Observe

$$p'^2 = B^2 - 1,$$

i.e. $B^2 \geq 1$ because of $p'^2 \geq 0$, and $q'^2 = 1 + A^2$, $p'q' = AB$. We now would like to prove

$$|A| < B.$$

Case $A \geq 0$. Here we get

$$(q-p)e\sinh t \leq |(q-p)e\sinh t| \leq \sqrt{(q-p)^2 e^2}\,\sinh|t|,$$

i.e. $(q-p)e\sinh t \leq \sqrt{1+p^2}\,\sinh|t| < \sqrt{1+p^2}\,\cosh t$.
Case $A < 0$. We must prove $-A < B$. Observe

$$(-q-p)e\sinh t \leq |(q+p)e\sinh t| \leq \sqrt{1+p^2}\,\sinh|t| < \sqrt{1+p^2}\,\cosh t.$$

Because of $B^2 \geq 1$ and $|A| < B$, we obtain $B \geq 1$ and

$$\left| -\frac{A}{B} \right| < 1,$$

i.e. $\tanh\alpha = -\frac{A}{B}$ determines $\alpha \in \mathbb{R}$ uniquely. Hence

$$\sinh\alpha = -\frac{A}{\sqrt{B^2 - A^2}}, \quad \cosh\alpha = \frac{B}{\sqrt{B^2 - A^2}}.$$

In view of $p'^2 = B^2 - 1$, $p'q' = AB$, $q'^2 = A^2 + 1$, we thus obtain

$$(q^*)^2 = (B^2 - 1) \frac{A^2}{B^2 - A^2} - 2AB \frac{AB}{B^2 - A^2} + (A^2 + 1) \frac{B^2}{B^2 - A^2} = 1,$$

$$p^*q^* = -(A^2 + B^2) \frac{AB}{B^2 - A^2} + AB \frac{A^2 + B^2}{B^2 - A^2} = 0.$$

Notice

$$T_t(p \cosh \xi + q \sinh \xi) = p' \cosh \xi + q' \sinh \xi,$$

and put $\eta := \xi - \alpha$. Then

$$p' \cosh \xi + q' \sinh \xi = p^* \cosh \eta + q^* \sinh \eta,$$

by $\cosh \xi = \cosh \alpha \cosh \eta + \sinh \alpha \sinh \eta$ and $\sinh \xi = \sinh \alpha \cosh \eta + \dots$. Hence

$$T_t(l) = \{p^* \cosh \eta + q^* \sinh \eta \mid \eta \in \mathbb{R}\}. \qquad \square$$

2.15 Hyperbolic coordinates

Let $n \geq 2$ be an integer and suppose that V is a subspace of dimension n of the vector space X. Let $b_1, \dots, b_n$ be a basis of V satisfying $b_i b_j = 0$ for $i \neq j$ and $b_i^2 = 1$ for all $i, j \in \{1, \dots, n\}$. If $p \in V$ and if

$$p = p_1 b_1 + \dots + p_n b_n$$

holds true with $p_1, \dots, p_n \in \mathbb{R}$, then $(p_1, \dots, p_n)$ will be called the *cartesian coordinates* of p, and $(x_1, \dots, x_n)$ with

$$\sqrt{1 + p_2^2 + \dots + p_n^2} \sinh x_1 = p_1,$$

$$\sqrt{1 + p_3^2 + \dots + p_n^2} \sinh x_2 = p_2,$$

$$\vdots$$

$$\sqrt{1 + p_n^2} \sinh x_{n-1} = p_{n-1},$$

$$\sinh x_n = p_n,$$

its *hyperbolic coordinates*. Designate by π the mapping which associates for every $p \in V$ to the cartesian coordinates $(p_1, \dots, p_n)$ of p its hyperbolic coordinates

$$\pi(p_1, \dots, p_n) = (x_1, \dots, x_n).$$

The mapping π is bijective, because $\pi^{-1}(x_1, \dots, x_n)$ is given by $(p_1, \dots, p_n)$ with

$$p_n = \sinh x_n,$$

$$p_{n-1} = \sinh x_{n-1} \cdot \cosh x_n,$$

$$p_{n-2} = \sinh x_{n-2} \cdot \cosh x_{n-1} \cdot \cosh x_n,$$

$$\vdots$$

$$p_1 = \sinh x_1 \cdot \cosh x_2 \cdot \cosh x_3 \cdots \cosh x_n.$$

Proposition 29. *Let* $e \in V$ *satisfy* $e^2 = 1$. *Extend* $e =: b_1$ *to a basis* $b_1, \ldots, b_n$ *of* V, *again with*

$$b_i b_j = \begin{cases} 0 & \text{for} \quad i \neq j \\ 1 & \text{for} \quad i = j \end{cases}$$

for all $i, j \in \{1, \ldots, n\}$. *Representing then the points of* V *by hyperbolic coordinates* $(x_1, \ldots, x_n)$,

$$T_t(x_1, \ldots, x_n) = (x_1 + t, x_2, \ldots, x_n)$$

holds true for all $t \in \mathbb{R}$ *and for all points of* V *in hyperbolic coordinates* $(x_1, \ldots, x_n)$, *where* T_t *are hyperbolic translations with axis* e.

Proof. Put $S_1 := \sinh x_1, C_1 := \cosh x_1$, and so on. Then

$$T_t(x_1, \ldots, x_n) = T_t(S_1 C_2 \ldots C_n b_1 + S_2 C_3 \ldots C_n b_2 + \cdots + S_n b_n)$$
$$= p + [pe(\cosh t - 1) + C_1 C_2 \ldots C_n \sinh t] e$$

with $p := S_1 C_2 \ldots C_n b_1 + \cdots + S_n b_n$. Put $S_t := \sinh t, C_t := \cosh t$. Then

$$
\begin{aligned}
T_t(x_1, \ldots, x_n) &= p + [S_1 C_2 \ldots C_n (C_t - 1) + C_1 C_2 \ldots C_n S_t] b_1 \\
&= p + (S_1 C_t - S_1 + C_1 S_t) C_2 C_3 \ldots C_n b_1 \\
&= \sinh(x_1 + t) C_2 \ldots C_n b_1 + S_2 C_3 \ldots C_n b_2 + \cdots + S_n b_n \\
&= (x_1 + t, x_2, x_3, \ldots, x_n). \qquad \square
\end{aligned}
$$

2.16 All isometries of $(X,\ \mathbf{eucl})$, $(X,\ \mathbf{hyp})$

The mapping $f : S \to S$ of a metric space (S, d) will be called an *isometry* of (S, d) provided

$$d\left(f(x), f(y)\right) = d(x, y) \tag{2.59}$$

holds true for all $x, y \in S$.

Isometries are injective mappings since $x \neq y$ for $x, y \in S$ implies

$$0 \neq d(x, y) = d\left(f(x), f(y)\right),$$

i.e. $f(x) \neq f(y)$. However, isometries need not be surjective. In chapter 1 we presented an example of an orthogonal mapping ω which is not surjective. Because of

$$d\left(\omega(x), \omega(y)\right) = d(x, y),$$

$d(x, y) := \|x - y\|$, for all x, y of the underlying real inner product space X, this ω hence represents an isometry of the metric space $(X,\ eucl)$ which is not surjective.

Surjective isometries of (S,d) are called *motions* of (S,d). Of course, the set of all motions of the metric space (S,d) is a group $M(S,d)$ under the permutation product. In view of I (see the proof of Theorem 7, chapter 1) we already know

$$M(X,d) = \{\alpha T_t\beta \mid \alpha,\beta \in O(X),\, t \in \mathbb{R}\} \tag{2.60}$$

for $(X,\text{ eucl})$ or $(X,\text{ hyp})$. Here and throughout section 2.3 T is the euclidean or hyperbolic translation group with a given axis $e \in X$, $e^2 = 1$, i.e.

$$T_t(x) = x + te \tag{2.61}$$

in the euclidean and

$$T_t(x) = x + [(xe)(\cosh t - 1) + \sqrt{1+x^2}\,\sinh t]\,e \tag{2.62}$$

in the hyperbolic case for all $x \in X$.

The following statement now presents the set of all isometries of (X,d) in the euclidean or hyperbolic case.

Proposition 30. *The set of all isometries of* (X,d) *is given by*

$$I(X,d) = \{\alpha T_t\beta \mid \alpha \in O(X),\, \beta \in \tilde{O}(X),\, t \in \mathbb{R}\}, \tag{2.63}$$

where $\tilde{O}(X)$ *designates the set of all orthogonal mappings of* X.

Proof. Suppose that δ is an isometry of (X,d) and that $\delta(0) =: p$. Because of A (see the proof of Theorem 7 in chapter 1) there exists $\gamma \in O(X)$ with

$$\gamma\delta(0) = \|p\|e.$$

In view of property (T 2) of a translation group, there exists $t \in \mathbb{R}$ satisfying

$$T_t\gamma\delta(0) = 0.$$

The mapping $\varphi := T_t\gamma\delta$ preserves distances and it satisfies $\varphi(0) = 0$.
Euclidean case. Hence for all $x,y \in X$,

$$\|x - y\| = \|\varphi(x) - \varphi(y)\|.$$

Thus $\varphi \in \tilde{O}(X)$, in view of Proposition 3 of chapter 1. This implies

$$\delta = \gamma^{-1}T_{-t}\varphi$$

with $\gamma^{-1} \in O(X)$.
Hyperbolic case. $\text{hyp}(x,y) = \text{hyp}(\varphi(x),\varphi(y))$ for all $x,y \in X$ implies

$$\sqrt{1+x^2}\,\sqrt{1+y^2} - xy = \sqrt{1+\xi^2}\,\sqrt{1+\eta^2} - \xi\eta \tag{2.64}$$

with $\xi := \varphi(x)$, $\eta := \varphi(y)$ and, especially for $x = 0$, $y = z$,

$$z^2 = [\varphi(z)]^2$$

for all $z \in X$, i.e., by (2.64), $xy = \varphi(x)\varphi(y)$ for all $x,y \in X$. Hence

$$\|x - y\| = \|\varphi(x) - \varphi(y)\|,$$

and thus $\varphi \in \tilde{O}(X)$, by Proposition 3, chapter 1.   □

2.17 Isometries preserving a direction

Let T be a translation group with axis $e \in X$, $e^2 = 1$. The following three statements hold true for hyperbolic as well as for euclidean geometry. The given proof of Lemma 31 is based on $(X,\ \mathrm{hyp})$.

Lemma 31. *Given $\alpha \in O(X)$ with $\alpha(e) = \varepsilon e$, $\varepsilon \in \mathbb{R}$. Then $\alpha T_t \alpha^{-1}(x) = T_{\varepsilon t}(x)$ for all $x \in X$ and $t \in \mathbb{R}$.*

Proof. $[\alpha(e)]^2 = e^2$ implies $\varepsilon^2 = 1$. With $\alpha^{-1}(e) = \varepsilon e$ and

$$x = h + x_0 e,\ h \in e^{\perp},\ x_0 \in \mathbb{R},$$

we obtain $\alpha^{-1}(h)\,\alpha^{-1}(e) = he = 0$ and $\alpha^{-1}(h) \in e^{\perp}$, and hence

$$\begin{aligned}
\alpha T_t \alpha^{-1}(x) &= \alpha T_t (\alpha^{-1}(h) + x_0 \varepsilon e) \\
&= \alpha \left(\alpha^{-1}(h) + [x_0 \varepsilon \cosh t + \sqrt{1 + h^2 + x_0^2}\ \sinh t]\, e \right) \\
&= x + [(xe)(\cosh \varepsilon t - 1) + \sqrt{1 + x^2}\ \sinh \varepsilon t]\, e = T_{\varepsilon t}(x),
\end{aligned}$$

by $\alpha^{-1}(h)\,\alpha^{-1}(h) = h^2$ and (2.62). □

Corollary. *Define $\chi(x) = h - x_0 e$ for $x = h + x_0 e$ with $h \in e^{\perp}$ and $x_0 \in \mathbb{R}$. Then*

$$\chi T_t = T_{-t}\,\chi$$

for all $t \in \mathbb{R}$.

Proof. Notice $\chi \in O(X)$ and $\chi(e) = -e$. □

Theorem 32. *Suppose that $f : X \to X$ is an isometry. Then*

$$f(x) - x \in \mathbb{R}e \text{ for all } x \in X \tag{2.65}$$

holds true if, and only if, $f \in T \cup T\chi$.

Proof. Obviously, $f \in T \cup T\chi$ satisfies (2.65). Let now $f : X \to X$ be an isometry satisfaying (2.65).

Case 1: $f \in \tilde{O}(X)$. Here $f = \mathrm{id}$ or $f = \chi$ holds true. In order to prove this statement observe

$$f(e) - e \in \mathbb{R}e,$$

i.e. $f(e) = \lambda e$ with a suitable $\lambda \in \mathbb{R}$. Hence $e^2 = [f(e)]^2$, i.e. $\lambda^2 = 1$. Because of

$$0 = he = f(h)\,f(e) = f(h) \cdot \lambda e$$

for $h \in e^{\perp}$, we obtain $f(h) \in e^{\perp}$, i.e.

$$f(h + x_0 e) = f(h) + x_0 \lambda e,\ f(h) \in e^{\perp}, \tag{2.66}$$

for $x = h + x_0 e$, $h \in e^\perp$, $x_0 \in \mathbb{R}$. By (2.65)

$$f(h + x_0 e) = h + x_0 e + \mu e \tag{2.67}$$

with a suitable $\mu \in \mathbb{R}$. Hence, by (2.66), (2.67), $f(h) = h$, i.e. by (2.66),

$$f(h + x_0 e) = h + x_0 \lambda e.$$

Thus $f = \mathrm{id}$ for $\lambda = 1$, and $f = \chi$ for $\lambda = -1$.

Case 2: $f \notin \tilde{O}(X)$. Since $f \in I(X, d)$,

$$f = \alpha T_t \beta$$

with $\alpha \in O(X)$, $\beta \in \tilde{O}(X)$ and $t \in \mathbb{R}$. Because of $f \notin \tilde{O}(X)$, we obtain $t \neq 0$. Hence $T_t(0) \neq 0$. By (2.65), $\alpha T_t \beta(0) = \lambda e$ with a suitable $\lambda \in \mathbb{R}$. Thus

$$0 \neq T_t(0) = \lambda \alpha^{-1}(e),$$

which implies $\alpha^{-1}(e) = \varepsilon e$, $\varepsilon \in \mathbb{R}$, in view of $T_t(0) \in \mathbb{R} e$. So we obtain $\alpha(e) = \varepsilon e$ with $\varepsilon^2 = 1$, i.e. by Lemma 31,

$$f = \alpha T_t \alpha^{-1} \cdot \alpha\beta = T_{\varepsilon t} \cdot \gamma$$

with $\gamma := \alpha\beta \in \tilde{O}(X)$. Since $T_{-\varepsilon t}$ and $f = T_{\varepsilon t} \cdot \gamma$ have property (2.65), hence also their product γ. This implies, by Case 1, $\gamma = \mathrm{id}$ or $\gamma = \chi$. Thus $f \in T \cup T\chi$. $\square$

2.18   A characterization of translations

The following Theorem 33 is essentially a corollary of Theorem 32.

Theorem 33. *An isometry f of (X, d) is a translation $\neq \mathrm{id}$ with axis e if, and only if,*

$$0 \neq f(x) - x \in \mathbb{R} e \tag{2.68}$$

holds true for all $x \in X$.

Proof. Suppose that the isometry $f : X \to X$ satisfies (2.68). Hence, by Theorem 32, $f \in T \cup T\chi$. We will show that $f = \mathrm{id}$ and also $f \in T\chi$ have at least one fixpoint, i.e. a point x with $f(x) = x$, i.e. with $0 = f(x) - x$, so that f must be a translation $\neq \mathrm{id}$ with axis e.

Hyperbolic case: Here $h + e\sqrt{1 + h^2} \sinh \frac{t}{2}$ with $h \in e^\perp$ is a fixpoint of $T_t\chi$.

Euclidean case: Here $h + \frac{t}{2} e$ is a fixpoint of $T_t\chi$.

Suppose, vice versa, that the translation $T_t \neq \mathrm{id}$ has axis e. Then, of course, $f(x) - x \in \mathbb{R} e$ holds true for all $x \in X$. Property (T 2) of a translation group implies that $t = 0$ is a consequence of $T_t(x_0) = x_0$ for a point x_0. $\square$

2.19 Different representations of isometries

Let again $e \in X$ be given with $e^2 = 1$ and suppose that T is the euclidean or hyperbolic translation group with axis e. Given isometries

$$\alpha T_t \beta, \ \gamma T_s \delta$$

of (X, d) with $\alpha, \gamma \in O(X)$, $\beta, \delta \in \tilde{O}(X)$, $t, s \in \mathbb{R}$, we would like to answer the question, when and only when $\alpha T_t \beta$ and $\gamma T_s \delta$ represent the same isometry.

Theorem 34. *Given* $(X, d) \in \{(X, \ \mathrm{eucl}), (X, \ \mathrm{hyp})\}$ *and*

$$\alpha, \gamma \in O(X), \ \beta, \delta \in \tilde{O}(X), \ t, s \in \mathbb{R}.$$

Then

$$\alpha T_t \beta = \gamma T_s \delta \tag{2.69}$$

holds true if, and only if,

$$t = s = 0, \ \alpha\beta = \gamma\delta \ \text{for} \ ts = 0,$$

or

$$0 \neq t = \varepsilon s, \ \varepsilon^2 = 1, \ \alpha\beta = \gamma\delta, \ \alpha(e) = \varepsilon\gamma(e) \ \text{for} \ ts \neq 0.$$

Proof. Of course, (2.69) holds true for $t = s = 0$, $\alpha\beta = \gamma\delta$, but also if all the presented conditions for $ts \neq 0$ are satisfied, by observing Lemma 31 and

$$\gamma^{-1}\alpha T_{\varepsilon s}\beta = \gamma^{-1}\alpha T_{\varepsilon s}(\gamma^{-1}\alpha)^{-1} \cdot \gamma^{-1}\alpha\beta = T_{\varepsilon^2 s} \cdot \delta.$$

Assume now (2.69), i.e. $\xi T_t \beta = T_s \delta$ with $\xi := \gamma^{-1}\alpha$. Because of

$$\xi T_t \beta(0) = T_s \delta(0)$$

we obtain

$$\xi(e) \cdot \sinh t = e \cdot \sinh s \tag{2.70}$$

and $\xi(e)\xi(e) = ee = 1$.

Case $ts = 0$. Hence, by (2.70), $t = s = 0$, and thus $\alpha\beta = \gamma\delta$, by (2.69).

Case $ts \neq 0$. Because of (2.70) and $(\xi(e))^2 = 1$, we obtain $\xi(e) = \varepsilon e$, $\varepsilon^2 = 1$, $0 \neq t = \varepsilon s$. A consequence of $\xi = \gamma^{-1}\alpha$ then is $\alpha(e) = \varepsilon\gamma(e)$. Finally observe, by (2.69) and Lemma 31,

$$T_s \delta = \xi T_t \beta = \xi T_t \xi^{-1} \cdot \xi\beta = T_{\varepsilon t} \cdot \xi\beta,$$

i.e. $T_s \delta = T_s \xi\beta$, i.e. $\delta = \xi\beta$. $\qquad \square$

2.20   A characterization of isometries

Suppose that (X, d) is one of the metric spaces (X, eucl) or (X, hyp). Then the following theorem holds true.

Theorem 35. *Let $\varrho > 0$ be a fixed real number and $N > 1$ be a fixed integer. If $f : X \to X$ is a mapping satisfying*

$$d\left(f\left(x\right), f\left(y\right)\right) \leq \varrho \text{ for all } x, y \in X \text{ with } d\left(x, y\right) = \varrho, \tag{2.71}$$
$$d\left(f\left(x\right), f\left(y\right)\right) \geq N\varrho \text{ for all } x, y \in X \text{ with } d\left(x, y\right) = N\varrho, \tag{2.72}$$

then f must be an isometry of (X, d).

Proof. Euclidean case: $d\left(x, y\right) = \|x - y\|$ for all $x, y \in X$. In view of Theorem 4, chapter 1, we obtain

$$f\left(x\right) = \omega\left(x\right) + t$$

for all $x \in X$, where $\omega \in \tilde{O}\left(X\right)$, and t a fixed element of X. Obviously, f satisfies (2.59).

Hyperbolic case: $d\left(x, y\right) = \text{hyp}\left(x, y\right)$ for all $x, y \in X$.

1. *The mapping f preserves hyperbolic distances ϱ and 2ϱ.*

Proof. Let p, q be points of (hyperbolic) distance ϱ, and

$$x\left(\xi\right) = a \cosh \xi + b \sinh \xi, \ \xi \in \mathbb{R},$$

with $a, b \in X$, $ab = 0$, $b^2 = 1$, be the line through p, q. If $p = x\left(\alpha\right)$, $q = x\left(\beta\right)$, then $|\beta - \alpha| = \varrho$. We may assume $\beta - \alpha = \varrho$, since otherwise we would work with $y\left(\xi\right) := x\left(-\xi\right)$ instead of $x\left(\xi\right)$, and $\alpha' := -\alpha$, $\beta' := -\beta$ instead of α, β. Hence

$$p = x\left(\alpha\right) \text{ and } q = x\left(\alpha + \varrho\right).$$

Define

$$x_\lambda := x\left(\alpha + \lambda\varrho\right), \ \lambda \in \{0, 1, \ldots, N\}.$$

Since $\text{hyp}\left(x_0, x_N\right) = N\varrho$, (2.72) implies

$$\text{hyp}\left(x_0', x_N'\right) \geq N\varrho$$

with $x' := f\left(x\right)$ for $x \in X$. Observe

$$\text{hyp}\left(x_\lambda', x_{\lambda+1}'\right) \leq \varrho$$

for $\lambda = 0, \ldots, N - 1$, in view of $\text{hyp}\left(x_\lambda, x_{\lambda+1}\right) = \varrho$. Hence

$$N\varrho \ \leq \text{hyp}\left(x_0', x_N'\right) \leq \text{hyp}\left(x_0', x_2'\right) + \sum_{\lambda=2}^{N-1} \text{hyp}\left(x_\lambda', x_{\lambda+1}'\right)$$
$$\leq \text{hyp}\left(x_0', x_1'\right) + \text{hyp}\left(x_1', x_2'\right) + \sum_{\lambda=2}^{N-1} \text{hyp}\left(x_\lambda', x_{\lambda+1}'\right) \leq N\varrho.$$

This yields hyp $(x'_\lambda, x'_{\lambda+1}) = \varrho$ and hyp $(x'_0, x'_2) = 2\varrho$. Hence hyp $(p', q') = \varrho$. If p, r are points of distance 2ϱ, we may write

$$p = x(\alpha) \text{ and } r = x(\alpha + 2\varrho).$$

Working now with $q := x(\alpha + \varrho)$, the proof above leads to $2\varrho = $ hyp $(x'_0, x'_2) = $ hyp (p', r'). $\qquad\square$

2. *If a, b, m are points with $a \neq b$ and*

$$\text{hyp } (a, m) = \text{hyp } (m, b) = \frac{1}{2} \text{ hyp } (a, b), \tag{2.73}$$

then m must be the hyperbolic midpoint of a, b.

Proof. If $x(\xi) = p \cosh \xi + q \sinh \xi$, $\xi \in \mathbb{R}$, with $pq = 0$, $q^2 = 1$, contains a, b, we may write $\alpha < \beta$,

$$a = x(\alpha) \text{ and } b = x(\beta).$$

Equation (2.73) implies

$$\text{hyp } (a, b) = \text{hyp } (a, m) + \text{hyp } (m, b),$$

and hence that a, b, m are collinear, i.e. on a common line (see the notion of a Menger line). Put $m = x(\gamma)$. By (2.73), we obtain

$$|\gamma - \alpha| = |\beta - \gamma|,$$

i.e. $\gamma = \frac{1}{2}(\alpha + \beta)$, in view of $\alpha < \beta$. $\qquad\square$

3. *Given points p, q of distance ϱ, we will write*

$$p = x(\alpha) \text{ and } q = x(\alpha + \varrho).$$

If $y(\eta), \eta \in \mathbb{R}$, is the line through $p' := f(p)$, $q' := f(q)$, we may write, by step 1,

$$p' = y(\beta), \ q' = y(\beta + \varrho).$$

Then

$$f(x(\alpha + \lambda\varrho)) = y(\beta + \lambda\varrho) \tag{2.74}$$

holds true for all integers $\lambda \geq 0$.

Proof. Clear for $\lambda \in \{0, 1\}$. Put $p_\lambda := x(\alpha + \lambda\varrho)$. Now

$$\varrho = \text{hyp } (p_{\lambda-1}, p_\lambda) = \text{hyp } (p_\lambda, p_{\lambda+1}) = \frac{1}{2} \text{ hyp } (p_{\lambda-1}, p_{\lambda+1})$$

and step 1 imply

$$\varrho = \text{hyp } (p'_{\lambda-1}, p'_\lambda) = \text{hyp } (p'_\lambda, p'_{\lambda+1}) = \frac{1}{2} \text{ hyp } (p'_{\lambda-1}, p'_{\lambda+1}) \tag{2.75}$$

for $\lambda = 1, 2, 3, \ldots$. Assume that (2.74) is proved up to $\lambda \geq 1$. Because of (2.75), the points $p'_{\lambda-1}, p'_{\lambda}, p'_{\lambda+1}$ must be collinear. Put $p'_{\lambda+1} = y\,(\gamma)$. Hence

$$\beta + \lambda \varrho = \frac{1}{2}\left(\beta + (\lambda - 1)\varrho + \gamma\right)$$

by (2.75) and step 2. Thus $\gamma = \beta + (\lambda + 1)\varrho$, i.e. (2.74) holds true also for $\lambda + 1$. $\quad\square$

A consequence of step 3 is that f preserves all (hyperbolic) distances $\lambda\varrho$ with $\lambda \in \{1, 2, 3, \ldots\}$.

4. *There exists a sequence $\alpha_1, \alpha_2, \alpha_3, \ldots$ of positive real numbers tending to 0 such that f preserves all hyperbolic distances α_i.*

Proof. Let $\mu > 1$ be an integer and A, B, C be points with

$$\mathrm{hyp}\,(A, B) = \mu\varrho = \mathrm{hyp}\,(A, C)$$

and $\mathrm{hyp}\,(B, C) = 2\varrho$. Such a triangle exists because of

$$\mathrm{hyp}\,(B, C) < \mathrm{hyp}\,(B, A) + \mathrm{hyp}\,(A, C).$$

We are now interested in the uniquely determined points B_μ, C_μ with

$$\mathrm{hyp}\,(A, B_\mu) = \varrho, \quad \mathrm{hyp}\,(B_\mu, B) = (\mu - 1)\varrho,$$
$$\mathrm{hyp}\,(A, C_\mu) = \varrho, \quad \mathrm{hyp}\,(C_\mu, C) = (\mu - 1)\varrho.$$

Since this configuration remains unaltered in its lengths under f, also the hyperbolic distance $\mathrm{hyp}\,(B_\mu, C_\mu)$ is preserved under f. Applying the hyperbolic cosine theorem twice, we get

$$\cosh \mathrm{hyp}\,(B_\mu, C_\mu) = \cosh^2 \varrho - \sinh^2 \varrho \cdot \frac{\cosh^2 \mu\varrho - \cosh 2\varrho}{\sinh^2 \mu\varrho},$$

i.e. $\sinh \mu\varrho \cdot \sinh \frac{1}{2}\,\mathrm{hyp}\,(B_\mu, C_\mu) = \sinh^2 \varrho$. The sequence

$$\alpha_{\mu-1} := \mathrm{hyp}\,(B_\mu, C_\mu) > 0, \quad \mu = 2, 3, \ldots,$$

hence tends to 0. All hyperbolic distances α_μ are preserved under f. $\quad\square$

5. *If $\alpha > 0$ and $x, y \in X$ satisfy $\mathrm{hyp}\,(x, y) = \alpha$, then*

$$\mathrm{hyp}\,\big(f\,(x),\, f\,(y)\big) \leq \alpha.$$

Proof. This is proved as soon as

$$\forall_{\varepsilon > 0}\ \forall_{x, y \in X}\ \mathrm{hyp}\,(x, y) = \alpha \Rightarrow \mathrm{hyp}\,\big(f\,(x),\, f\,(y)\big) < \alpha + \varepsilon$$

is shown. Let x, y be elements of X with $\mathrm{hyp}\,(x, y) = \alpha$ and let $x\,(\xi)$, $\xi \in \mathbb{R}$, be the line joining x, y with $x = x\,(\sigma)$ and $y = x\,(\sigma + \alpha)$ for a suitable σ. Suppose that $\varepsilon > 0$ is given. Take an element γ of the sequence $\alpha_1, \alpha_2, \alpha_3, \ldots$ of step 4 with $2\gamma < \varepsilon$ and elements $\gamma_1, \ldots, \gamma_n$ of $\{\alpha_1, \alpha_2, \ldots\}$ satisfying

$$0 < \alpha - (\gamma_1 + \cdots + \gamma_n) < 2\gamma.$$

The γ_i's need not be pairwise distinct. Then

$$0 < \alpha - (\gamma_1 + \cdots + \gamma_n) < 2\gamma < [\alpha - (\gamma_1 + \cdots + \gamma_n)] + \varepsilon \qquad (2.76)$$

holds true. Define $x_1 = x\,(\sigma + \gamma_1), \ldots, x_n = x\,(\sigma + \gamma_1 + \cdots + \gamma_n)$. Take $p \in X$ with

$$\mathrm{hyp}\,(x_n, p) = \gamma = \mathrm{hyp}\,(p, y).$$

The triangle x_n, p, y exists, because of

$$\mathrm{hyp}\,(x_n, y) < \mathrm{hyp}\,(x_n, p) + \mathrm{hyp}\,(p, y),$$

i.e. because of $\mathrm{hyp}\,(x_n, y) = \alpha - (\gamma_1 + \cdots + \gamma_n) < 2\gamma$. If we designate $f(z)$ by z' for $z \in X$, then the triangle inequality implies

$$\mathrm{hyp}\,(x', y') \leq \mathrm{hyp}\,(x', x_1') + \cdots + \mathrm{hyp}\,(x_{n-1}', x_n') + \mathrm{hyp}\,(x_n', p') + \mathrm{hyp}\,(p', y').$$

Since distances $\gamma_1, \ldots, \gamma_n, \gamma$ are preserved under f, we get

$$\gamma_1 = \mathrm{hyp}\,(x, x_1) = \mathrm{hyp}\,(x', x_1'), \ldots, \gamma_n = \mathrm{hyp}\,(x_{n-1}, x_n) = \mathrm{hyp}\,(x_{n-1}', x_n')$$

and $\gamma = \mathrm{hyp}\,(x_n, p) = \mathrm{hyp}\,(x_n', p'), \ldots$. Hence

$$\mathrm{hyp}\,(x', y') \leq \gamma_1 + \cdots + \gamma_n + \gamma + \gamma < \alpha + \varepsilon,$$

in view of (2.76). $\qquad\square$

6. *If r is a positive rational number, then f preserves the hyperbolic distance $r\varrho$.*

Proof. Let $n > 1$ be an integer. Then step 5 implies

$$\forall_{x,y \in X}\ \mathrm{hyp}\,(x, y) = \frac{\varrho}{n} \Rightarrow \mathrm{hyp}\,(f(x), f(y)) \leq \frac{\varrho}{n}.$$

Since distance ϱ is preserved, we get

$$\forall_{x,y \in X}\ \mathrm{hyp}\,(x, y) = n \cdot \frac{\varrho}{n} \Rightarrow \mathrm{hyp}\,(f(x), f(y)) = n \cdot \frac{\varrho}{n},$$

i.e. we get (2.71), (2.72) for $\frac{\varrho}{n}$ instead of ϱ and for n instead of N. Hence steps 1 and 3, carried out for the present values $\frac{\varrho}{n}$ and n, imply that all distances $\lambda \cdot \frac{\varrho}{n}$ with $\lambda \in \{1, 2, 3, \ldots\}$ are preserved. $\qquad\square$

7. *If t is a positive rational number and if x, y are points satisfying $\mathrm{hyp}\,(x, y) < t\varrho$, then $\mathrm{hyp}\,\big(f(x), f(y)\big) \leq t\varrho$.*

Proof. We shall write again $v' := f(v)$ for $v \in X$. Take $z \in X$ with

$$\mathrm{hyp}\,(x, z) = \frac{1}{2}\, t\varrho = \mathrm{hyp}\,(z, y).$$

Step 6 implies $\mathrm{hyp}\,(x', z') = \frac{1}{2}\, t\varrho = \mathrm{hyp}\,(z', y')$. Hence

$$\mathrm{hyp}\,(x', y') \le \mathrm{hyp}\,(x', z') + \mathrm{hyp}\,(z', y') = t\varrho. \qquad \square$$

8. *If r, s are positive rational numbers and x, y are points satisfying*

$$r\varrho < \mathrm{hyp}\,(x, y) < s\varrho,$$

then $r\varrho \le \mathrm{hyp}\,(x', y') \le s\varrho$ holds true.

Proof. Let $x(\tau)$, $\tau \in \mathbb{R}$, be the line joining x, y with $x = x(\xi)$, $y = x(\eta)$, $\xi < \eta$. Hence $\mathrm{hyp}\,(x, y) = \eta - \xi$ and thus $r\varrho < \eta - \xi < s\varrho$. Notice $\mathrm{hyp}\,(x', y') \le s\varrho$, by step 7. Define $p := x(\xi + s\varrho)$. Then

$$\mathrm{hyp}\,(y, p) = \xi + s\varrho - \eta,$$

i.e. $\mathrm{hyp}\,(y, p) = s\varrho - (\eta - \xi) < (s - r)\,\varrho$. Hence $\mathrm{hyp}\,(y', p') \le (s - r)\,\varrho$, by step 7. Moreover, $\mathrm{hyp}\,(x, p) = s\varrho$ implies $\mathrm{hyp}\,(x', p') = s\varrho$, on account of step 6. Hence, by the triangle inequality,

$$\mathrm{hyp}\,(x', y') \ge \mathrm{hyp}\,(x', p') - \mathrm{hyp}\,(y', p') \ge s\varrho - (s - r)\,\varrho = r\varrho. \qquad \square$$

9. $\mathrm{hyp}\,(x, y) = \mathrm{hyp}\,\big(f(x),\, f(y)\big)$ *holds true for all $x, y \in X$.*

Proof. If $\mathrm{hyp}\,(x, y) > 0$, take sequences r_ν, s_ν $(\nu = 1, 2, \ldots)$ of positive rational numbers satisfying

$$r_\nu \varrho < \mathrm{hyp}\,(x, y) < s_\nu \varrho, \quad \nu = 1, 2, \ldots,$$

and $\lim r_\nu = \frac{1}{\varrho}\,\mathrm{hyp}\,(x, y) = \lim s_\nu$. Hence

$$r_\nu \varrho \le \mathrm{hyp}\,\big(f(x),\, f(y)\big) \le s_\nu \varrho, \quad \nu = 1, 2, \ldots,$$

by step 8, i.e. $\mathrm{hyp}\,(x, y) = \mathrm{hyp}\,\big(f(x),\, f(y)\big)$. $\qquad \square$

Because of step 9 the mapping $f : X \to X$ must be a hyperbolic isometry. This finally proves Theorem 35. $\qquad \square$

Remark. If the dimension of X is finite, then

$$\forall_{x, y \in X}\ d(x, y) = \varrho \Rightarrow d\big(f(x),\, f(y)\big) = \varrho, \qquad (2.77)$$

d the euclidean or the hyperbolic distance function, for a fixed $\varrho > 0$ characterizes the isometries (F.S. Beckman, D.A. Quarles [1], B. Farrahi [1], A.V. Kuz'minyh [1]). In other words, if X is finite dimensional, then also $N = 1$ is allowed in Theorem 35. The euclidean part of Theorem 35 was proved in the context of strictly convex linear spaces by W. Benz, H. Berens [1], in the context of a more general $N \in \mathbb{R}$ by F. Radó, D. Andreescu, D. Valcán [1]. The theory beyond the Beckman–Quarles result started with the important contribution of E.M. Schröder [1]. The hyperbolic part of Theorem 35 was proved by W. Benz [8].

2.21 A counterexample

The following examples show that (2.77) does not generally characterize the isometries in the infinite dimensional case. The special example $(X, \text{ eucl})$ was given by Beckman, Quarles [1], the one concerning hyperbolic geometry by W. Benz [8].

Let X be the set of all sequences

$$a = (a_1, a_2, a_3, \ldots)$$

of real numbers $a_1, a_2, a_3, \ldots$ such that almost all a_i are zero. Define

$$\begin{aligned}
a + b &:= (a_1 + b_1, a_2 + b_2, \ldots), \\
\lambda a &:= (\lambda a_1, \lambda a_2, \ldots), \\
a \cdot b &:= a_1 b_1 + a_2 b_2 + \cdots
\end{aligned}$$

for all a, b in X and all real λ. This is a real inner product space which, in other terms, we already introduced in chapter 1. Let X_{rat} be the set of all $a \in X$ such that the a_i's of a are rational. Since X_{rat} is countable, let

$$\omega : \mathbb{N} \to X_{\text{rat}}$$

with $\mathbb{N} = \{1, 2, 3, \ldots\}$ be a fixed bijection. Moreover, suppose that $\varrho > 0$ is a fixed real number. Define

$$\psi\left(\omega\left(i\right)\right) := (x_{i1}, x_{i2}, \ldots) \text{ for } i = 1, 2, \ldots$$

with

(*euclidean case*) $x_{ii} = \frac{\varrho}{\sqrt{2}}$ and $x_{ij} = 0$ for $i \neq j$,

(*hyperbolic case*) $x_{ii} = \sqrt{2} \sinh \frac{\varrho}{2}$ and $x_{ij} = 0$ for $i \neq j$.

We hence get a mapping $\psi : X_{\text{rat}} \to X$. Another mapping $\varphi : X \to X_{\text{rat}}$ will play a role: For every $a \in X$ choose an element $\varphi\left(a\right)$ in X_{rat} such that

$$d\left(a, \varphi\left(a\right)\right) < \frac{\varrho}{2}. \tag{2.78}$$

It is now easy to show that

$$f : X \to X$$

with $f\left(x\right) := \psi\left(\varphi\left(x\right)\right)$ for $x \in X$ preserves distance ϱ, but no other positive distance. In fact, if $\varphi\left(x\right) = \varphi\left(y\right)$ for $x, y \in X$, we then obtain $d\left(f\left(x\right), f\left(y\right)\right) = 0$. If $\varphi\left(x\right) \neq \varphi\left(y\right)$ for $x, y \in X$, then $d\left(f\left(x\right), f\left(y\right)\right) = \varrho$. What we finally have to show is that $d\left(x, y\right) = \varrho$ implies $\varphi\left(x\right) \neq \varphi\left(y\right)$. But $\varphi\left(x\right) = \varphi\left(y\right)$ would lead, in the case $d\left(x, y\right) = \varrho$, to the contradiction

$$\varrho = d\left(x, y\right) \leq d\left(x, \varphi\left(x\right)\right) + d\left(\varphi\left(y\right), y\right) < \frac{\varrho}{2} + \frac{\varrho}{2},$$

in view of (2.78).

2.22   An extension problem

Let again (X, d) be one of the metric spaces (X, eucl), (X, hyp).

Lemma 36. *Let $a_1 \neq a_2$ and $b_1 \neq b_2$ be points with*

$$d(a_1, a_2) = d(b_1, b_2). \tag{2.79}$$

Then there exists a motion $\mu \in M(X, d)$ satisfying

$$\mu(a_1) = b_1 \text{ and } \mu(a_2) = b_2.$$

Proof. Because of step D.a of the proof of Theorem 7, chapter 1, there exist motions μ_1, μ_2 with $\mu_1(a_1) = 0$, $\mu_1(a_2) = \lambda_1 e$ and $\mu_2(b_1) = 0$, $\mu_2(b_2) = \lambda_2 e$ where λ_1, λ_2 are suitable positive real numbers. Now

$$
\begin{aligned}
d(a_1, a_2) &= d\big(\mu_1(a_1), \mu_1(a_2)\big) &= d(0, \lambda_1 e), \\
d(b_1, b_2) &= d\big(\mu_2(b_1), \mu_2(b_2)\big) &= d(0, \lambda_2 e)
\end{aligned}
$$

and (2.79) imply $\lambda_1 = \lambda_2$, in view of $\lambda_1, \lambda_2 > 0$. Hence

$$\mu_2^{-1}\mu_1(a_i) = b_i \text{ for } i = 1, 2. \qquad \square$$

Lemma 37. *Let m be a positive integer, b an element of X, and suppose that $a_1, \ldots, a_m, a_{m+1}$ are $m + 1$ linearly independent elements of X. If*

$$a_{m+1}^2 = b^2 \text{ and } a_{m+1}a_i = ba_i \tag{2.80}$$

hold true for $i = 1, \ldots, m$, there exists $\omega \in O(X)$ with $\omega = \omega^{-1}$,

$$\omega(a_{m+1}) = b \text{ and } \omega(a_i) = a_i \tag{2.81}$$

for $i = 1, \ldots, m$.

Proof. Take an orthogonal basis $c_1, \ldots, c_m$ with $c_i^2 = 1$, $i = 1, \ldots, m$, of the vector space V spanned by $a_1, \ldots, a_m$. Two cases are now important. If $a_{m+1} + b \in V$, put $c_{m+1} = 0$, and if $a_{m+1} + b \notin V$ put

$$r := \frac{a_{m+1} + b}{2} - \sum_{i=1}^{m} \left(\frac{a_{m+1} + b}{2} c_i \right) c_i, \tag{2.82}$$

and, moreover, since $r \neq 0$,

$$c_{m+1} := \frac{r}{\|r\|}.$$

In the second case $c_1, \ldots, c_{m+1}$ must be an orthogonal basis of the vector space spanned by $a_1, \ldots, a_m, a_{m+1} + b$. Define $\omega : X \to X$ by

$$\omega(x) = -x + 2 \sum_{i=1}^{m+1} (x c_i) c_i. \tag{2.83}$$

Since a_j is in V, we obtain $a_j c_{m+1} = 0$, and hence

$$w(a_j) = -a_j + 2 \sum_{i=1}^{m} (a_j c_i) c_i = a_j$$

for $j = 1, \ldots, m$. If we write $c_i = \varrho_{i1} a_1 + \cdots + \varrho_{im} a_m$ for suitable real numbers ϱ_{ij}, we get, by (2.80),

$$a_{m+1} c_i = b c_i \tag{2.84}$$

for $i = 1, \ldots, m$. If $a_{m+1} + b \in V$, then

$$\frac{a_{m+1} + b}{2} = \sum_{i=1}^{m} \left(\frac{a_{m+1} + b}{2} c_i \right) c_i$$

holds true, i.e., by (2.84), $a_{m+1} + b = 2 \sum (a_{m+1} c_i) c_i$, i.e., by (2.83), $w(a_{m+1}) = b$. If $a_{m+1} + b \notin V$, we obtain, by (2.82) and (2.80),

$$(a_{m+1} - b) r = -\sum_{i=1}^{m} \left(\frac{a_{m+1} + b}{2} c_i \right) (a_{m+1} c_i - b c_i),$$

i.e., (by (2.84), $(a_{m+1} - b) c_{m+1} = 0$. Hence from (2.84)

$$\frac{a_{m+1} + b}{2} = \sum_{i=1}^{m+1} \left(\frac{a_{m+1} + b}{2} c_i \right) c_i = \sum_{i=1}^{m+1} (a_{m+1} c_i) c_i,$$

i.e. $w(a_{m+1}) = b$.

Since w is linear and an involution, and since it satisfies $[w(x)]^2 = x^2$ for all $x \in X$, it must be in $O(X)$. □

Remark. If one of the elements a_{m+1}, b is in the vector space W spanned by $a_1, \ldots, a_m, a_{m+1} + b$, then also is the other one. In this case $V \neq W$ holds true. We then get

$$a_{m+1} = \sum_{i=1}^{m+1} (a_{m+1} c_i) c_i = \sum_{i=1}^{m+1} (b c_i) c_i = b.$$

The subspaces of (X, d) are given, by Proposition 13, by the subspaces Y of the vector space X and their images under motions of (X, d). If Y has dimension $n \in \{0, 1, 2, \ldots\}$, then the dimension of $\mu(Y)$ for every $\mu \in M(X, d)$ will also be defined by n. In order to show that this dimension of $\mu(Y)$ is well-defined we consider another subspace Y' of the vector space X such that there exists a motion ν with $\mu(Y) = \nu(Y')$. Observe $Y' = \sigma(Y)$ for the motion $\sigma := \nu^{-1}\mu$ which also can be written, in view of Proposition 30, in the form $\sigma = \alpha T_t \beta$ with suitable $\alpha, \beta \in O(X)$ and a suitable translation with respect to an axis e. Hence

$$Y' = \alpha T_t \beta(Y), \quad R' := \alpha^{-1}(Y') = T_t(R)$$

with $R := \beta(Y)$. Since α, β are linear and bijective, R and R' are subspaces of the vector space X with $n = \dim Y = \dim R$. We will show that the equation $R' = T_t(R)$ implies $R' = R$, i.e. $\dim Y' = \dim R' = \dim R = n$. There is nothing to prove for $t = 0$. So assume $t \neq 0$. As subspaces of the vector space X, both spaces R and R' contain $0 \in X$. Hence $0, T_t(0) \in R'$ implies $\mathbb{R}e \subseteq R'$, and $0, T_{-t}(0) \in R$, obviously, $\mathbb{R}e \subseteq R$. Assume now $z \in R \backslash \mathbb{R}e$. Hence $T_t(z) \in R'$ and R' contains the subspace W_z of X spanned by $0, e, T_t(z)$. Thus $z \in W_z \subseteq R'$. Similarly $R' \subseteq R$, because of $R = T_{-t}(R')$.

The following theorem will now be proved.

Theorem 38. *Let $S \neq \emptyset$ be a (finite or infinite) subset of a finite-dimensional subspace of (X, d), and let $f : S \to X$ satisfy*

$$d(x, y) = d(f(x), f(y))$$

for all $x, y \in S$. Then there exists $\varphi \in M(X, d)$ with $f(x) = \varphi(x)$ for all $x \in S$.

Proof. 1. If $S = \{a_1, a_2\}$ contains exactly two elements, define $b_i := f(a_i)$, $i = 1, 2$. Then Lemma 36 proves our theorem in this special case. If $S = \{a\}$, put $b := f(a)$. Because of D.a (see the proof of Theorem 7 in chapter 1), there exists a motion μ_1 such that $\mu_1(a) = 0$, and also a motion μ_2 with $\mu_2(f(a)) = 0$. Hence $\varphi = \mu_2^{-1}\mu_1$ is a motion transforming a into $f(a)$. So we may assume that S contains at least three distinct points. Let $a \neq p$ be elements of S and take $a_1 \in X$ with

$$d(0, a_1) = d(a, p),$$

and, in view of step 1, $\alpha \in M(X, d)$ such that $\alpha(a) = 0$, $\alpha(p) = a_1$. Because of

$$d(0, a_1) = d(a, p) = d(f(a), f(p)),$$

take $\beta \in M(X, d)$ satisfying $\beta(f(a)) = 0$, $\beta(f(p)) = a_1$. Instead of S we would like to work with $\alpha(S)$ containing $0, a_1$, and instead of f with

$$\beta f \alpha^{-1} : \alpha(S) \to X.$$

Notice that $\alpha(S) \ni 0$ implies that $\alpha(S)$ is a subspace of the vector space X. It is hence sufficient to prove Theorem 38 in the following form.

2. *Let $S \ni 0$, a_1 with $a_1 \neq 0$ be a subset of a finite-dimensional subspace Σ of the vector space X, and let $f : S \to X$ satisfy $f(0) = 0$, $f(a_1) = a_1$ and*

$$d(x, y) = d(f(x), f(y))$$

for all $x, y \in S$. Then there exists $\varphi \in M(X, d)$ with $f(x) = \varphi(x)$ for all $x \in S$.

3. *The euclidean case.* If $\dim \Sigma = 1$, $\Sigma = \mathbb{R}a_1$ holds true. Since

$$\|x - y\| = \|f(x) - f(y)\| \tag{2.85}$$

must be satisfied for all $x, y \in S$, we obtain, by $f(0) = 0$, $f(a_1) = a_1$,

$$\|\lambda a_1 - 0\| = \|f(\lambda a_1) - 0\|$$

and $\|\lambda a_1 - a_1\| = \|f(\lambda a_1) - a_1\|$ for all real λ with $\lambda a_1 \in S$. If

$$\{0, a_1, \lambda a_1\} =: \{x, y, z\},$$

we get for a suitable order

$$\|z - x\| = \|z - y\| + \|y - x\|.$$

This carries over to the f-images, and these must hence be collinear. Moreover, we obtain $f(\lambda a_1) = \lambda a_1$, i.e. $f(s) = s$ for all $s \in S$. Put $\varphi = \mathrm{id}$.

Assume $\dim \Sigma \geq 2$ and that statement 2 holds true for all subspaces Π of X with $\dim \Pi \leq m$ where m is a positive integer. We will show that then statement 2 holds true also in the case $\dim \Pi = m + 1$, provided $\dim X \geq m + 1$. Besides a_1 take elements $a_2, \ldots, a_{m+1}$ in S such that $a_1, \ldots, a_{m+1}$ are linearly independent. If they do not exist, S must already be contained in a subspace Π_0 with $\dim \Pi_0 \leq m$, and there is nothing to prove. Apply 2 for $S_0 = \{0, a_1, \ldots, a_m\}$, and there hence exists $\varphi_1 \in M(X, \mathrm{eucl})$ with $\varphi_1(0) = 0 = f(0)$, i.e. $\varphi_1 \in O(X)$, and

$$\varphi_1(a_i) = f(a_i), \ i = 1, \ldots, m.$$

Instead of f we will work with $f_1 := \varphi_1^{-1} f : S \to X$. Observe

$$\|x - y\| = \|f_1(x) - f_1(y)\| \tag{2.86}$$

for all $x, y \in S$, and $f_1(a_i) = a_i$, $i = 1, \ldots, m$. Put $b := f_1(a_{m+1})$. If we apply (2.86) for $x = 0$, $y = a_{m+1}$, and also for $x = a_i$, $y = a_{m+1}$, $i = 1, \ldots, m$, we get (2.80). In view of Lemma 37, there hence exists $\varphi_2 \in O(X)$ with $\varphi_2(a_i) = a_i$, $\varphi_2(a_{m+1}) = b$ for $i = 1, \ldots, m$. Put $f_2 := \varphi_2^{-1} f_1$ and observe $f_2(a_i) = a_i$ for $i = 1, \ldots, m + 1$, and

$$\|x - y\| = \|f_2(x) - f_2(y)\| \tag{2.87}$$

for all $x, y \in S$.

Let now s be an arbitrary element of S and define $t := f_2(s)$. Suppose that $e_1, \ldots, e_{m+1}$ is an orthogonal basis of the vector space V spanned by $a_1, \ldots, a_{m+1}$ with $e_i^2 = 1$, $i = 1, \ldots, m + 1$. Define $e = 0$ for $t \in V$, and otherwise such that $e_1, \ldots, e_{m+1}, e$ is an orthogonal basis, $e^2 = 1$, for the vector space W spanned by $a_1, \ldots, a_{m+1}, t$. From (2.87) we get, by $f_2(0) = 0 \in S$,

$$s^2 = t^2 \text{ and } sa_i = ta_i, \ i = 1, \ldots, m + 1.$$

If $e_i = \varrho_{i,1} a_1 + \cdots + \varrho_{i,m+1} a_{m+1}$, we hence obtain $se_i = te_i$, $i = 1, \ldots, m + 1$. Observe $s \in S \subseteq \Pi$ and therefore

$$s = (se_1) e_1 + \cdots + (se_{m+1}) e_{m+1}$$

and $t = (te_1) e_1 + \cdots + (te_{m+1}) e_{m+1} + (te) e$, i.e. $t = s + (te) e$. Since $se = 0$, we obtain

$$s^2 = t^2 = s^2 + (te)^2,$$

i.e. $te = 0$, i.e. $s = t$. Hence $f_2 = \mathrm{id}$ on S, and the identity mapping of $O(X)$ extends f_2 on X. Thus $\varphi_1 \varphi_2(x) = f(x)$ for all $x \in S$. This implies that if statement 2 holds true for all $\Pi, \dim \Pi \le m$, then also for Π with $\dim \Pi = m + 1$ provided $\dim X \ge m + 1$.

4. *The hyperbolic case.* If $\dim \Sigma = 1$, again $\Sigma = \mathbb{R}a_1$ holds true. Since

$$\mathrm{hyp}\,(x, y) = \mathrm{hyp}\,\big(f(x),\, f(y)\big) \tag{2.88}$$

must be satisfied for all $x, y \in S$, we obtain, by $f(0) = 0$, $f(a_1) = a_1$,

$$(\lambda a_1)^2 = \big(f(\lambda a_1)\big)^2$$

and $\lambda a_1^2 = a_1 f(\lambda a_1)$ for all real λ with $\lambda a_1 \in S$. If

$$\{0, a_1, \lambda a_1\} =: \{x, y, z\},$$

we get for a suitable order

$$\mathrm{hyp}\,(z, x) = \mathrm{hyp}\,(z, y) + \mathrm{hyp}\,(y, x), \tag{2.89}$$

since $0, a_1, \lambda a_1$ are on a common hyperbolic line. This carries over to the f-images implying collinearity for the image points, i.e. for $0, a_1, f(\lambda a_1)$. Since (2.89) holds also true for the image points

$$f(x),\, f(y),\, f(z),$$

we obtain $f(\lambda a_1) = \lambda a_1$, i.e. $f(s) = s$ for all $s \in S$.

Assume $\dim \Sigma \ge 2$ and that statement 2 holds true for all subspaces Π of X with $\dim \Pi \le m$ where m is a positive integer. We now will proceed as in step 3 up till formula (2.86), which must be replaced by

$$\mathrm{hyp}\,(x, y) = \mathrm{hyp}\,\big(f_1(x),\, f_1(y)\big) \tag{2.90}$$

for all $x, y \in S$. It is important to note that the stabilizer of $M(X, \mathrm{hyp})$ in the point 0 is given by $O(X)$, i.e. that $\gamma \in M(X, \mathrm{hyp})$ and $\gamma(0) = 0$ imply $\gamma \in O(X)$, so that $\varphi_1 \in M(X, \mathrm{hyp})$ must be in $O(X)$, because of $\varphi_1(0) = 0$. As in step 3 we put $b := f_1(a_{m+1})$. Applying (2.90) for $x = 0$, $y = a_{m+1}$, and also for $x = a_i, y = a_{m+1}$, $i = 1, \ldots, m$, we get $a_{m+1}^2 = b^2$ and

$$\sqrt{1 + a_i^2}\,\sqrt{1 + a_{m+1}^2} - a_i a_{m+1} = \sqrt{1 + a_i^2}\,\sqrt{1 + b^2} - a_i b,$$

i.e. (2.81). Proceeding as in step 3, we arrive at

$$\mathrm{hyp}\,(x, y) = \mathrm{hyp}\,\big(f_2(x),\, f_2(y)\big) \tag{2.91}$$

for all $x, y \in S$, instead of (2.87), with $f_2(a_i) = a_i$, $i = 1, \ldots, m + 1$. With the further definitions of step 3, we obtain from (2.91), by $f_2(0) = 0 \in S$,

$$\text{hyp } (0, s) = \text{hyp } (0, t), \ \text{hyp } (a_i, s) = \text{hyp } (a_i, t)$$

for $i = 1, \ldots, m + 1$, i.e.

$$s^2 = t^2 \text{ and } sa_i = ta_i, \ i = 1, \ldots, m + 1.$$

This leads to $s = t$ as in step 3, and finally to $\varphi_1 \varphi_2(x) = f(x)$ for all $x \in S$.

This finishes the proof of Theorem 38. $\qquad\square$

2.23 A mapping which cannot be extended

We already know an example of an orthogonal mapping $\omega : X \to X$ which is not surjective (see chapter 1 where orthogonal mappings are defined). Of course, in this special case there cannot exist $\varphi \in M(X, d)$ with $\omega(x) = \varphi(x)$ for all $x \in S := X$, since $\varphi : X \to X$ is bijective. Here S is not contained in a finite-dimensional subspace of X.

In order to present a mapping $f : S \to X$ which cannot be extended and where S is a proper subset of X, take as X all sequences

$$(a_1, a_2, a_3, \ldots)$$

of real numbers such that almost all of the a_i's are 0. Define, as usual,

$$
\begin{aligned}
a + b &:= (a_1 + b_1, a_2 + b_2, \ldots), \\
\lambda a &:= (\lambda a_1, \lambda a_2, \ldots), \\
a \cdot b &:= \sum_{i=1}^{\infty} a_i b_i
\end{aligned}
$$

for $a, b \in X$, $\lambda \in \mathbb{R}$. A basis of this real inner product space is

$$e_1 = (1, 0, 0, \ldots), \ e_2 = (0, 1, 0, \ldots), \ \ldots.$$

Define $S = \{0, e_2, e_3, e_4, \ldots\}$, $f(0) = 0$ and $f(e_i) = e_{i-1}$ for $i = 2, 3, \ldots$. Then

$$\|x - y\| = \|f(x) - f(y)\| \text{ and hyp } (x, y) = \text{hyp } (f(x), f(y))$$

hold true for all $x, y \in S$. The smallest subspace Σ of X containing S is spanned by $e_2, e_3, \ldots$. Hence Σ is infinite-dimensional. If there existed $\varphi \in M(X, d)$ with $\varphi(s) = f(s)$ for all $s \in S$, we would obtain $\varphi \in O(X)$, in view of $\varphi(0) = 0$. Assuming now

$$\varphi(e_1) =: \lambda_1 e_{e_1} + \cdots + \lambda_n e_{i_n}$$

with $\lambda_j \in \mathbb{R}$ and $1 \leq i_1 < i_2 < \cdots < i_n$, would imply

$$0 = e_1 e_{i_j + 1} = \varphi(e_1) \varphi(e_{i_j + 1}) = \varphi(e_1) e_{i_j} = \lambda_j$$

for $j = 1, \ldots, n$, i.e. $\varphi(e_1) = 0$, contradicting

$$1 = e_1 e_1 = \varphi(e_1) \varphi(e_1).$$

There hence does not exist $\varphi \in M(X, d)$ extending f.

Chapter 3

Sphere Geometries of Möbius and Lie

Also in this chapter X denotes a real inner product space of arbitrary (finite or infinite) dimension ≥ 2.

3.1 Möbius balls, inversions

The elements of $X' := X \cup \{\infty\}$ are called *points*. A *Möbius ball* (M-ball) is a set

$$B(c, \varrho) \text{ with } c \in X \text{ and } 0 < \varrho \in \mathbb{R},$$

or a set

$$H'(a, \alpha) := H(a, \alpha) \cup \{\infty\} \text{ with } 0 \neq a \in X \text{ and } \alpha \in \mathbb{R},$$

where $B(c, \varrho)$ is a ball of $(X, \text{ eucl})$ (see section 4, chapter 2) and $H(a, \alpha)$ a euclidean hyperplane of X. The bijections of X' such that images and inverse images of M-balls are M-balls are called *Möbius transformations* (M-transformations) of X. The set of all these M-transformations is a group under the usual product of bijections, the so-called *Möbius group* $\mathbb{M}(X)$ of X. The geometry $(X', \mathbb{M}(X))$ (see section 9 of chapter 1) is defined to be the *Möbius sphere geometry* over X.

If $\omega \in O(X)$ (see section 5 of chapter 1), $\gamma \in \mathbb{R}$ with $\gamma \neq 0$, and $a \in X$, then, obviously,

$$f(x) := \gamma\omega(x) + a \text{ for } x \in X, \text{ and } f(\infty) = \infty \tag{3.1}$$

is an M-transformation, called the *similitude* (γ, ω, a). Since also $-\omega$ is in $O(X)$, $(-\omega)(x) := -\omega(x)$ for $x \in X$, we always will assume $\gamma > 0$ in (3.1). Hence

$$(\gamma, \omega, a) = \text{id} \Leftrightarrow \gamma = 1, \omega = \text{id}, a = 0, \tag{3.2}$$

because $x = \gamma\omega(x) + a$ implies $a = 0$ for $x = 0$ and

$$x^2 = \gamma^2[\omega(x)]^2 = \gamma^2 x^2$$

implies $\gamma^2 = 1$ for $x \neq 0$, i.e. $\gamma = 1$, by $\gamma > 0$.

If $H(a, \alpha)$ is a euclidean hyperplane of X, then there exist to every $x \in X$ uniquely determined $h \in H(a, \alpha) = \{x \in X \mid ax = \alpha\}$ and $\lambda \in \mathbb{R}$ with

$$x = h + \lambda a. \tag{3.3}$$

In fact, if h, λ exist as described, then $ax = ah + \lambda a^2$ and (3.3) imply

$$\lambda = \frac{ax - \alpha}{a^2} \text{ and } h = x - \frac{ax - \alpha}{a^2}a. \tag{3.4}$$

Vice versa, h, λ of (3.4) satisfy (3.3), $h \in H(a, \alpha)$, $\lambda \in \mathbb{R}$.

The *reflection* in the euclidean hyperplane $H(a, \alpha)$ is the mapping φ from X into X with

$$\varphi(h + \lambda a) = h - \lambda a \tag{3.5}$$

for all $h \in H(a, \alpha)$ and $\lambda \in \mathbb{R}$. The mapping φ is an involution of X,

$$\varphi \neq \mathrm{id} = \varphi^2,$$

a bijection of X, and it leaves invariant exactly the points of $H(a, \alpha)$. By (3.4), (3.5),

$$\varphi(x) = x + 2\frac{\alpha - ax}{a^2}a \tag{3.6}$$

for all $x \in X$. Moreover, $\varphi(x) = \omega(x) + \frac{2\alpha a}{a^2}$, where

$$\omega(x) := x - \frac{2ax}{a^2}a, \tag{3.7}$$

$x \in X$, is in $O(X)$ because of $[\omega(x)]^2 = x^2$ for all $x \in X$, the linearity of ω and the bijectivity of φ. The corresponding similitude

$$f = \left(1, \omega, \frac{2\alpha a}{a^2}\right)$$

is called the *inversion* in the M-ball $H'(a, \alpha)$. It is an involutorial M-transformation with

$$\mathrm{Fix}(f) := \{x \in X' \mid f(x) = x\} = H'(a, \alpha).$$

The mapping $\iota : X' \to X'$ is defined by

$$\iota(x) := \tfrac{x}{x^2} \text{ for } 0 \neq x \neq \infty,$$
$$\iota(0) := \infty \text{ and } \iota(\infty) := 0.$$

It is an involution. In fact, putting $y := \iota(x)$ for $0 \neq x \neq \infty$ we obtain $x^2 y^2 = 1$, i.e.

$$x = yx^2 = \frac{y}{y^2}.$$

Hence ι is a bijection of X'. Moreover, ι is an M-transformation. The image

(1) of $B(c, \varrho)$ with $c^2 \neq \varrho^2$ is $B\left(\frac{c}{c^2 - \varrho^2}, \frac{\varrho}{|c^2 - \varrho^2|}, \right)$,

(2) of $B(c, \varrho)$ with $c^2 = \varrho^2$ is $H'(2c, 1)$,

(3) of $H'(a, \alpha)$ with $\alpha \neq 0$ is $B\left(\frac{a}{2\alpha}, \left\|\frac{a}{2\alpha}\right\|\right)$,

(4) of $H'(a, 0)$ is $H'(a, 0)$.

The *inversion* $f = \iota(c, \varrho)$ in the M-ball $B(c, \varrho)$ is defined by

$$f(x) = c + \varrho^2 \cdot \frac{x - c}{(x - c)^2} \text{ for } c \neq x \neq \infty \tag{3.8}$$

and by $f(c) = \infty$, $f(\infty) = c$. The mapping ι is hence the inversion in the *unit ball* $B(0, 1)$. If φ is the similitude

$$\varphi = (\varrho, \text{ id}, c), \tag{3.9}$$

then, obviously,

$$f = \varphi \iota \varphi^{-1}, \tag{3.10}$$

so that the inversion in $B(c, \varrho)$ is also an M-transformation. Moreover, we obtain Fix $(f) = B(c, \varrho)$. The inversion in $B(c, \varrho)$ can be characterized as a mapping which interchanges c and ∞, and which carries the point $x \notin \{c, \infty\}$ into $y \in X$ such that

$$y - c = \lambda(x - c),$$

$$\|x - c\| \cdot \|y - c\| = \varrho^2$$

hold with $0 < \lambda \in \mathbb{R}$. The points c, x, y are hence on a common euclidean line.

Proposition 1. *If b_1, b_2 are M-balls, there exists $\mu \in \mathbb{M}(X)$ with*

$$b_2 = \mu(b_1) := \{\mu(x) \mid x \in b_1\}.$$

Proof. a) $(\varrho, \text{ id}, c)$ carries $B(0, 1)$ onto $B(c, \varrho)$ for given $\varrho > 0$ and $c \in X$.

b) The inversion in $B(0, 1)$ carries $B(c, \|c\|)$ onto $H'\left(c, \frac{1}{2}\right)$ for all $c \neq 0$.

c) $\left(1, \text{ id}, \frac{c}{2c^2}\right)$ carries $H'(c, 0)$ onto $H'\left(c, \frac{1}{2}\right)$ for all $0 \neq c \in X$. $\square$

Let e be a fixed element of X with $e^2 = 1$. If $H'(a, \alpha)$ is an M-ball, we may assume $a^2 = 1$ without loss of generality, since

$$H(a, \alpha) = H\left(\frac{a}{\|a\|}, \frac{\alpha}{\|a\|}\right).$$

Take $\omega \in O(X)$ with $\omega(e) = a$ (see step A of the proof of Theorem 7, chapter 1) and put

$$\varphi = (1, \omega, \alpha a). \tag{3.11}$$

Then $\varphi(H'(e, 0)) = H'(a, \alpha)$, since $ex = 0$ is equivalent with $\omega(e)\,\omega(x) = 0$, i.e. with

$$0 = a\left(\varphi(x) - \alpha a\right) = a\varphi(x) - \alpha.$$

Let j be the inversion in $H'(e, 0)$ and f be that one in $H'(a, \alpha)$. Then

$$f = \varphi j \varphi^{-1}. \tag{3.12}$$

We will prove this by applying (3.6) for $H'(e, 0)$,

$$j(x) = x - 2(ex)\,e,$$

and for $H'(a, \alpha)$,

$$f(x) = x + 2(\alpha - ax)\,a.$$

Equation (3.11) implies $\varphi^{-1}(x) = \omega^{-1}(x - \alpha a)$. Hence

$$j\varphi^{-1}(x) = \omega^{-1}(x - \alpha a) - 2a\,(x - \alpha a) \cdot e,$$

on account of $e\omega^{-1}(x - \alpha a) = \omega(e) \cdot (x - \alpha a)$ and $\omega(e) = a$. Thus

$$\varphi j \varphi^{-1}(x) = (x - \alpha a) - 2(ax - \alpha)\,\omega(e) + \alpha a = f(x).$$

For the following proposition we do not ask for $a^2 = 1$, but, of course, for $a \neq 0$.

Proposition 2. *The inversion in* $H'(a, \alpha)$, $\alpha \neq 0$, *can be written in the form* $f \iota f^{-1}$, *where* $f := \iota \varphi$ *with*

$$\varphi = \left(\left\|\frac{a}{2\alpha}\right\|, \mathrm{id}, \frac{a}{2\alpha}\right)$$

carries $B(0, 1)$ *onto* $H'(a, \alpha)$. *Similarly, the inversion in* $H'(a, 0)$ *is given by* $g \iota g^{-1}$, *where* $g := \psi \iota \varphi$ *with*

$$\varphi = (\|a\|, \mathrm{id}, a), \quad \psi = \left(1, \mathrm{id}, -\frac{a}{2a^2}\right),$$

carries $B(0, 1)$ *onto* $H'(a, 0)$.

3.2   An application to integral equations

If S is a subset of X, put $\lambda S := \{\lambda x \mid x \in S\}$ for $\lambda \in \mathbb{R}$, and

$$a + S := \{a + x \mid x \in S\}$$

for $a \in X$.

A parametric representation of the M-ball $B\left(c, \varrho\right)$ is given by

$$B\left(c, \varrho\right) = \left\{ c + \frac{\varrho x}{\|x\|} \,\middle|\, x \in X \backslash \{0\} \right\} = c + \varrho B\left(0, 1\right),$$

and a parametric representation for the euclidean hyperplane $H\left(a, \alpha\right)$, $a^2 = 1$, by

$$H\left(a, \alpha\right) = \alpha a + \left\{ \frac{v - \left(va\right)a}{1 - va} \,\middle|\, v \in B\left(0, 1\right) \backslash \{a\} \right\}. \tag{3.13}$$

In order to prove (3.13), observe $1 \neq va$, since otherwise $\left(va\right)^2 = v^2 a^2$, i.e. $v = \pm a$, i.e. $v = -a$, i.e. $1 = va = -1$. Of course, the right-hand side of (3.13) is a subset of $H\left(a, \alpha\right)$. If, vice versa, x is in $H\left(a, \alpha\right)$, put

$$v := a + \frac{2z}{z^2}, \; z := x - \left(\alpha + 1\right)a,$$

and observe $z \neq 0$, since otherwise $\alpha = ax = a\left(\alpha + 1\right)a = \alpha + 1$. Hence $v^2 = 1$, $v \neq a$, and

$$\alpha a + \frac{v - \left(va\right)a}{1 - va} = x.$$

Equation (3.13) and $B\left(0, 1\right) = \left\{ \frac{x}{\|x\|} \,\middle|\, x \in X \backslash \{0\} \right\}$ imply

$$H\left(a, \alpha\right) = \alpha a + \left\{ \frac{v - \left(va\right)a}{\|v\| - va} \,\middle|\, v \in X \backslash \{0\} \text{ with } \frac{v}{\|v\|} \neq a \right\}. \tag{3.14}$$

If we take the example b) of section 2, chapter 1, as space X, then formula (3.14) solves certain integral equations:

Corollary. *Let $\alpha < \beta$ be real numbers and $h : I \to \mathbb{R}$, $I := \left[\alpha, \beta\right]$, a function continuous in I, satisfying $h\left(t\right) > 0$ for all $t \in I \backslash T$, where T is a finite subset of I. Suppose that ϱ is a real number and $a : I \to \mathbb{R}$ a real function also continuous in I. All functions $x\left(t\right)$ continuous in I satisfying*

$$\int_\alpha^\beta h\left(t\right) a\left(t\right) x\left(t\right) dt = \varrho \tag{3.15}$$

are then given by (3.14), i.e. by

$$x = \varrho a + \left\{ \frac{v - \left(va\right)a}{\|v\| - va} \,\middle|\, v \in X \backslash \{0\} \text{ with } \frac{v}{\|v\|} \neq a \right\}. \tag{3.16}$$

The proof is obvious, since (3.15) asks for all points of the hyperplane $H\left(a, \varrho\right) \subset X$. As a matter of fact (3.14) solves even more integral equations. Let $J \neq \emptyset$ be a subset of $\mathbb{R}$ and

$$a : J \to X \backslash \{0\}, \; \varrho : J \to \mathbb{R}$$

be functions. The problem then is to solve

$$a\left(s\right) \cdot x\left(s\right) = \varrho\left(s\right)$$

for all $s \in J$, i.e. to solve

$$\int_{\alpha}^{\beta} h\left(t\right) a\left(s,t\right) x\left(s,t\right) dt = \varrho\left(s\right) \tag{3.17}$$

for all $s \in J$. The solution of (3.17) is again given by (3.16), however, for each single $s \in J$, so that also v in (3.16) must be a function of s. Of course, instead of $J \subseteq \mathbb{R}$ we could consider $J \subseteq \mathbb{R}^n$ as well, n a positive integer, i.e. functions $a\left(s_1, \ldots, s_n, t\right)$, $\varrho\left(s_1, \ldots, s_n\right)$, $x\left(s_1, \ldots, s_n, t\right)$.

3.3   A fundamental theorem

If a, b are elements of X', there exists $\mu \in M\left(X\right)$ with $\mu\left(a\right) = b$ and $\mu^2 = \mathrm{id}$. This is clear for $a = b$. So assume $a \neq b$. If $\infty \in \{a, b\}$, take the M-ball $B\left(c, 1\right)$ with $\{c, \infty\} = \{a, b\}$. The inversion in this M-ball interchanges a and b. If $a, b \in X$, take the inversion in

$$H'\left(a - b, \frac{a^2 - b^2}{2}\right),$$

which also interchanges a and b (apply (3.6) for the present situation). We now would like to prove the following theorem which we call the fundamental theorem of Möbius sphere geometry.

Theorem 3. *Let f be an M-transformation of X. If $f\left(\infty\right) = \infty$ then f is a similitude. Otherwise there exist similitudes α, β with $f = \alpha\iota\beta$.*

Proof. a) Case: $f\left(\infty\right) = \infty$. The restriction φ of f on X is then a bijection of X, and images and inverse images under f of M-balls through ∞ are M-balls containing ∞. Hence images and inverse images under φ of euclidean hyperplanes of X are euclidean hyperplanes. Let p, q be distinct elements of X and let l be the euclidean line passing through p, q. In view of Proposition 23 and its proof, chapter 2, the intersection of all euclidean hyperplanes H through p and q, must be l. Hence

$$\varphi\left(l\right) = \varphi\left(\bigcap_{p,q \in H} H\right) = \bigcap_{p,q \in H} \varphi\left(H\right).$$

If J is a hyperplane through $\varphi\left(p\right)$, $\varphi\left(q\right)$, then $f^{-1}(J \cup \{\infty\})$ is an M-ball through ∞, p, q. Thus $\varphi^{-1}(J)$ is a hyperplane through p, q. Hence

$$\varphi\left(l\right) = \bigcap_{p,q \in H} \varphi\left(H\right) = \bigcap_{\varphi\left(p\right), \varphi\left(q\right) \in J} J$$

is a line, and φ thus maps euclidean lines of X onto such lines.

b) Based on this mapping φ we now would like to prove that

$$\gamma(x) := \varphi(x) - \varphi(0), \ x \in X, \tag{3.18}$$

is a bijective linear mapping of X. Since φ is a bijection of X, so must be γ. If l is a (euclidean) line of X, so is $\varphi(l)$ and therefore $\gamma(l)$ as well. If g is a line, take distinct points p, q of $\gamma^{-1}(g)$. Denote by h the line through p, q. Then g and $\gamma(h)$ are lines through $\gamma(p)$ and $\gamma(q)$, i.e. $\gamma^{-1}(g)$ must be a line as well. Hence images and inverse images of lines under γ must be lines.

A set $T \subset X$ is called *collinear* provided there exists a line $l \supseteq T$. Let v_1, v_2 be linearly independent elements of X. Define $w_i = \gamma(v_i)$, $i = 1, 2$. Then w_1, w_2 must also be linearly independent. Otherwise, $\{0, w_1, w_2\}$ would be collinear, and hence

$$\gamma^{-1}(\{0, w_1, w_2\}) = \{0, v_1, v_2\},$$

since the inverse image under γ of the line through $0, w_1, w_2$ must be a line. But the collinearity of $\{0, v_1, v_2\}$ contradicts the fact that v_1, v_2 are linearly independent. Define P, Q by

$$\begin{aligned} P &= \{\alpha_1 v_1 + \alpha_2 v_2 \mid \alpha_1, \alpha_2 \in \mathbb{R}\}, \\ Q &= \{\beta_1 w_1 + \beta_2 w_2 \mid \beta_1, \beta_2 \in \mathbb{R}\}, \end{aligned}$$

respectively. $\gamma : P \to Q$ is a bijection. Here γ also denotes the restriction of the original mapping γ to P. In order to prove this statement, we only need to show that

$$\forall_{p \in P} \ \gamma(p) \in Q \text{ and } \forall_{q \in Q} \ \gamma^{-1}(q) \in P, \tag{3.19}$$

since γ is bijective. The image of the line ξ through $0, v_1$ is the line ξ' through $0, w_1$, and the image of the line η through $0, v_2$ is the line η' through $0, w_2$. For a given $p \in P$ choose $a \in \xi$ and $b \in \eta$ with $a \neq b$ such that a, b, p are collinear. Now $\gamma(a), \gamma(b), \gamma(p)$ are also collinear and since $\gamma(a) \in \xi'$, $\gamma(b) \in \eta'$ hold true, we get $\gamma(p) \in Q$. For the remaining statement of (3.19) observe that also γ^{-1} maps lines onto lines.

Write (α_1, α_2) instead of $\alpha_1 v_1 + \alpha_2 v_2$ and (β_1, β_2) instead of the element $\beta_1 w_1 + \beta_2 w_2$. The mapping $\gamma : P \to Q$ may then be considered to be a bijection δ of $\mathbb{R}^2$ by defining

$$\delta(\alpha_1, \alpha_2) = (\beta_1, \beta_2)$$

if, and only if, $\gamma(\alpha_1 v_1 + \alpha_2 v_2) = \beta_1 w_1 + \beta_2 w_2$. Images and inverse images under δ of lines of $\mathbb{R}^2$,

$$\{(\lambda_1, \lambda_2) + \mu(\varrho_1, \varrho_2) \mid \mu \in \mathbb{R}\} =: (\lambda_1, \lambda_2) + \mathbb{R}(\varrho_1, \varrho_2), \ (\varrho_1, \varrho_2) \neq (0, 0),$$

must be lines, since

$$\gamma(a + \mathbb{R}t) = b + \mathbb{R}s$$

for

$$a = \alpha_1 v_1 + \alpha_2 v_2, \quad t = \tau_1 v_1 + \tau_2 v_2 \neq 0,$$
$$b = \beta_1 w_1 + \beta_2 w_2, \quad s = \sigma_1 w_1 + \sigma_2 w_2 \neq 0,$$

is equivalent to

$$\delta\left((\alpha_1, \alpha_2) + \mathbb{R}\left(\tau_1, \tau_2\right)\right) = (\beta_1, \beta_2) + \mathbb{R}\left(\sigma_1, \sigma_2\right).$$

A basic theorem of geometry (see, for instance, Proposition 2 of section 1.2 of W. Benz [3]) now says

$$\delta\left(\alpha_1, \alpha_2\right) = (\alpha_1 a_{11} + \alpha_2 a_{21}, \; \alpha_1 a_{12} + \alpha_2 a_{22})$$

for fixed real numbers a_{ij}, in view of $\delta\left(0, 0\right) = (0, 0)$. In particular,

$$\delta\left(1, 0\right) = (a_{11}, a_{12}), \quad \delta\left(0, 1\right) = (a_{21}, a_{22}).$$

Hence $(a_{11}, a_{12}) = (1, 0)$ and $(a_{21}, a_{22}) = (0, 1)$ since

$$w_i = \gamma\left(v_i\right) = a_{i1} w_1 + a_{i2} w_2$$

for $i = 1, 2$. Thus $\delta\left(\alpha_1, \alpha_2\right) = (\alpha_1, \alpha_2)$, i.e.

$$\gamma\left(\alpha_1 v_1 + \alpha_2 v_2\right) = \alpha_1 w_1 + \alpha_2 w_2 = \alpha_1 \lambda\left(v_1\right) + \alpha_2 \lambda\left(v_2\right). \tag{3.20}$$

Let now p_1, p_2 be linearly dependent elements of X. In this case, we would also like to show that

$$\gamma\left(\alpha_1 p_1 + \alpha_2 p_2\right) = \alpha_1 \gamma\left(p_1\right) + \alpha_2 \gamma\left(p_2\right)$$

for $\alpha_1, \alpha_2 \in \mathbb{R}$. Since there is nothing to prove for $p_1 = 0 = p_2$, we may assume that $p_1 \neq 0$ without loss of generality. Then $p_2 = \varrho p_1$ with $\varrho \in \mathbb{R}$. Since the dimension of X is at least 2, there exists $q \in X$ such that p_1, q are linearly independent. Hence, in view of (3.20),

$$\gamma\left(\alpha_1 p_1 + \alpha_2 p_2\right) = \gamma\left((\alpha_1 + \varrho \alpha_2) p_1 + 0 \cdot q\right) = (\alpha_1 + \varrho \alpha_2) \gamma\left(p_1\right)$$

and

$$\begin{aligned} \alpha_1 \gamma\left(p_1\right) + \alpha_2 \gamma\left(p_2\right) &= \alpha_1 \gamma\left(p_1 + 0 \cdot q\right) + \alpha_2 \gamma\left(\varrho p_1 + 0 \cdot q\right) \\ &= (\alpha_1 + \varrho \alpha_2) \gamma\left(p_1\right). \end{aligned}$$

c) The mapping φ of step a) can hence be written, by (3.18), as

$$\varphi\left(x\right) = \gamma\left(x\right) + \varphi\left(0\right)$$

for all $x \in X$, where γ is a bijective linear mapping of X. This mapping φ was the restriction of an M-transformation f of X satisfying $f\left(\infty\right) = \infty$. Since $B\left(0, 1\right)$ is an M-ball not containing ∞, we obtain

$$\varphi\left(B\left(0, 1\right)\right) = f\left(B\left(0, 1\right)\right) = B\left(c, \varrho\right)$$

with suitable $c \in X$ and $\varrho > 0$. Hence

$$\left(\varphi \left(\frac{x}{\|x\|} \right) - c \right)^2 = \varrho^2$$

for all $x \neq 0$ in X, i.e.

$$\left(\frac{1}{\|x\|} \gamma(x) + a \right)^2 = \varrho^2 \tag{3.21}$$

with $a := \varphi(0) - c$ for all $x \neq 0$ in X. Applying (3.21) for x and $-x$ yields $\gamma(x) \cdot a = 0$ for all $x \neq 0$ in X. If a were $\neq 0$, then

$$a^2 = \gamma \left(\gamma^{-1}(a) \right) \cdot a = 0.$$

Hence $c = \varphi(0)$, i.e. $\left(\gamma(x) \right)^2 = \varrho^2 x^2$ for all $x \in X$. Since

$$\omega(x) := \frac{1}{\varrho} \gamma(x), \ x \in X,$$

is linear and satisfies $\|x\| = \|\omega(x)\|$ for all $x \in X$, it must be orthogonal (see section 5 of chapter 1). Hence

$$\varphi(x) = \varrho \omega(x) + \varphi(0).$$

Thus f is a similitude.

d) Case: $f(\infty) = c \in X$. Define $g := \iota(c, 1) \cdot f$. Hence g is an M-transformation with $g(\infty) = \infty$ and thus a similitude according to step c). In view of (3.10) there exists a similitude α with

$$\iota(c, 1) = \alpha \iota \alpha^{-1}.$$

Hence $f = \iota(c, 1) \cdot g = \alpha \iota \cdot (\alpha^{-1} g) = \alpha \iota \beta$ with the similitude $\beta := \alpha^{-1} g$ $\qquad \square$

Proposition 4. *If $\alpha, \beta, \gamma, \delta$ are similitudes, then $\alpha \iota \beta = \gamma \iota \delta$ holds true if, and only if, $\alpha(0) = \gamma(0)$ and*

$$\delta(x) = \frac{1}{\varrho^2} \cdot \varphi \beta(x), \tag{3.22}$$

where $\varphi := \gamma^{-1} \alpha =: \varrho \omega$ with $0 < \varrho \in \mathbb{R}$ and $\omega \in O(X)$.

Proof. Put $\psi := \delta \beta^{-1}$. Then we have to prove that $\varphi \iota = \iota \psi$ holds true provided $\varphi(0) = 0$ and $\psi = \frac{1}{\varrho^2} \varphi$.

a) Assume $\varphi \iota = \iota \psi$. Applying this for $x \in \{0, \infty\}$, we obtain $\varphi(0) = 0$ and $\psi(0) = 0$. Since φ, ψ are similitudes leaving invariant 0, we get

$$\varphi =: \varrho \omega, \ \psi =: \sigma \pi$$

with $\omega, \pi \in O(X)$ and positive reals ϱ, σ, without loss of generality. If $x \notin \{0, \infty\}$, then, by $x^2 = (\pi(x))^2$,

$$\frac{\varrho}{x^2} \omega(x) = \varphi \iota(x) = \iota \psi(x) = \frac{1}{\sigma x^2} \pi(x),$$

i.e. $\varrho \sigma \omega(x) = \pi(x)$. Hence $(\varrho \sigma)^2 = 1$, i.e.

$$\psi = \sigma \pi = \varrho \sigma^2 \omega = \frac{1}{\varrho} \omega = \frac{1}{\varrho^2} \varphi,$$

i.e. (3.22).

b) Assume $\varphi(0) = 0$ and $\psi = \frac{1}{\varrho^2} \varphi$. Hence $\psi(0) = 0$ and thus $\varphi \iota(x) = \iota \psi(x)$ for $x \in \{0, \infty\}$. For $x \notin \{0, \infty\}$ we get

$$\iota \psi(x) = \iota \left(\frac{\varphi(x)}{\varrho^2} \right) = \varrho \frac{\omega(x)}{x^2} = \varphi \iota(x). \qquad \square$$

Remark. Suppose that α, β are similitudes. All similitudes γ, δ satisfying $\alpha \iota \beta = \gamma \iota \delta$ are then, by Proposition 4, determined as follows. Choose γ arbitrarily such that $\gamma(0) = \alpha(0)$ holds true and define δ to be the mapping (3.22). Of course, there are $\alpha \iota \beta$ which cannot be written in the form $\iota \delta$, for instance when $\alpha(0) \neq \mathrm{id}(0) = 0$. There are also $\alpha \iota \beta$ which are not of the form $\gamma \iota$, since $\alpha \iota \beta = \gamma \iota$ is equivalent to $\beta^{-1} \iota \alpha^{-1} = \iota \gamma^{-1}$. Clearly, $\alpha \iota \beta$ cannot be a similitude γ if α, β are similitudes, since otherwise $\iota = \alpha^{-1} \gamma \beta^{-1}$ would be a similitude, contradicting $\iota(\infty) \neq \infty$.

3.4   Involutions

A Möbius transformation f is called an *involution* provided $f^2 = \mathrm{id} \neq f$.

Theorem 5. *The similitude $f(x) = \varrho \omega(x) + a$ with $\varrho > 0$, $\omega \in O(X)$ and $a \in X$ is an involution if, and only if, $\varrho = 1$, $\omega(a) = -a$ and $\omega^2 = \mathrm{id} \neq \omega$. All involutions f of $\mathrm{M}(X)$ which are not similitudes are given by*

$$f = g \cdot \iota(c, \tau) \tag{3.23}$$

with $c \in X$, $\tau > 0$, $\varphi \in O(X)$ such that $\varphi^2 = \mathrm{id}$ and

$$g(x) = \varphi(x) + [c - \varphi(c)]. \tag{3.24}$$

Proof. a) Obviously, if $\varrho = 1$, $\omega(a) = -a$ and ω is an involution, then

$$f^2 = \mathrm{id} \neq f \tag{3.25}$$

for $f(x) = \varrho \omega(x) + a$. Assume now (3.25). Then

$$f^2(x) = \varrho^2 \omega^2(x) + [\varrho \omega(a) + a] = x \tag{3.26}$$

for all $x \in X$. For $x = 0$ we obtain $\varrho \omega (a) + a = 0$. Since $\varphi := \omega^2$ is in $O(X)$, we get $\varphi (x) \varphi (x) = x^2$, i.e. $\varrho = 1$, by $\varrho^2 \varphi (x) = x$ and $\varrho > 0$. Hence, by (3.26),

$$\omega^2(x) = x$$

for all $x \in X$ and $\omega (a) = -a$. If ω were equal to id, then

$$-a = \omega (a) = a,$$

i.e. $a = 0$, i.e. $f = \text{id}$, a contradiction.

b) A mapping (3.23) cannot be a similitude, because otherwise the inversion $\iota (c, \tau) = g^{-1} f$ would be a similitude. Observe $g^2 = \text{id}$ and $g \iota (c, \tau)(c) = \infty$, $g \iota (c, \tau)(\infty) = c$. For $x \in X \backslash \{c\}$ we get

$$g \iota (c, \tau)(x) = \tau^2 \varphi \left(\frac{x - c}{(x - c)^2} \right) + c = \iota (c, \tau) g (x)$$

and hence $[g \iota (c, \tau)]^2 = \text{id}$.

c) Suppose now that $f \in M(X)$ is an involution, but not a similitude. Then, by Theorem 3, $f = \alpha \iota \beta$ with suitable similitudes α, β. Because of $f^2 = \text{id}$, we get

$$\alpha \iota \beta = \beta^{-1} \iota \alpha^{-1}.$$

Hence, by Proposition 4, $\beta \alpha (0) = 0$ and $\alpha^{-1} = \frac{1}{\varrho^2} \beta \alpha \cdot \beta$ with $\beta \alpha =: \varrho \omega$, $\varrho > 0$ and $\omega \in O(X)$. Thus

$$\varrho^2 \, \text{id} = (\beta \alpha)^2 = \varrho^2 \omega^2,$$

i.e. $\omega^2 = \text{id}$, i.e.

$$f = \alpha \iota \beta = \alpha \iota \varrho \omega \alpha^{-1} \tag{3.27}$$

with $\varrho \in \mathbb{R}$, $\omega \in O(X)$ satisfying $\varrho > 0$, $\omega^2 = \text{id}$ and where ϱ also designates the similitude $\varrho (x) = \varrho \cdot x$, $x \in X$, $\varrho (\infty) = \infty$. Put

$$\alpha (x) =: \sigma \nu (x) + c$$

with $c \in X$, $\sigma > 0$ and $\nu \in O(X)$. Hence

$$\alpha^{-1}(x) = \frac{1}{\sigma} \nu^{-1}(x - c).$$

Observe $\iota \varrho \omega (0) = \infty$, $\iota \varrho \omega (\infty) = 0$ and

$$\iota \varrho \omega (x) = \frac{\omega (x)}{\varrho x^2}$$

for $0 \neq x \in X$, on account of $\omega (x) \omega (x) = xx$. This implies with (3.27) and $\varphi := \nu \omega \nu^{-1}$,

$$f (x) = \frac{\sigma^2}{\varrho} \frac{\varphi (x - c)}{(x - c)^2} + c \tag{3.28}$$

for $x \in X \backslash \{c\}$. With $g(x) := \varphi(x - c) + c$ we hence get for $x \neq c$ in X,

$$f(x) = \alpha \iota \varrho \omega \alpha^{-1}(x) = g\iota(c, \tau)(x), \tag{3.29}$$

where $\tau \in \mathbb{R}$ is defined by $\sqrt{\varrho} \cdot \tau = \sigma$. The formula (3.29) is also correct for $x \in \{c, \infty\}$, by observing $\alpha(0) = c$ and that $\iota \varrho \omega$ interchanges 0 and ∞. $\qquad \square$

Theorem 6. *If $s \in M(X)$, then $s\iota s^{-1} = \iota(c, \varrho)$ in the case*

$$s(B(0,1)) = B(c, \varrho), \ \varrho > 0.$$

Moreover, $s\iota s^{-1}$ is the inversion in $H'(a, \alpha)$ in the case

$$s(B(0,1)) = H'(a, \alpha).$$

If j is the inversion in the M-ball b and if $s \in M(X)$, then sjs^{-1} is the inversion in $s(b)$.

Proof. a) Observe $s\iota s^{-1} \neq \mathrm{id} = (s\iota s^{-1})^2$. Hence $s\iota s^{-1}$ is an involution. Applying again the definition
$$\mathrm{Fix}(\lambda) := \{x \in X' \mid \lambda(x) = x\}$$
for $\lambda \in M(X)$, we obtain $\mathrm{Fix}(s\iota s^{-1}) = s\big(\mathrm{Fix}(\iota)\big) = s\big(B(0,1)\big)$.

b) Case 1: $s(B(0,1)) = B(c, \varrho)$.
If $f := s\iota s^{-1}$ is a similitude, then, by Theorem 5,

$$f(x) = \omega(x) + a$$

with $a \in X$, $\omega \in O(X)$, $\omega \neq \mathrm{id} = \omega^2$, $\omega(a) = -a$. Hence

$$c + \frac{\varrho x}{\|x\|} = f\left(c + \frac{\varrho x}{\|x\|}\right) = \omega(c) + \frac{\varrho \omega(x)}{\|x\|} + a \tag{3.30}$$

for all $x \neq 0$ in X, by $\mathrm{Fix}(f) = B(c, \varrho)$ and the parametric representation of the M-ball $B(c, \varrho)$ (see section 2). If we apply (3.30) for x and $-x$, we obtain $\omega(x) = x$ for all $x \neq 0$ in X, a contradiction. So f cannot be a similitude. Hence, by Theorem 5,

$$f := s\iota s^{-1} = g \cdot \iota(a, \tau) \tag{3.31}$$

with $a \in X$, $\tau > 0$, $\varphi \in O(X)$ such that $\varphi^2 = \mathrm{id}$ and

$$g(x) = \varphi(x) + a - \varphi(a).$$

Observe $f(a) \neq a$, since otherwise $a = g\iota(a, \tau)(a) = \infty$. Hence

$$c + \frac{\varrho x}{\|x\|} = f\left(c + \frac{\varrho x}{\|x\|}\right) \tag{3.32}$$

for all $x \neq 0$ in X, by Fix $(f) = B(c, \varrho)$, implies

$$a \neq c + \frac{\varrho x}{\|x\|} \tag{3.33}$$

for all $x \in X \backslash \{0\}$. By (3.31), (3.32) and $b := c - a$, we obtain $0 \neq b + \frac{\varrho x}{\|x\|}$ and

$$b + \frac{\varrho x}{\|x\|} = \tau^2 \cdot \frac{\varphi(b)\|x\|^2 + \varrho \varphi(x)\|x\|}{(b\|x\| + \varrho x)^2} \tag{3.34}$$

for all $x \neq 0$ in X. Applying this equation for x and $-x$ in the case $x \neq 0$ and $xb = 0$ implies

$$x = \frac{\tau^2}{b^2 + \varrho^2} \varphi(x)$$

for $x = b$ and all $x \in X$ with $x \perp b$, i.e. $xb = 0$. Hence $\tau^2 = b^2 + \varrho^2$, by $x^2 = [\varphi(x)]^2$ and since there exist $x \neq 0$ in X with $xb = 0$. Thus $b = \varphi(b)$ and $x = \varphi(x)$ for all $x \perp b$. If $b = 0$, then $c = a$, $\tau = \varrho$ from $\tau^2 = b^2 + \varrho^2$, and $g = \mathrm{id}$ because of $x \perp b$ for all $x \in X$, i.e. because of $\varphi(x) = x$ for all $x \in X$. Hence $f = \iota(a, \tau) = \iota(c, \varrho)$ for $b = 0$ from (3.31). We now will prove that the case $b \neq 0$ does not occur. So assume $b \neq 0$. Observe

$$y = \frac{yb}{b^2} b + \left(y - \frac{yb}{b^2} b\right)$$

with $y - \frac{yb}{b^2} b \perp b$ for all $y \in X$. Hence

$$\varphi(y) = \frac{yb}{b^2} \varphi(b) + \varphi\left(y - \frac{yb}{b^2} b\right),$$

i.e. $\varphi(y) = y$ for all $y \in X$, in view of $\varphi(b) = b$ and $\varphi(x) = x$ for all $x \perp b$. Thus $g = \mathrm{id}$, i.e. $f = \iota(a, \tau)$, by (3.31). Now (3.34) and $\varphi = \mathrm{id}$ imply

$$\left(b + \frac{\varrho x}{\|x\|}\right)\left(1 - \frac{\tau^2}{\left(b + \frac{\varrho x}{\|x\|}\right)^2}\right) = 0,$$

i.e. $\left(b + \frac{\varrho x}{\|x\|}\right)^2 = \tau^2$, by (3.33), for all $x \neq 0$ in X. Hence $bx = 0$ for all $x \neq 0$ in X, by $b^2 + \varrho^2 = \tau^2$. Since $b \neq 0$, we get $b^2 = 0$ for $x = b$, a contradiction.

b) Case 2: $s(B(0,1)) = H'(a, \alpha)$.

Since $s(B(0,1)) = s(\mathrm{Fix}\, \iota) = \mathrm{Fix}(s\iota s^{-1})$, by step a), we get $f(\infty) = \infty$ for $f := s\iota s^{-1}$. Hence, by Theorem 5,

$$f(x) = \omega(x) + b$$

with $\omega \in O(X)$, $\omega \neq \mathrm{id} = \omega^2$, $\omega(b) = -b$. Without loss of generality we will assume $a^2 = 1$. From (3.14) we get

$$H(a, \alpha) = \alpha a + \left\{ \frac{v - (va)\,a}{\|v\| - va} \mid 0 \neq v \in X \text{ with } \frac{v}{\|v\|} \neq a \right\}.$$

Because of $\mathrm{Fix}(f) = H'(a, \alpha)$, we obtain

$$\alpha a + \frac{v - (va)\,a}{\|v\| - va} = b + \alpha \omega(a) + \frac{\omega(v) - (va)\,\omega(a)}{\|v\| - va}$$

for all $v \neq 0$ satisfying $v \neq a \cdot \|v\|$. Applying this for v and $-v$ for $0 \neq v \perp a$ yields

$$\alpha a = b + \alpha \omega(a) \text{ and } v = \omega(v) \qquad (3.35)$$

for all $v \neq 0$ with $v \perp a$. Since $x = (xa)\,a + [x - (xa)\,a]$ for all $x \in X$, we get from (3.35)

$$\omega(x) = (xa)\,\omega(a) + [x - (xa)\,a] \qquad (3.36)$$

for all $x \in X$, by observing $[x - (xa)\,a] \perp a$. Note $\omega(a) \neq a$ because otherwise $\omega = \mathrm{id}$ from (3.36). Hence (3.36) implies for $x = \omega(a)$, by $\omega^2 = \mathrm{id}$,

$$0 = \big(\omega(a) - a\big)\big(\omega(a)\,a + 1\big),$$

i.e. $\omega(a) \cdot a = -1$. Thus $\big(\omega(a)\,a\big)^2 = 1 = [\omega(a)]^2 \cdot a^2$ and $\omega(a)$, a must be linearly dependent by Lemma 1, chapter 1. Hence $\omega(a) = -a$ from $\omega(a) \neq a$. Thus

$$\omega(x) = x - 2(xa)\,a$$

from (3.36), and $b = 2\alpha a$ from (3.35). This implies that f is the inversion in $H'(a, \alpha)$ (see section 1).

d) Let j be the inversion in the M-ball b and take, by Proposition 1, $\mu \in \mathbb{M}(X)$ with $\mu\big(B(0, 1)\big) = b$. Then, according to steps b), c) of the proof of the present theorem, $j = \mu \iota \mu^{-1}$. Hence

$$sjs^{-1} = (s\mu)\,j\,(s\mu)^{-1}.$$

Again, by steps b), c), $(s\mu)\,j\,(s\mu)^{-1}$ must be the inversion in

$$(s\mu)\big(B(0, 1)\big) = s\big[\mu\big(B(0, 1)\big)\big] = s(b). \qquad \square$$

Remark. If b is an M-ball, the inversion in b will be denoted by inv_b. The last statement of Theorem 6 can hence be expressed as follows: If b is an M-ball and s an M-transformation, then

$$s \cdot \mathrm{inv}_b \cdot s^{-1} = \mathrm{inv}_{s(b)}. \qquad (3.37)$$

3.5 Orthogonality

Let a, b be M-balls. a will be called *orthogonal* to b, $a \perp b$, provided $a \neq b$ and $\mathrm{inv}_b(a) = a$.

Proposition 7. *Let a, b be M-balls with $a \perp b$. Then $\mu \in \mathbb{M}(X)$ implies $\mu(a) \perp \mu(b)$.*

Proof. From $\mathrm{inv}_{\mu(b)} = \mu \cdot \mathrm{inv}_b \cdot \mu^{-1}$, by (3.37), and $\mathrm{inv}_b(a) = a$, we obtain

$$\mathrm{inv}_{\mu(b)}\big(\mu(a)\big) = \mu \cdot \mathrm{inv}_b(a) = \mu(a). \qquad \square$$

Lemma 8. *If a, b are M-balls with $a \perp b$, then $a \cap b$ contains at least two distinct points.*

Proof. In view of Proposition 1 there exists $\mu \in \mathbb{M}(X)$ with $\mu(b) = B(0,1)$. Hence, by Proposition 7, $\mu(a) \perp B(0,1)$, and thus $\iota\big((\mu(a)\big) = \mu(a)$.

Case 1: $\mu(a) = B(c, \varrho)$.

Since the image of $\mu(a)$ under ι is $\mu(a)$, we obtain from the list of images of M-balls under ι (see section 1),

$$B(c, \varrho) = B\left(\frac{c}{c^2 - \varrho^2}, \frac{\varrho}{|c^2 - \varrho^2|}\right) \text{ with } c^2 \neq \varrho^2.$$

Proposition 11, chapter 2, yields

$$c = \frac{c}{c^2 - \varrho^2} \text{ and } \varrho = \frac{\varrho}{|c^2 - \varrho^2|}. \tag{3.38}$$

The second equation implies $|c^2 - \varrho^2| = 1$, because of $\varrho > 0$. Hence, if $c = 0$, then $\varrho = 1$, i.e. $\mu(b) = \mu(a)$. But $b \neq a$. Thus $c \neq 0$, i.e. $c^2 - \varrho^2 = 1$, by (3.38). Take $p \in X$ with $pc = 0$, $p^2 = 1$ and observe

$$\frac{c}{c^2} \pm \frac{\varrho}{\|c\|} p \in B(0,1) \cap B(c, \varrho) = \mu(b) \cap \mu(a).$$

Case 2: $\mu(a) = H'(p, \alpha)$ with $p^2 = 1$.

The list of images of M-balls under ι yields $\alpha = 0$, since $\iota\big(\mu(a)\big) = \mu(a)$. Take $q \in X$ with $pq = 0$, $q^2 = 1$. Then $\pm q \in \mu(a) \cap \mu(b)$. $\qquad \square$

Proposition 9. *If a, b are M-balls with $a \perp b$, then also $b \perp a$ holds true.*

Proof. Take r in $a \cap b$ and $\mu_1 \in \mathbb{M}(X)$ with $\mu_1(r) = \infty$ (see the beginning of section 3). Hence $\mu_1(a) \perp \mu_1(b)$, by Proposition 7, and

$$\mu_1(a) =: H'(p, \alpha), \; \mu_1(b) =: H'(q, \beta)$$

with $p^2 = 1$, $q^2 = 1$.

Because of Lemma 8 there exists a point $s \neq r$ on $a \perp b$. Define the similitude

$$\mu_2(x) := x - \mu_1(s)$$

and $\mu := \mu_2 \mu_1$. Hence $\mu(a) \perp \mu(b)$ and

$$\mu(a) = H'(p, 0), \ \mu(b) = H'(q, 0).$$

Thus $\mathrm{inv}_{\mu(b)}(\mu(a)) = \mu(a)$ and, by (3.7),

$$\mathrm{inv}_{\mu(b)}(x) = x - 2(qx)q, \ x \in X.$$

Consequently, $t \in p^\perp = H(p, 0)$ implies $\mathrm{inv}_{\mu(b)}(t) \in p^\perp$, i.e.

$$\left(t - 2(qt)q \right) p = 0,$$

i.e. $(qt)(qp) = 0$ for all $t \in p^\perp$. If $qt = 0$ would hold true for all $t \in p^\perp$, then $H(p, 0) \subseteq H(q, 0)$, contradicting $\mu(a) \neq \mu(b)$, by Proposition 12, chapter 2. Hence $pq = 0$, which implies

$$\mu(b) \perp \mu(a),$$

i.e. $b \perp a$, by Proposition 7.   □

Proposition 10.   1) $H'(t, \alpha) \perp H'(s, \beta) \Leftrightarrow ts = 0$.

2) $B(c, \varrho) \perp H'(t, \alpha) \Leftrightarrow tc = \alpha \Leftrightarrow c \in H(t, \alpha)$.

3) $B(c, \varrho) \perp B(d, \sigma) \Leftrightarrow (c - d)^2 = \varrho^2 + \sigma^2$.

Proof. 1) Put $a := H'(t, \alpha)$, $b := H'(s, \beta)$ and assume $ts = 0$, $t^2 = 1$, $s^2 = 1$. Because of

$$\mathrm{inv}_b(x) = x - 2(sx)s + 2\beta s, \tag{3.39}$$

we obtain $\mathrm{inv}_b(a) \subseteq a$, i.e. $\mathrm{inv}_b(a) = a$, by Proposition 12, chapter 2, i.e. $a \perp b$. Vice versa assume $a \perp b$. Since $a \neq b$ there exists $x_0 \in X$ with $tx_0 = \alpha$ and $sx_0 \neq \beta$, because otherwise $a \subseteq b$ would imply $a = b$. Hence, by (3.39) and $a \perp b$,

$$\alpha = t \cdot \mathrm{inv}_b(x_0) = t \cdot (x_0 - 2sx_0 \cdot s + 2\beta s),$$

i.e. $(\beta - sx_0) ts = 0$, i.e. $ts = 0$.

2) Put $a := B(c, \varrho)$, $b := H'(t, \alpha)$ and assume $a \perp b$, i.e. $b \perp a$, by Proposition 9, i.e.

$$\iota(c, \varrho)(b) = b.$$

Hence $c = \iota(c, \varrho)(\infty) \in b$, i.e. $tc = \alpha$. Vice versa assume $c \in H(t, \alpha)$. If $x \neq c, \infty$ is on b, then

$$\iota(c, \varrho)(x) = c + \varrho^2 \frac{x - c}{(x - c)^2}$$

is also on b, since $tx = \alpha$, i.e. $t(x - c) = 0$ yields

$$t \cdot [\iota(c, \varrho)(x) - c] = 0.$$

Now $\mathrm{inv}_a(b) \subseteq b$ implies $b = \mathrm{inv}_a(\mathrm{inv}_a(b)) \subseteq \mathrm{inv}_a(b)$, i.e. $b \perp a$.

3) Assume $a := B(c, \varrho) \perp B(d, \sigma) =: b$. If c were equal to d, then $a \cap b \neq \emptyset$ would imply $\varrho = \sigma$, i.e. $a = b$. Put $c - d =: p$, take $t \in X$ with $t^2 = 1$, $tp = 0$, and observe

$$x := c + \varrho t \in B(c, \varrho). \tag{3.40}$$

Hence $\mathrm{inv}_b(x) \in B(c, \varrho)$, by $a \perp b$, i.e.

$$\left(c - d - \sigma^2 \frac{c + \varrho t - d}{(c + \varrho t - d)^2} \right)^2 = \varrho^2, \tag{3.41}$$

in view of (3.8). Thus, by $tp = 0$, $|p^2 - \sigma^2| = \varrho^2$. Interchanging the role of a and b, we also obtain $|p^2 - \varrho^2| = \sigma^2$. Hence $p^2 = \varrho^2 + \sigma^2$. Vice versa assume $(c - d)^2 = \varrho^2 + \sigma^2$. This implies $p := c - d \neq 0$. Observe $d \notin B(c, \varrho)$, since otherwise $p^2 = \varrho^2$, i.e. $\sigma = 0$. The arbitrary point of $B(c, \varrho)$ is given by (3.40) with $t^2 = 1$. In order to prove $a \perp b$, we will show (3.41) for all $t \in X$ satisfying $t^2 = 1$, by noticing $p + \varrho t \neq 0$, since otherwise $p = -\varrho t$, i.e. $p^2 = \varrho^2$. In fact,

$$\left(p - \sigma^2 \frac{p + \varrho t}{(p + \varrho t)^2} \right)^2 - \varrho^2 = \sigma^2 - 2\sigma^2 \frac{p^2 + \varrho p t}{(p + \varrho t)^2} + \frac{\sigma^4}{(p + \varrho t)^2} = 0$$

because of $p^2 - \varrho^2 = \sigma^2$. □

Remark. If $B(c, \varrho) \perp B(d, \sigma)$, then $(x - c)(x - d) = 0$ for $x \in B(c, \varrho) \cap B(d, \sigma)$. This follows from

$$\varrho^2 + \sigma^2 = (c - d)^2 = \left((c - x) - (d - x) \right)^2 = \varrho^2 + \sigma^2 - 2(x - c)(x - d).$$

Remark. If $B(c, \varrho) \perp B(d, \sigma)$, then $(x - y)(c - d) = 0$ for $x, y \in B(c, \varrho) \cap B(d, \sigma)$. This follows from

$$(x - c)^2 = \varrho^2 = (y - c)^2, \quad (x - d)^2 = \sigma^2 = (y - d)^2,$$

i.e. from $-2xc + 2yc = y^2 - x^2 = -2xd + 2yd$.

Proposition 11. *Suppose that a is an M-ball and that $p \notin a$ is a point. Then $\{p, q := \mathrm{inv}_a(p)\}$ is the intersection I of all M-balls b satisfying $p \in b \perp a$.*

Proof. Without loss of generality we may assume $p = \infty$, and, by applying a suitable similitude, even $a = B(0, 1)$. The balls b with $\infty \in b \perp B(0, 1)$ are given by all $H'(t, 0)$, $t \neq 0$. Observe now $q = 0$ and

$$\{\infty, 0\} = \bigcap_{0 \neq t \in X} H'(t, 0) =: R$$

since $r \in X' \backslash \{\infty, 0\}$ implies $r \notin R$, in view of $r \notin H'(r, 0)$. □

Proposition 12.   a) *A line cuts $B(c, \varrho)$ in at most two distinct points.*

b) $H(a, \alpha) \not\subseteq B(c, \varrho) \not\subseteq H(b, \beta)$.

c) *If $p \in B(c, \varrho)$, there is exactly one hyperplane $H(a, \alpha)$ with $B(c, \varrho) \cap H(a, \alpha) = \{p\}$, namely $H\left(c - p, (c - p) p\right)$, the so-called tangent hyperplane of $B(c, \varrho)$ in p.*

Proof. a) If $p + \mathbb{R}v$, $v^2 = 1$, is a line, the points of intersection $p + \lambda v$ with $B(c, \varrho)$ are determined by

$$\varrho^2 = (p + \lambda v - c)^2 = (p - c)^2 + 2\lambda v (p - c) + \lambda^2.$$

There are hence at most two solutions.

b) $H(a, \alpha)$ contains a line, but not $B(c, \varrho)$. Both points

$$c \pm \varrho \cdot \frac{b}{\|b\|}$$

are in $B(c, \varrho)$, but they are not both in $H(b, \beta)$.

c) Let p be a point on the M-ball $B(c, \varrho)$. Hence $(p - c)^2 = \varrho^2$, i.e. $p \neq c$. If $H(a, \alpha)$ contains p, then $\alpha = ap$. Observe

$$2c - p - \frac{2a(c - p)}{a^2} a \in H(a, ap) \cap B(c, \varrho).$$

Asking for $\{p\} = H(a, ap) \cap B(c, \varrho)$, we obtain

$$2c - p - \frac{2a(c - p)}{a^2} a = p,$$

i.e. $c - p = \lambda a$ with a suitable real $\lambda \neq 0$, in view of $p \neq c$. Hence

$$H(a, ap) = H\left(c - p, (c - p) p\right).$$

Assuming now $q \in H\left(c - p, (c - p) p\right) \cap B(c, \varrho)$, we get

$$(c - p) q = (c - p) p \text{ and } (q - c)^2 = \varrho^2.$$

This implies $(c - p)(q - p) = 0$ and $\left((q - p) - (c - p)\right)^2 = \varrho^2$, i.e.

$$\varrho^2 = (q - p)^2 + (c - p)^2 = (q - p)^2 + \varrho^2,$$

i.e. $q = p$. $\square$

3.6 Möbius circles, M_N- and M^N-spheres

Proposition 13. *If p, q, r are three distinct points and also p', q', r', then there exists $f \in \mathbb{M}(X)$ with $p' = f(p)$, $q' = f(q)$, $r' = f(r)$.*

Proof. Case 1: $p = \infty = p'$.

If $s, t \in X$ satisfy $s^2 = t^2$, then there exists $\omega \in O(X)$ with $\omega(s) = t$. In the case $s \neq t$ take the mapping (1.7) with $a := s - t$,

$$\omega(x) = x - \frac{2(s-t)x}{(s-t)^2}(s-t),$$

and observe $\omega(s) = t$ by $2(s-t)s = s^2 - 2ts + t^2$. Now choose $\varrho > 0$ with

$$\varrho^2(q-r)^2 = (q'-r')^2$$

and put $s := \varrho(q-r)$, $t := q' - r'$. There hence exists $\omega \in O(X)$ with $q' - r' = \varrho\omega(q-r)$. Define now

$$f(x) := \varrho\omega(x-q) + q'.$$

If $\varrho(q-r) = s = t = q' - r'$, put $f(x) = \varrho(x-q) + q'$.

Case 2: $p = \infty \neq p'$.

Take $\tau \in \mathbb{M}(X)$ with $\tau(p') = \infty$ and put $\tau(p') =: p'_0$, $\tau(q') =: q'_0$, $\tau(r') = : r'_0$. Because of case 1, there exists $\varphi \in \mathbb{M}(X)$ with $p'_0 = \varphi(p)$, $q'_0 = \varphi(q)$, $r'_0 = \varphi(r)$. Now put $f := \tau^{-1}\varphi$.

Case 3: $p \neq \infty$.

Take $\sigma \in \mathbb{M}(X)$ with $\sigma(p) = \infty$ and put $\sigma(p) =: p_0$, $\sigma(q) =: q_0$, $\sigma(r) =: r_0$. Because of cases 1 or 2, there exist $\varphi \in \mathbb{M}(X)$ with $p' = \varphi(p_0)$, $q' = \varphi(q_0)$, $r' = \varphi(r_0)$. Now put $f := \varphi\sigma$. $\qquad\square$

If p, q, r are three distinct points, then the intersection of all M-balls containing p, q, r will be called a *Möbius circle*, or also an *M-circle*, of X.

Proposition 14. *If l is a fixed euclidean line of X, then $\mu(l \cup \{\infty\})$ is an M-circle for all $\mu \in \mathbb{M}(X)$. There are no other M-circles.*

Proof. If l is a (euclidean) line, we will designate the set $l \cup \{\infty\}$ by l'. Let $p \neq q$ be points of l and put $r = \infty$. All the M-balls through p, q, r are then given by all $H'(a, \alpha)$ with $p, q \in H(a, \alpha)$. In view of step a) of the proof of Theorem 3, we hence get that l' is the intersection of all M-balls through p, q, r. Let $\mu \in \mathbb{M}(X)$ be given. $\mu(l')$ must then be the intersection of all M-balls through $\mu(p)$, $\mu(q)$, $\mu(r)$, i.e. $\mu(l')$ must be an M-circle itself. If, finally, c is the M-circle through the pairwise distinct points p', q', r', we take $\mu \in \mathbb{M}(X)$, by Proposition 13, satisfying $p' = \mu(p)$, $q' = \mu(q)$, $r' = \mu(r)$. Hence $c = \mu(l')$. $\qquad\square$

Let n be a positive integer with $\dim X > n$. If $p \in X$ and if $s_1, \ldots, s_n \in X$ are linearly independent, then

$$\{p + \xi_1 s_1 + \cdots + \xi_n s_n \mid \xi_1, \ldots \xi_n \in \mathbb{R}\}$$

will be called an *n-plane* of X. An M_n-*sphere* is a set $P'_n = P_n \cup \{\infty\}$ where P_n is an n-plane or a set $P_{n+1} \cap B(c, \varrho)$ where P_{n+1} is an $(n+1)$-plane containing $c \in X$.

Instead of P'_n we also will speak of an extended n-plane.

Proposition 15. *All M-circles of X which are not of type l', l a line, are given by*

$$\{x \in P_2 \mid (x - c)^2 = \varrho^2\}, \tag{3.42}$$

where P_2 is an arbitrary 2-plane, $\varrho > 0$ a real number, and c a point in P_2.

Proof. Let P_2 be given by

$$P_2 = \{c + \xi r + \eta t \mid \xi, \eta \in \mathbb{R}\}$$

with $r^2 = 1 = t^2$ and $rt = 0$. Hence, by (3.42),

$$P_2 \cap B(c, \varrho) = \{c + \xi r + \eta t \mid \xi^2 + \eta^2 = \varrho^2\}, \tag{3.43}$$

or also

$$P_2 \cap B(c, \varrho) = \{c + \varrho r \cos \varphi + \varrho t \sin \varphi \mid 0 \leq \varphi < 2\pi\}. \tag{3.44}$$

The image of (3.43) under the inversion $\iota(c + \varrho r, \varrho)$ is, by (3.8), l' where l is the line through $a, a + t$ with $a := c + \frac{1}{2}\varrho r$. Hence, by Proposition 14, (3.43) is an M-circle. Finally we must show that every M-circle $\not\ni \infty$ is of type (3.42). If

$$\alpha(x) = \sigma \omega(x) + b, \ \sigma > 0, \ \omega \in O(X), \ b \in X, \tag{3.45}$$

is a similitude, then

$$B(c, \varrho) = \left\{c + \frac{\varrho x}{\|x\|} \,\Big|\, x \neq 0\right\}$$

(see section 2) implies

$$\alpha(B(c, \varrho)) = B(\alpha(c), \sigma\varrho).$$

Hence $\alpha(P_2 \cap B(c, \varrho)) = Q_2 \cap B(\alpha(c), \sigma\varrho)$ where Q_2 is the 2-plane $\alpha(P_2)$, must again be of type (3.42) since $\alpha(c) \in \alpha(P_2)$. We will refer to this result later on by (R).

In view of Proposition 14, let now $\mu(l')$ be an arbitrary M-circle which does not contain ∞. Then μ cannot be a similitude, since otherwise $\infty = \mu(\infty) \in \mu(l')$. Hence, by Theorem 3, $\mu = \alpha \iota \beta$ with similitudes α, β. From $\alpha \iota \beta(l') \not\ni \infty$ we get

$$\iota \beta(l') \not\ni \alpha^{-1}(\infty) = \infty,$$

i.e. $\iota(h') \not\ni \infty$ where $h' := \beta(l')$. Because of $h' \not\ni \iota^{-1}(\infty) = 0$, we may write

$$h = \{p + \lambda q \mid \lambda \in \mathbb{R}\}$$

with suitable elements $p \neq 0$ and q of X with $pq = 0$, $q^2 = 1$. If $\iota(h')$ is of type (3.42), then also, by (R),

$$\alpha[\iota(h')] = \alpha\iota\beta(l') = \mu(l').$$

But

$$\iota(h') = \{\xi p + \eta q \mid \xi, \eta \in \mathbb{R}\} \cap B\left(\frac{p}{2p^2}, \frac{1}{2\|p\|}\right),$$

since $\iota(h') = \{0\} \cup \left\{\frac{p + \lambda q}{p^2 + \lambda^2} \mid \lambda \in \mathbb{R}\right\}$. $\qquad\square$

Remark. As a consequence of Proposition 15 we obtain that the M-circles of X are exactly the M_1-spheres.

Proposition 16. *Let n be an integer satisfying $1 \leq n < \dim X$. If P_n is a fixed n-plane, then $\mu(P'_n)$ is an M_n-sphere for all $\mu \in \mathbb{M}(X)$. There are no other M_n-spheres.*

Proof. We already proved this result separately for the (for us) more important case $n = 1$ (see Proposition 14). If μ is a similitude, then $\mu(P'_n)$ is again an extended n-plane. So we have to show, by Theorem 3, that $\alpha\iota\beta(P'_n)$ is an M_n-sphere, where α, β are similitudes. Let $\beta(P'_n)$ be the extended n-plane Q'_n with

$$Q_n = \{q + \eta_1 t_1 + \cdots + \eta_n t_n \mid \eta_i \in \mathbb{R}\}$$

such that $t_i^2 = 1$ and $t_i t_j = 0$ for $i \neq j$. If $0 \in Q'_n$, then $\iota(Q'_n) = Q'_n$ and $\alpha\iota(Q'_n) = \mu(P'_n)$ is an extended n-plane. So assume $0 \notin Q'_n$, i.e. $\infty \notin \iota(Q'_n)$. Since $\infty \in Q'_n$, we get $0 \in \iota(Q'_n)$. We may write

$$q = \gamma_1 t_1 + \cdots + \gamma_n t_n + \gamma t$$

with $t^2 = 1$, $t_i t = 0$ and $\gamma \neq 0$, by $0 \notin Q'_n$. Hence

$$\iota(Q'_n) = \{0\} \cup \left\{\frac{\Sigma(\eta_i + \gamma_i)t_i + \gamma t}{\Sigma(\eta_i + \gamma_i)^2 + \gamma^2} \,\Big|\, \eta_i \in \mathbb{R}\right\}.$$

Thus $\iota(Q'_n) \subseteq Q_{n+1} := \{\Sigma\eta_i t_i + \eta t \mid \eta_i, \eta \in \mathbb{R}\}$. Moreover,

$$\iota(Q'_n) = Q_{n+1} \cap B\left(\frac{t}{2\gamma}, \frac{1}{2|\gamma|}\right),$$

i.e. $\mu(P'_n) = \alpha[\iota(Q'_n)] = \alpha(Q_{n+1}) \cap B\left(\alpha\left(\frac{t}{2\gamma}\right), \frac{\sigma}{2|\gamma|}\right)$ is an M_n-sphere, where α is defined by (3.45).

Let now Σ_n be an arbitrary M_n-sphere.

Case 1: Σ_n is an extended n-plane R'_n with

$$R_n = \{p + \xi_1 s_1 + \cdots + \xi_n s_n \mid \xi_i \in \mathbb{R}\}$$

such that $s_i^2 = 1$ and $s_i s_j = 0$ for $i \neq j$. Put

$$P_n = \{a + \lambda_1 v_1 + \cdots + \lambda_n v_n \mid \lambda_i \in \mathbb{R}\}$$

with $v_i^2 = 1$ and $v_i v_j = 0$ for $i \neq j$. Define

$$S = \{a, a + v_1, \ldots, a + v_n\}$$

and $f : S \to X$ by $f(a) = p$, $f(a + v_1) = p + s_1, \ldots, f(a + v_n) = p + s_n$. Applying Theorem 38 of chapter 2 for (X, eucl), we obtain that there exists $\varphi \in M(X, \text{eucl})$ with $f(x) = \varphi(x)$ for all $x \in S$. Hence

$$\varphi(x) = \omega(x) + d, \ x \in X,$$

with $\omega \in O(X)$ and a fixed $d \in X$. Thus

$$\varphi(a) = p \text{ and } \omega(v_i) = s_i,$$

i.e. $\varphi(a + \lambda_1 v_1 + \cdots + \lambda_n v_n) = \varphi(a) + \lambda_1 s_1 + \cdots + \lambda_n s_n$. Now observe that φ, extended by $\varphi(\infty) = \infty$, is in $M(X)$, and that $\varphi(P_n) = R_n$, i.e. $\varphi(P'_n) = R'_n$.

Case 2: $\Sigma_n = P_{n+1} \cap B(c, \varrho)$, $c \in P_{n+1}$.

Put $P_{n+1} = \{c + \xi_1 s_1 + \cdots + \xi_{n+1} s_{n+1} \mid \xi_i \in \mathbb{R}\}$ with $s_i^2 = 1$ and $s_i s_j = 0$ for $i \neq j$. Now observe

$$\iota(c + \varrho s_1, \varrho)(\Sigma_n) = Q'_n$$

with $Q_n = \{c + \frac{\varrho}{2} s_1 + \sum_{i=2}^{n+1} \eta_i s_i \mid \eta_i \in \mathbb{R}\}$. Because of case 1 there exists $\varphi \in M(X)$ with $\varphi(P'_n) = Q'_n$. Hence

$$\Sigma_n = \iota(c + \varrho s_1, \varrho)(Q'_n) = \iota(c + \varrho s_1, \varrho) \varphi(P'_n). \qquad \square$$

Remark. If Σ_n is a fixed M_n-sphere of X, then all M_n-spheres of X are given by $\mu(\Sigma_n)$, $\mu \in M(X)$. This follows immediately from Proposition 16.

Let n be a positive integer with $\dim X > n$. The points $p_1, \ldots p_{n+2}$ are called *spherically independent*, if, and only if, there is no M_{n-1}-sphere containing these points, where every subset T of X' with $\#T = 2$ is said to be an M_0-sphere.

Obviously, three distinct points are always spherically independent.

Proposition 17. *If n is a positive integer and $p_1, \ldots, p_{n+2}$ are spherically independent, there is exactly one M_n-sphere through these points.*

Proof. We may assume $p_{n+2} = \infty$, without loss of generality. Define

$$P := \{p_1 + \xi_1(p_2 - p_1) + \cdots + \xi_n(p_{n+1} - p_1) \mid \xi_i \in \mathbb{R}\}$$

and observe $p_1, \ldots, p_{n+1}, p_{n+2} \in P'$. If $p_2 - p_1, \ldots, p_{n+1} - p_1$ were linearly dependent, say,

$$p_{n+1} - p_1 = \beta_1(p_2 - p_1) + \cdots + \beta_{n-1}(p_n - p_1),$$

then $p_1, \ldots, p_{n+2} \in \{p_1 + \eta_1(p_2 - p_1) + \cdots + \eta_{n-1}(p_n - p_1) \mid \eta_i \in \mathbb{R}\} \cup \{\infty\}$ would be contained in an M_{n-1}-sphere. There hence exists an M_n-sphere, namely P', through $p_1, \ldots, p_{n+2}$. If $Q \cup \{\infty\}$ is another M_n-sphere through $p_1, \ldots, p_{n+2}$, then, obviously, $P \subseteq Q$, i.e. $P = Q$ since the underlying vector spaces of P and Q are both of dimension n. □

Remark. If $p_1, \ldots, p_{n+2}$ are spherically independent and also $q_1, \ldots, q_{n+2}$, there does not necessarily exist $\mu \in \mathbb{M}(X)$ with $\mu(p_i) = q_i$, $i = 1, \ldots, n + 2$. Assume $\dim X > 2$ and that $v, w \in X$ are linearly independent. Take

$$p_1 = \infty, \quad p_2 = 0, \quad p_3 = v, \quad p_4 = w,$$

$$q_1 = \infty, \quad q_2 = 0, \quad q_3 = v, \quad q_4 = 2w.$$

Observe that both quadruples are spherically independent. If there were $\mu \in \mathbb{M}(X)$ with $\mu(p_i) = q_i$, then $\mu \in O(X)$ which contradicts $\|w\| = \|w(w)\| = 2\|w\|$.

Let n be a positive integer and assume $\dim X > n$. Suppose that μ is a Möbius transformation and that

$$H(a_1, \alpha_1), \ldots, H(a_n, \alpha_n)$$

are hyperplanes such that $a_1, \ldots, a_n$ are linearly independent. Then

$$\mu\big(H'(a_1, \alpha_1) \cap \cdots \cap H'(a_n, \alpha_n)\big)$$

will be called an M^n-sphere of X.

Obviously, if X is finite-dimensional, $\dim X = N$, then every M^n-sphere is an M_{N-n}-sphere as well.

Remark. The M^1-spheres of X are exactly its M-balls.

Proposition 18. *Suppose* $\dim X > n$. *If* A, B *are* M^n-spheres *of* X, *there exists* $\mu \in \mathbb{M}(X)$ *with* $\mu(A) = B$.

Proof. According to the definition there exist representations

$$A = \mu_1\big(H'(a_1, \alpha_1) \cap \cdots \cap H'(a_n, \alpha_n)\big),$$

$$B = \mu_1\big(H'(b_1, \beta_1) \cap \cdots \cap H'(b_n, \beta_n)\big).$$

So we have to prove the existence of $\varphi \in \mathbb{M}(X)$ with

$$\bigcap_{\nu=1}^{n} H'(b_\nu, \beta_\nu) = \varphi \left(\bigcap_{\nu=1}^{n} H'(a_\nu, \alpha_\nu) \right). \tag{3.46}$$

Take

$$v_i = \sum_{\nu=1}^{n} \gamma_{i\nu} a_\nu, \ i = 1, \ldots, n,$$

with $v_i^2 = 1$ and $v_i v_j = 0$ for $i \neq j$, where $\gamma_{i\nu}$ are suitable reals. Then

$$\bigcap_{i=1}^{n} H(a_i, \alpha_i) = \bigcap_{i=1}^{n} H(v_i, \alpha_i')$$

with $\alpha_i' := \sum_{\nu=1}^{n} \gamma_{i\nu} \alpha_\nu$, $i = 1, \ldots, n$. So we may assume, without loss of generality,

$$a_i^2 = 1 = b_i^2 \text{ and } a_i a_j = 0 = b_i b_j \text{ for } i \neq j$$

in (3.46). Define $a := \Sigma \alpha_i a_i$, $b := \Sigma \beta_i b_i$ and observe

$$a \in \bigcap H(a_i, \alpha_i), \ b \in \bigcap H(b_i, \beta_i).$$

Define $S = \{a, a + a_1, \ldots, a + a_n\}$ and $f : S \to X$ by $f(a) = b$, $f(a + a_i) = b + b_i$. Applying again Theorem 38 of chapter 2, as in Proposition 16, we obtain a similitude $\varphi = (1, \omega, d)$ with $\varphi(a) = b$ and $\omega(a_i) = b_i$. It is easy to verify that φ satisfies (3.46): $a_i x = \alpha_i$ implies $a_i(x - a) = 0$, i.e. $\omega(a_i) \omega(x - a) = 0$. Hence $b_i(\varphi(x) - b) = 0$, i.e. $b_i \varphi(x) = \beta_i$. $\qquad \square$

Lemma 19. *Let c be an M-circle and b be an M-ball. Then $\#(c \cap b) \geq 3$ implies $c \subseteq b$.*

Proof. We may assume $\infty \in c \cap b$, without loss of generality. Hence $l := c \backslash \{\infty\}$ is a euclidean line and $b \backslash \{\infty\}$ a euclidean hyperplane H. But $\#(l \cap H) \geq 2$ implies $l \subseteq H$. $\qquad \square$

If c is an M-circle and b an M-ball, we will say that c is *orthogonal* to b, or b is *orthogonal* to c provided every M-ball through c is orthogonal to b. We shall write $c \perp b$ or $b \perp c$.

Proposition 20. *If the M-circle c is orthogonal to the M-ball b, then $\#(c \cap b) = 2$.*

Proof. $c \not\subseteq b$, because otherwise $b \perp b$. Hence, by Lemma 19, c and b have at most two points in common. Assume $c \cap b = \{p\}$ and, without loss of generality, $p = \infty$. Hence $c \backslash \{p\}$ is a line

$$l = \{r + \lambda v \mid \lambda \in \mathbb{R}\},$$

and $b\backslash\{p\}$ a hyperplane $H(t, \alpha)$. Since $l \cap H = \emptyset$, we get $tv = 0$. Hence $c = l'$ is contained in $H'(t, tr)$. Thus, by $c \perp b$,

$$H'(t, \alpha) = b \perp H'(t, tr),$$

which contradicts Proposition 10, statement 1. Assume, finally, $c \cap b = \emptyset$. We may write $b = H'(v, 0)$ and, by (3.44),

$$c = \{m + \varrho r \cos\varphi + \varrho t \sin\varphi \mid 0 \le \varphi < 2\pi\} \tag{3.47}$$

with $r^2 = 1 = t^2$ and $rt = 0$. Because of $c \cap b = \emptyset$, we obtain

$$(m + \varrho r \cos\varphi + \varrho t \sin\varphi) \cdot v \ne 0$$

for all $\varphi \in [0, 2\pi[$. Since there exists $\varphi \in [0, 2\pi[$ with

$$\cos\varphi \cdot \varrho r v + \sin\varphi \cdot \varrho t v = 0,$$

we obtain $m \cdot v \ne 0$. Hence, by Proposition 10, 2),

$$b = H'(v, 0) \not\perp B(m, \varrho),$$

which contradicts $c \subseteq B(m, \varrho)$. $\qquad\square$

Proposition 21. *If $p \ne q$ are points, and if b is an M-ball such that q is not the image of p under the inversion in b, then there exists exactly one M-circle c through p, q which is orthogonal to b.*

Proof. If at least one of the points p, q is on b, we may assume $q = \infty \in b$, i.e. $b =: H'(t, \alpha)$. Then $\{p + \lambda t \mid \lambda \in \mathbb{R}\} \cup \{\infty\}$ is the only M-circle through p, q and orthogonal to b. If $\{p, q\} \cap b = \emptyset$, we also may assume $q = \infty$. Hence b is in this case of the form $B(c, \varrho)$. Since p is assumed not to be the image of q under the inversion in b, we get $p \ne c$. Then

$$\{c + \lambda(p - c) \mid \lambda \in \mathbb{R}\} \cup \{\infty\}$$

is the only M-circle through p, q and orthogonal to b. $\qquad\square$

We already introduced in section 12, chapter 2, the notion of the cross ratio of an ordered quadruple z_1, z_2, z_3, z_4 of four distinct points of a euclidean line l. Now we will allow that the points are even elements of l', or more general, of an arbitrary M-circle.

Lemma 22. *Let c be an M-circle, and let p_1, p_2, p_3, p_4 be four distinct points on c. If q_1, q_2, q_3 are three distinct points, and μ_1, μ_2 M-transformations with*

$$\mu_i(p_j) = q_j \text{ for all } i \in \{1, 2\} \text{ and } j \in \{1, 2, 3\},$$

then $\mu_1(p_4) = \mu_2(p_4)$.

Proof. Take a fixed element $e \neq 0$ of X. Then, by Proposition 13, there exist $f, g \in \mathrm{M}(X)$ with

$$f(\infty) = p_1, \; f(0) = p_2, \; f(e) = p_3,$$

and $g(\infty) = q_1, \; g(0) = q_2, \; g(e) = q_3$. Put $g^{-1}\mu_i f =: \nu_i$ for $i \in \{1, 2\}$. Hence

$$\nu_i(\infty) = \infty, \; \nu_i(0) = 0, \; \nu_i(e) = e \qquad (3.48)$$

for $i \in \{1, 2\}$, and ν_i must be a similitude

$$\nu_i(x) = \sigma_i \omega_i(x) + a_i, \; \sigma_i > 0, \; i \in \{1, 2\},$$

in view of Theorem 3. Thus, by (3.48), $\nu_i = \omega_i$, $i \in \{1, 2\}$. Since p_1, p_2, p_3, p_4 are on a common M-circle, so must be their images under f^{-1}. Hence $f^{-1}(p_4) = \alpha e$ with a suitable real $\alpha \notin \{1, 2\}$. Thus

$$\nu_i f^{-1}(p_4) = \nu_i(\alpha e) = \omega_i(\alpha e) = \alpha \omega_i(e) = \alpha e,$$

i.e. $\mu_1(p_4) = \mu_2(p_4)$. □

If p_1, p_2, p_3, p_4 is an ordered quadruple consisting of four distinct points on a common M-circle, we define its *cross ratio*. Take arbitrarily $e \neq 0$ in X and take arbitrarily $\mu \in \mathrm{M}(X)$ with $\mu(p_1) = \infty$, $\mu(p_2) = 0$, $\mu(p_3) = e$. If then $\mu(p_4) = \alpha e$, $\alpha \in \mathbb{R}$, put

$$\{p_1, p_2; p_3, p_4\} := \alpha.$$

Since $\mu(p_1)$, $\mu(p_2)$, $\mu(p_3)$, $\mu(p_4)$ must be on a common M-circle, there exists in fact $\alpha \in \mathbb{R}$ with $\mu(p_4) = \alpha e$. Obviously, $\alpha \neq 0$ and $\alpha \neq 1$. Moreover, $\{p_1, p_2; p_3, p_4\}$ does not depend on the chosen e and μ, as we would like to verify now.

Lemma 23. *If $e \neq 0$ and $k \neq 0$ are elements of X, and if $\mu \in \mathrm{M}(X)$ satisfies $\mu(\infty) = \infty$, $\mu(0) = 0$, $\mu(e) = k$, then $\mu(\alpha e) = \alpha k$ for all $\alpha \in \mathbb{R}$.*

Proof. $\mu(\infty) = \infty$, $\mu(0) = 0$ imply, by Theorem 3, that μ is a similitude of the form $\mu(x) = \sigma \omega(x)$, $\sigma > 0$, $\omega \in O(X)$. Hence

$$\mu(\alpha e) = \sigma \omega(\alpha e) = \sigma \alpha \omega(e) = \alpha k.$$ □

In order to show that $\{p_1, p_2; p_3, p_4\}$ does not depend on e and μ, take $e_1 \neq 0$ in X and $\mu_1 \in \mathrm{M}(X)$ with

$$\mu_1(p_1) = \infty, \; \mu_1(p_2) = 0, \; \mu_1(p_3) = e_1. \qquad (3.49)$$

We then have to show $\mu_1(p_4) = \alpha e_1$. In view of Proposition 13 there exists $\tau \in \mathrm{M}(X)$ with $\tau(\infty) = \infty$, $\tau(0) = 0$, $\tau(e) = e_1$. Lemma 23 yields $\tau(\alpha e) = \alpha e_1$. Since

$$\tau\mu(p_1) = \infty, \; \tau\mu(p_2) = 0, \; \tau\mu(p_3) = e_1$$

holds true, we get, by Lemma 22 and (3.49),

$$\mu_1(p_4) = \tau\mu(p_4),$$

i.e. $\mu_1(p_4) = \tau(\alpha e) = \alpha e_1$. □

Proposition 24. *Given four distinct points*

$$p_i = r + \alpha_i v, \ i \in \{1, 2, 3, 4\},$$

on the line $l = r + \mathbb{R}v$, $v \neq 0$, then

$$\{p_1, p_2; p_3, p_4\} = \frac{\alpha_1 - \alpha_3}{\alpha_1 - \alpha_4} : \frac{\alpha_2 - \alpha_3}{\alpha_2 - \alpha_4}. \tag{3.50}$$

Moreover,

$$\{\infty, p_2; p_3, p_4\} = \frac{\alpha_2 - \alpha_4}{\alpha_2 - \alpha_3}, \quad \{p_1, \infty; p_3, p_4\} = \frac{\alpha_1 - \alpha_3}{\alpha_1 - \alpha_4},$$

$$\{p_1, p_2; \infty, p_4\} = \frac{\alpha_2 - \alpha_4}{\alpha_1 - \alpha_4}, \quad \{p_1, p_2; p_3, \infty\} = \frac{\alpha_1 - \alpha_3}{\alpha_2 - \alpha_3}.$$

If $p_i = m + \varrho r \cos \varphi_i + \varrho t \sin \varphi_i$, $i \in \{1, 2, 3, 4\}$ are four distinct points on (3.47) with $0 \leq \varphi_i < 2\pi$, then

$$\{p_1, p_2; p_3, p_4\} = \frac{\sin \frac{1}{2}(\varphi_1 - \varphi_3)}{\sin \frac{1}{2}(\varphi_1 - \varphi_4)} : \frac{\sin \frac{1}{2}(\varphi_2 - \varphi_3)}{\sin \frac{1}{2}(\varphi_2 - \varphi_4)}. \tag{3.51}$$

Proof. Given four distinct points $p_1, p_2, p_3, p_4 \in X$, we define $\varphi \in \mathbb{M}(X)$ by

$$\varphi(x) = \iota(p_1, 1)(x) - \iota(p_1, 1)(p_2).$$

Put $e := \varphi(p_3)$. Hence $\varphi(p_1) = \infty$, $\varphi(p_2) = 0$, $\varphi(p_3) = e$. If

$$p_i = r + \alpha_i v,$$

then $\varphi(p_4) = \{p_1, p_2; p_3, p_4\} \cdot e$. This implies (3.50). If

$$p_i = m + \varrho r \cos \varphi_i + \varrho t \sin \varphi_i,$$

$0 \leq \varphi_i < 2\pi$, then $\varphi(p_4) = \{p_1, p_2; p_3, p_4\} \cdot e$ yields (3.51). $\square$

Proposition 25. *Given an M-circle c and four distinct points p_1, p_2, p_3, p_4 on c. Then*

$$\{p_1, p_2; p_3, p_4\} = \{p_4, p_3; p_2, p_1\},$$

$$\{p_1, p_2; p_3, p_4\} = \{p_2, p_1; p_4, p_3\},$$

$$\{p_1, p_2; p_3, p_4\} \cdot \{p_1, p_2; p_4, p_3\} = 1,$$

$$\{p_1, p_2; p_3, p_4\} + \{p_1, p_3; p_2, p_4\} = 1.$$

Proof. Apply Proposition 24. $\square$

Proposition 26. *If $\mu \in \mathbb{M}(X)$ and if p_1, p_2, p_3, p_4 are four distinct points on a common M-circle, then*

$$\{\mu(p_1), \mu(p_2); \mu(p_3), \mu(p_4)\} = \{p_1, p_2; p_3, p_4\}.$$

Proof. Take $e \neq 0$ in X and $\varphi \in \mathbb{M}(X)$ with $\varphi(p_1) = \infty$, $\varphi(p_2) = 0$, $\varphi(p_3) = e$. Hence $\varphi(p_4) = \{p_1, p_2; p_3, p_4\} \cdot e$. Moreover with $\psi := \varphi\mu^{-1}$,

$$\psi\left(\mu(p_1)\right) = \infty, \ \psi\left(\mu(p_2)\right) = 0, \ \psi\left(\mu(p_3)\right) = e.$$

Thus $\{\mu(p_1), \mu(p_2); \mu(p_3), \mu(p_4)\} \cdot e = \psi\left(\mu(p_4)\right) = \varphi(p_4)$.   □

Let $\Gamma_4(X)$ be the set of all ordered quadruples (p_1, p_2, p_3, p_4) consisting of four distinct elements of X' which are on a common M-circle. We are then interested in the following problem. Find all functions $f : \Gamma_4 \to \mathbb{R}$ such that

$$f(p_1, p_2, p_3, p_4) = f\left(\mu(p_1), \mu(p_2), \mu(p_3), \mu(p_4)\right) \qquad (3.52)$$

holds true for all $(p_1, p_2, p_3, p_4) \in \Gamma_4(X)$ and all $\mu \in \mathbb{M}(X)$.

Theorem 27. *All solutions of the functional equation (3.52) are given as follows. Take an arbitrary function $\varphi : \mathbb{R} \backslash \{0, 1\} \to \mathbb{R}$ and put*

$$f(p_1, p_2, p_3, p_4) = \varphi\left(\{p_1, p_2; p_3, p_4\}\right) \qquad (3.53)$$

for all $(p_1, p_2, p_3, p_4) \in \Gamma_4(X)$.

Proof. In view of Proposition 26, (3.53) is a solution of (3.52). Let now vice versa $f : \Gamma_4(X) \to \mathbb{R}$ solve (3.52). Take a fixed $e \neq 0$ in X and define

$$\varphi(\alpha) := f(\infty, 0, e, \alpha e)$$

for every real $\alpha \notin \{0, 1\}$. For four distinct points p_1, p_2, p_3, p_4 on a common M-circle we hence get for suitable $\mu \in \mathbb{M}(X)$ with

$$\mu(p_1) = \infty, \ \mu(p_2) = 0, \ \mu(p_3) = e,$$

by (3.52), $f(p_1, p_2, p_3, p_4) = f(\infty, 0, e, \alpha e) = \varphi(\alpha)$ with $\alpha = \{p_1, p_2; p_3, p_4\}$.   □

3.7   Stereographic projection

Besides our real inner product space X, dim ≥ 2, we also will consider the real vector space

$$Y := X \oplus \mathbb{R},$$

equipped with the inner product

$$(a, \alpha) \cdot (b, \beta) := ab + \alpha\beta \qquad (3.54)$$

for $(a, \alpha), (b, \beta) \in Y$. Obviously, Y itself is a real inner product space under the scalar product (3.54). We hence may apply to Y everything we developed for X. We are now interested in the unit ball of Y, namely in

$$U := \{(x, \xi) \in Y \mid (x, \xi)^2 = 1\}.$$

Call $N := (0, 1)$ the North Pole of U, and put $U_0 := U \backslash \{N\}$. We identify $x \in X$ with $(x, 0) \in Y$, so that X is a proper subset of Y. The *stereographic projection* ψ associates to $(x, \xi) \in U_0$ the point of intersection of the line

$$N + \mathbb{R}\big((x, \xi) - N\big) \tag{3.55}$$

of Y, with X, and it associates to N the point ∞.

Proposition 28. *The stereographic projection* $\psi : U \to X'$ *is a bijection.*

Proof. There is only one point $(x, 1)$, $x \in X$, in U, namely N. This follows from $x^2 + 1 = 1$. Hence $(x, \xi) \in U_0$ implies $\xi \neq 1$. The line (3.55) cuts X exactly in

$$\psi\,(x, \xi) = \frac{x}{1 - \xi}, \tag{3.56}$$

where $x^2 + \xi^2 = 1$, in view of $\xi \neq 1$. If $z \in X$, then

$$\left(\frac{2z}{z^2 + 1}, \frac{z^2 - 1}{z^2 + 1} \right) \tag{3.57}$$

is the only point p on U_0 with $\psi\,(p) = z$. $\square$

Also in the case of the space Y we distinguish between hyperplanes and quasi-hyperplanes of Y. The hyperplanes are given by all

$$H\,\big((a, \alpha),\, \gamma\big) = \{(x, \xi) \in Y \mid ax + \alpha\xi = \gamma\}$$

with $(a, \alpha) \neq (0, 0) =: 0$.

We will call a hyperplane H of Y a *suitable hyperplane* of Y, if it cuts U in more than one point.

Lemma 29. *The hyperplane* $H\,\big((b, \beta), \alpha\big)$ *of* Y *is suitable if, and only if,* $\alpha^2 < b^2 + \beta^2$.

Proof. Assume $\alpha^2 < b^2 + \beta^2$ and take $(c, \gamma) \perp (b, \beta)$ with $(c, \gamma)^2 = 1$. Then the elements of Y,

$$\alpha \frac{(b, \beta)}{(b, \beta)^2} \pm \lambda\,(c, \gamma), \quad \lambda := \sqrt{\frac{b^2 + \beta^2 - \alpha^2}{(b, \beta)^2}},$$

are both in $H \cap U$. Suppose now, vice versa, $\#(H \cap U) > 1$ and $(x_0, \xi_0) \in H \cap U$. We obtain

$$[(b, \beta)(x_0, \xi_0)]^2 \leq (b, \beta)^2 (x_0, \xi_0)^2,$$

by the inequality of Cauchy–Schwarz, i.e. $\alpha^2 \leq (b, \beta)^2 = b^2 + \beta^2$, since $(x_0, \xi_0) \in U$ implies $(x_0, \xi_0)^2 = 1$, and since $(x_0, \xi_0) \in H$, obviously, $(b, \beta)(x_0, \xi_0) = \alpha$. We must exclude $\alpha^2 = (b, \beta)^2$. But

$$[(b, \beta)(x_0, \xi_0)]^2 = \alpha^2 = (b, \beta)^2 (x_0, \xi_0)^2$$

implies, by Lemma 1, chapter 1, $(b, \beta) = \lambda (x_0, \xi_0)$, i.e. $\lambda = \alpha$ from

$$(b, \beta)(x_o, \xi_0) = \alpha.$$

The tangent hyperplane of $U = B((0,0),1)$ in $(x_0, \xi_0) \in U$ is hence, by Proposition 12, c),

$$H\left(-(x_0, \xi_0), -(x_0, \xi_0)^2\right),$$

i.e. $H((b, \beta), \alpha)$. Thus $\#(H \cap U) = 1$, contradicting $\#(H \cap U) > 1$. □

Theorem 30. *If H is a suitable hyperplane of Y, then $\psi(H \cap U)$ is an M-ball of X. If b is an M-ball of X, then there exists a uniquely determined suitable hyperplane H of Y with $b = \psi(H \cap U)$.*

Proof. Let $H((b, \beta), \alpha)$ be a suitable hyperplane. The points (y, η) on $H \cap U$ are given by the equations

$$by + \beta\eta = \alpha, \ y^2 + \eta^2 = 1. \tag{3.58}$$

Assume $\eta \neq 1$ and put $x := \psi(y, \eta)$. Hence, by (3.56),

$$bx(1 - \eta) + \beta\eta = \alpha, \ x^2(1 - \eta)^2 + \eta^2 = 1,$$

i.e. $(\beta - \alpha) x^2 + 2bx = \beta + \alpha$. From (3.58) we get $N \in H$ if, and only if, $\alpha = \beta$. In this case $b^2 > 0$, i.e. $b \neq 0$, holds true, by Lemma 29. We hence obtain

$$bx = \alpha$$

for $(y, \eta) \in H \cap U \backslash \{N\}$, $N \in H$, and thus

$$\psi(H \cap U) \subseteq H(b, \alpha) \cup \{\infty\}.$$

If $x \in H(b, \alpha) \cup \{\infty\}$, then $\psi^{-1}(x) = N$ for $x = \infty$, and otherwise, by (3.57)

$$\psi^{-1}(x) = \left(\frac{2x}{x^2 + 1}, \frac{x^2 - 1}{x^2 + 1} \right),$$

i.e. $\psi^{-1}(x) \in H((b, \alpha), \alpha)$. Hence $\psi(H \cap U) = H'(b, \alpha)$. In the case $\alpha \neq \beta$ we get

$$\psi\left(H((b, \beta), \alpha) \cap U\right) \subseteq B\left(\frac{b}{\alpha - \beta}, \frac{\sqrt{b^2 + \beta^2 - \alpha^2}}{|\alpha - \beta|} \right) \tag{3.59}$$

from $(\beta - \alpha) x^2 + 2bx = \beta + \alpha$ and $x = \psi(y, \eta)$. If x is an element of the right-hand side of (3.59), then $\psi^{-1}(x)$ is on $H((b, \beta), \alpha)$. Hence equality holds true in (3.59). If the M-ball $H'(b, \alpha)$ is given, we already know that its inverse image is given by $U \cap H((b, \alpha), \alpha)$. What is the inverse image of $B(m, \varrho)$? Since equality holds true in (3.59), we must solve

$$m = \frac{b}{\alpha - \beta}, \ \varrho^2 = \left(\frac{b}{\alpha - \beta} \right)^2 + \frac{\beta + \alpha}{\beta - \alpha}$$

with respect to b, β, α, by observing $\psi(N) \notin B(m, \varrho)$, i.e. $\alpha \neq \beta$. Instead of b, β, α we determine

$$\frac{b}{\alpha - \beta} = m, \quad \frac{\beta}{\alpha - \beta} = \frac{\alpha}{\alpha - \beta} - 1, \quad \frac{\alpha}{\alpha - \beta} = \frac{1 + m^2 - \varrho^2}{2}.$$

The inverse image of $B(m, \varrho)$ is hence

$$U \cap H\left((2m, m^2 - \varrho^2 - 1), m^2 - \varrho^2 + 1\right). \qquad \square$$

3.8 Poincaré's model of hyperbolic geometry

Let b be an M-ball. We will define the *two sides* of b.

Case 1: $b = H'(a, \alpha)$.

The two sides of b are here

$$\{x \in X \mid ax > \alpha\} \text{ and } \{x \in X \mid ax < \alpha\}.$$

Observe $X' = b_1 \cup H'(a, \alpha) \cup b_2$, where b_1, b_2 are the two sides of b, and, moreover,

$$b_1 \cap b = b \cap b_2 = b_2 \cap b_1 = \emptyset. \tag{3.60}$$

Case 2: $b = B(c, \varrho)$.

The two sides are defined by

$$\{x \in X \mid (x - c)^2 > \varrho^2\} \cup \{\infty\} \text{ and } \{x \in X \mid (x - c)^2 < \varrho^2\}.$$

Here we have $X' = b_1 \cup b \cup b_2$, too, and (3.60) for the sides of b.

Proposition 31. *If b is an M-ball, μ an M-transformation, and if b_1, b_2 are the two sides of b, then $\mu(b_1), \mu(b_2)$ are the two sides of the M-ball $\mu(b)$.*

Proof. Case 1. $b = H'(a, \alpha)$ and μ is a similitude.

If $\mu(x) = \sigma w(x) + t$, $\sigma > 0$, then $\mu(b) = H'\left(w(a), \alpha\sigma + tw(a)\right)$. Hence

$$\mu(\{x \in X \mid ax > \alpha\}) = \{y \in X \mid w(a)\, y > \alpha\sigma + tw(a)\},$$
$$\mu(\{x \in X \mid ax < \alpha\}) = \{y \in X \mid w(a)\, y < \alpha\sigma + tw(a)\},$$

since, for instance, $ax > \alpha$ is equivalent with $w(a)\, w(x) > \alpha$, i.e. with

$$w(a)\left(\frac{\mu(x) - t}{\sigma}\right) > \alpha.$$

Case 2. $b = B(c, \varrho)$ and μ is a similitude, say, of the form as above.

Here we get $\mu\left(b\right) = B\left(\mu\left(c\right), \sigma\varrho\right)$ and

$$
\begin{aligned}
\mu\left(\{x \in X \mid (c-x)^2 > \varrho^2\} \cup \{\infty\}\right) &= \{y \in X \mid \left(\mu\left(c\right) - y\right)^2 > (\varrho\sigma)^2\} \cup \{\infty\}, \\
\mu\left(\{x \in X \mid (c-x)^2 < \varrho^2\}\right) &= \{y \in X \mid \left(\mu\left(c\right) - y\right)^2 < (\varrho\sigma)^2\},
\end{aligned}
$$

since $(c-x)^2 > \varrho^2$ is equivalent with $\left(w\left(c\right) - w\left(x\right)\right)^2 = (c-x)^2 > \varrho^2$, i.e. with

$$
\left([\sigma w\left(c\right) + t] - [\sigma w\left(x\right) + t]\right)^2 > (\sigma\varrho)^2.
$$

Case 3. $b = H'(a, \alpha)$ and $\mu = \iota$.

From section 1 we know

$$
\iota\left(H'(a, \alpha)\right) = B\left(\frac{a}{2\alpha}, \left\|\frac{a}{2\alpha}\right\|\right)
$$

for $\alpha \neq 0$. We may assume $\alpha > 0$, because otherwise we would work with $H\left(-a, -\alpha\right)$. If $x \in X$ satisfies $ax > \alpha$, we get $x \neq 0$, i.e. $\iota\left(x\right) \in X$, and

$$
\left(\iota\left(x\right) - \frac{a}{2\alpha}\right)^2 < \frac{a^2}{4\alpha^2}.
$$

Vice versa,

$$
\left(y - \frac{a}{2\alpha}\right)^2 < \frac{a^2}{4\alpha^2}, \, y \in X,
$$

implies $y \neq 0$ and $a \cdot \iota\left(y\right) > \alpha$.

In the case $\alpha = 0$, we know from section 1 that

$$
\iota\left(H'(a, 0)\right) = H'(a, 0).
$$

Here, of course, $ax > 0$ is equivalent with $a\iota\left(x\right) > 0$ for $x \in X$.

Case 4. $b = B\left(c, \varrho\right)$ and $\mu = \iota$.

In the case $c^2 = \varrho^2$ we get $\iota\left(B\right) = H'(2c, 1)$. Since ι is involutorial, we also get $\iota\left(H'\right) = B$ and we may apply part 1 of Case 3. So assume $c^2 \neq \varrho^2$. From section 1 we get

$$
\iota\left(B\left(c, \varrho\right)\right) = B\left(\frac{c}{c^2 - \varrho^2}, \frac{\varrho}{|c^2 - \varrho^2|}\right).
$$

If $c^2 > \varrho^2$, then $(c-x)^2 < \varrho^2$ is equivalent with

$$
\left(\frac{c}{c^2 - \varrho^2} - \iota\left(x\right)\right)^2 < \frac{\varrho^2}{(c^2 - \varrho^2)}
$$

for all $x \in X \backslash \{0\}$, and if $c^2 > \varrho^2$, then the image of

$$
\{x \in X \mid (c-x)^2 > \varrho^2\} \cup \{\infty\}
$$

under ι is

$$\left\{ y \in X \mid \left(\frac{c}{c^2 - \varrho^2} - y \right)^2 < \left(\frac{\varrho}{c^2 - \varrho^2} \right)^2 \right\}.$$

Because of Theorem 3 no other cases need to be considered. □

Let B be an M-ball, and let Σ be one of its sides. Of course, B could be, for instance, the M-ball $B(c, \varrho)$ and Σ the side

$$\{x \in X \mid (c - x)^2 > \varrho\} \cup \{\infty\}$$

of B. We now would like to define the *hyperbolic geometry* (B, Σ), or (in shorter form) the *hyperbolic geometry* Σ. This will be the geometry (see section 9, chapter 1)

$$(\Sigma, G(\Sigma)),$$

where the group $G(\Sigma)$ is defined as follows. Take the subgroup

$$\Gamma(\Sigma) := \{\mu \in \mathbb{M}(X) \mid \mu(B) = B \text{ and } \mu(\Sigma) = \Sigma\}$$

of $\mathbb{M}(X)$ and put $G(\Sigma) := \{\mu|\Sigma \mid \mu \in \Gamma(\Sigma)\}$. The points of Σ are called *hyperbolic points* (*h*-points) of Σ. If c is an M-circle orthogonal to B, then $c \cap \Sigma$ is said to be a *hyperbolic line* (*h*-line) of Σ. The elements of $G(\Sigma)$ are called *hyperbolic transformations* of Σ.

Lemma 32. *Given $\mu_1, \mu_2 \in \Gamma(\Sigma)$ with $\mu_1|\Sigma = \mu_2|\Sigma$, then $\mu_1 = \mu_2$.*

Proof. 1) *If b is an M-ball and $p \notin b$ a point, then $p, \mathrm{inv}_b(p)$ are on different sides of b.*

Because of Proposition 31 we may assume $p = \infty$, i.e. $b = B(c_0, \varrho_0)$. Obviously, ∞, c_0 are on different sides of b.

2) Take $p \in \Sigma'$, where Σ, Σ' are the two sides of $B(c, \varrho)$. Define $q := \mathrm{inv}_B(p)$ and observe

$$\{p, q\} = \bigcap_{q \in b \perp B} b,$$

in view of Proposition 11. Moreover, we obtain $q \in \Sigma$, by 1). Now

$$\{\mu_i(p), \mu_i(q)\}, \ i \in \{1, 2\},$$

is the intersection of all M-balls $b \ni \mu_i(q)$ satisfying $b \perp \mu_i(B) = B$. Since $\mu_1(q) = \mu_2(q)$, by $q \in \Sigma$, we obtain

$$\{\mu_1(p), \mu_1(q)\} = \{\mu_1(p), \mu_2(q)\},$$

i.e. $\mu_1(p) = \mu_2(p)$. Hence $\mu_1 = \mu_2$. □

Remark. Because of Lemma 32 we may and we will identify $\Gamma(\Sigma)$ and $G(\Sigma)$.

Proposition 33. *Let also B' be an M-ball, and let Σ' be one of its sides. Then the two geometries $\left(\Sigma, G\left(\Sigma\right)\right)$ and $\left(\Sigma', G\left(\Sigma'\right)\right)$ are isomorphic.*

Proof. By Proposition 1 there exists $\mu_0 \in \mathbb{M}\left(X\right)$ with $\mu_0(B) = B'$. From Proposition 31 we get $\Sigma' \in \{\mu_0(\Sigma), \mu_0(\Sigma_1)\}$, where Σ, Σ' are the two sides of B. In the case $\Sigma' = \mu_0(\Sigma)$ put $\mu := \mu_0$, and otherwise $\mu := \alpha\mu_0$, where α denotes the inversion in B'. Hence

$$\mu\left(B\right) = B', \, \mu\left(\Sigma\right) = \Sigma'.$$

Define $\sigma : \Sigma \to \Sigma'$ by $\sigma\left(x\right) := \mu\left(x\right)$ for all $x \in \Sigma$, and

$$\tau : G\left(\Sigma\right) \to G\left(\Sigma'\right)$$

by $\tau\left(g\right) := \mu g \mu^{-1}$ for all $g \in G\left(\Sigma\right)$. Since the equations (1.15) hold true, we get $\left(\Sigma, G\left(\Sigma\right)\right) \cong \left(\Sigma', G\left(\Sigma'\right)\right)$. $\qquad\square$

Remark. In Theorem 37 we will show that the geometry $\left(\Sigma, G\left(\Sigma\right)\right)$, based on X, is isomorphic to the hyperbolic geometry over X (see section 10, chapter 1). $\left(\Sigma, G\left(\Sigma\right)\right)$ is called a *Poincaré model of hyperbolic geometry.*

Through two distinct h-points of Σ there is exactly one h-line. This follows from Proposition 21, and the fact that two distinct points $p, q \in X'$ must be on different sides of B in the case $q = \mathrm{inv}_B(p)$.

Let x, y be distinct h-points of Σ and let $c \ni x, y$ be the, by Proposition 21, uniquely determined M-circle orthogonal to B. Because of Proposition 20, we obtain $\#(c \cap B) = 2$. If a, b are the points of intersection of c and B, then

$$\delta\left(x, y\right) := |\ln\{a, b; x, y\}| \tag{3.61}$$

is called the *hyperbolic distance (h-distance)* of x, y. This expression is well-defined in view of Proposition 25. Moreover, put $\delta\left(x, x\right) = 0$ for $x \in \Sigma$. Observe that $\ln \xi = \eta$ is defined by $\xi = \exp\left(\eta\right)$ for real ξ, η with $\xi > 0$. So we have to show that

$$\{a, b; x, y\} > 0 \tag{3.62}$$

holds true for the described points a, b, x, y. Note that a, b separate c into two parts and, moreover, that x, y belong to the same part, since they are on the same side of B. Because of Propositions 13 and 26 we may assume $a = \infty$ and $b = 0$ and, moreover, that $y = \lambda x$, $\lambda > 0$. But then, by Proposition 24, (3.62) holds true.

Remark. If l is a hyperbolic line of (B, Σ), then

$$\mu\left(l\right) = \{\mu\left(x\right) \mid x \in l\}$$

must be an h-line of $\left(\mu\left(B\right), \mu\left(\Sigma\right)\right)$ for $\mu \in \mathbb{M}\left(X\right)$, since $l = c \cap \Sigma$, c an M-circle orthogonal to B, implies

$$\mu\left(l\right) = \mu\left(c\right) \cap \mu\left(\Sigma\right), \, \mu\left(c\right) \perp \mu\left(B\right).$$

In the case $\mu \in G(\Sigma)$, i.e. in the case $\mu(B) = B$ and $\mu(\Sigma) = \Sigma$, $\mu(l)$ is again an h-line of (B, Σ). If $x, y \in \Sigma$ and $\mu \in M(X)$, then

$$\delta(x, y) \text{ with respect to } (B, \Sigma)$$

is equal to

$$\delta(\mu, (x), \mu(y)) \text{ with respect to } (\mu(B), \mu(\Sigma)).$$

This will be shown as follows: assume $x \neq y$ and that c is the M-circle orthogonal to B with $x, y \in c \cap \Sigma$. Define $\{a, b\} = c \cap B$. Hence

$$\mu(x), \mu(y) \in \mu(c) \cap \mu(\Sigma)$$

and $\{\mu(a), \mu(b)\} = \mu(c) \cap \mu(B)$, $\mu(c) \perp \mu(B)$. Thus, by Proposition 26,

$$\delta(\mu(x), \mu(y)) = |\ln\{\mu(a), \mu(b); \mu(x), \mu(y)\}|.$$

On the basis of Proposition 33 it is sufficient to study now, more intensively, the following special situation of an M-ball B and one of its sides Σ.

Let t be a fixed element of X with $t^2 = 1$. Define

$$B := H'(t, 0)$$

and $\Sigma := \{x \in X \mid x_0 > 0\}$ where we applied the decomposition

$$X = t^\perp \oplus \mathbb{R}t$$

with $x = \bar{x} + x_0 t$, $\bar{x} \in t^\perp = H(t, 0)$, $x_0 \in \mathbb{R}$ (see the beginning of section 7, chapter 1).

Proposition 34. *In this present geometry* (B, Σ) *the h-distance* (3.61) *is given by*

$$\cosh \delta(x, y) = 1 + \frac{(x - y)^2}{2x_0 y_0} \tag{3.63}$$

for all $x, y \in \Sigma$, *and hence also by*

$$2 \sinh \frac{\delta(x, y)}{2} = \frac{\|x - y\|}{\sqrt{x_0 y_0}}. \tag{3.64}$$

Proof. We may assume $x \neq y$. By l denote the h-line through x, y, and by c the M-circle containing l.

Case 1. $\infty \in c$.

Hence $\bar{x} = \bar{y} =: a$ and, by $x = a + x_0 t$ and $y := a + y_0 t$ with Proposition 24,

$$\delta(x, y) = |\ln\{a, \infty; x, y\}| = \left|\ln \frac{x_0}{y_0}\right|,$$

i.e.

$$\cosh \delta\left(x, y\right) = \frac{e^{\delta} + e^{-\delta}}{2} = \frac{1}{2}\left(\frac{x_0}{y_0} + \frac{y_0}{x_0}\right)$$

$$= 1 + \frac{(x_0 - y_0)^2}{2x_0 y_0} = 1 + \frac{(x - y)^2}{2x_0 y_0}.$$

Case 2. $\infty \notin c$.

Define $\{a, b\} := c \cap B$. Hence $\delta\left(x, y\right) = |\ln\{a, b; x, y\}|$. Obviously

$$\varphi := \iota\left(b, \|a - b\|\right) \in G\left(\Sigma\right)$$

and $\varphi\left(a\right) = a$. For arbitrary $z \in \Sigma$ define $z_1 := z - b$ and observe $(z_1)_0 = z_0$. Hence $z_1 \in \Sigma$. Put $\|a - b\| =: \varrho$. With $v := \varphi\left(x\right)$ and $w := \varphi\left(y\right)$ we obtain

$$v_1 = \varrho^2 \cdot \frac{x_1}{x_1^2} \text{ and } w_1 = \varrho^2 \frac{y_1}{y_1^2},$$

and, especially, $v_0 = v_1 t = \varrho^2 \frac{x_0}{x_1^2}$, $w_0 = \varrho^2 \frac{y_0}{y_1^2}$. Since $\varphi\left(c\right)$ contains $\varphi\left(a\right) = a$ and $\varphi\left(b\right) = \infty$, the h-line $\varphi\left(l\right)$ is part of a euclidean line of X. With the last Remark before Proposition 34 and in view of Case 1, we hence get

$$\cosh \delta\left(x, y\right) = \cosh \delta\left(v, w\right) = 1 + \frac{(v - w)^2}{2 v_0 w_0}.$$

Observe now

$$\frac{(v_1 - w_1)^2}{v_0 w_0} - \frac{(x_1 - y_1)^2}{x_0 y_0} = \frac{1}{v_0 w_0}\left(\frac{v_0}{x_0} x_1 - \frac{w_0}{y_0} y_1\right)^2 - \frac{(x_1 - y_1)^2}{x_0 y_0} = 0,$$

by applying $v_0 x_1^2 = \varrho^2 x_0$ and $w_0 y_1^2 = \varrho^2 y_0$. Hence

$$\cosh \delta\left(x, y\right) = 1 + \frac{(v - w)^2}{2 v_0 w_0} = 1 + \frac{(x - y)^2}{2 x_0 y_0},$$

in view of $v_1 - w_1 = v - w$ and $x_1 - y_1 = x - y$. $\qquad\square$

Theorem 35. *All bijections ψ of Σ with*

$$\delta\left(\psi\left(x\right), \psi\left(y\right)\right) = \delta\left(x, y\right) \tag{3.65}$$

for all $x, y \in \Sigma$ are products of the restrictions on Σ of the following M-transformations:

(α) the similitudes $f\left(x\right) = kx$ with $0 < k \in \mathbb{R}$,

(β) the inversion ι,

(γ) the mappings $f\left(x\right) = \omega\left(x\right) + a$ with $\omega \in O\left(x\right)$, $\omega\left(t\right) = t$ and $a \in H\left(t, 0\right)$.

Proof. a) Observe that every $\mu \in G(\Sigma)$ satisfies (3.65) and, moreover, that all mappings (α), (β), (γ) belong to $G(\Sigma)$. The essential result of Theorem 35 is hence that $G(\Sigma)$ is the group of all h-distance preserving bijections of Σ.

b) $\omega \in O(X)$ with $\omega(t) = t$ implies

$$\omega(\overline{x} + x_0 t) = \omega(\overline{x}) + x_0 t, \ \omega(\overline{x}) t = 0,$$

since $0 = \overline{x}t = \omega(\overline{x})\omega(t) = \omega(\overline{x}) t$. Hence we get for a mapping (γ)

$$\overline{f(x)} = \omega(\overline{x}) + a \text{ and } [f(x)]_0 = x_0,$$

in view of $a = \overline{a}$. The inverse mapping of (γ) is

$$f^{-1}(x) = \omega^{-1}(x) - \omega^{-1}(a).$$

Observing $\omega^{-1} \in O(X)$, $\omega^{-1}(t) = t$, $0 = ta = \omega^{-1}(t)\omega^{-1}(a) = t\omega^{-1}(a)$, it must be also of type (γ). The inverse mapping of $f(x) = kx$, $k > 0$, is $f^{-1}(x) = \frac{1}{k}x$, i.e. it is of type (α) again.

c) Let the bijection ψ of Σ satisfy (3.65). Hence $\psi(t) \in \Sigma$, i.e.

$$\psi(t) =: \overline{b} + b_0 t, \ b_0 > 0, \tag{3.66}$$

and thus $g(x) = \frac{1}{b_0} x - \frac{1}{b_0} \overline{b}$ is the product of a mapping (α) and a mapping (γ):

$$x \to x - \overline{b} \to \frac{1}{b_0}(x - \overline{b}).$$

Note $g\psi(t) = t$. Put $g\psi(2t) =: c$. With (3.65) we get

$$\delta(t, 2t) = \delta(t, c),$$

i.e., by $c \in \Sigma$ and (3.63), $0 < \frac{1}{2}c_0 = \overline{c}^2 + (c_0 - 1)^2$. If $c_0 = 2$, then $\overline{c} = 0$, i.e. $c = 2t$, i.e. $g\psi(2t) = 2t$. If $c_0 \neq 2$, define $d \in H(t, 0)$ by

$$d := \frac{2\overline{c}}{2 - c_0}.$$

Put $h := \sigma^{-1}\lambda\iota\sigma$ with $\sigma(x) := x - d$, $\lambda(x) := (1 + d^2) x$. The mapping h is hence a product of mappings $(\alpha), (\beta), (\gamma)$. We obtain

$$hg\psi(t) = t, \ hg\psi(2t) = 2t,$$

in view of $\frac{1}{2}c_0 = \overline{c}^2 + (c_0 - 1)^2$ and $c_0 \neq 2$. There hence exists a product π of mappings $(\alpha), (\beta), (\gamma)$ with $\pi\psi(t) = t$ and $\pi\psi(2t) = 2t$. For every $x \in \Sigma$ we get

$$\delta(t, x) = \delta(t, \pi\psi(x)), \ \delta(2t, x) = \delta(2t, \pi\psi(x)),$$

i.e., by $y := \pi\psi(x)$, $x_0 = y_0$ and $x^2 = y^2$.

d) Put $\tau := \pi\psi$. If $\tau\left(\overline{x}+x_0 t\right) =: \varphi\left(\overline{x}, x_0\right) + x_0 t$, by observing $\tau\left(x\right) = y = \overline{y} + y_0 t = \overline{y} + x_0 t$, with $\varphi\left(\overline{x}, x_0\right) \in H\left(t, 0\right)$, then $\varphi\left(\overline{x}, x_0\right) = \varphi\left(\overline{x}, \xi\right)$ for all positive reals x_0 and ξ. In fact,

$$\delta\left(\overline{x} + x_0 t, \overline{x} + \xi t\right) = \delta\left(\tau\left(\overline{x} + x_0 t\right), \tau\left(\overline{x} + \xi t\right)\right)$$

implies $\left(x_0 - \xi\right)^2 = \left(\varphi\left(\overline{x}, x_0\right) - \varphi\left(\overline{x}, \xi\right)\right)^2 + \left(x_0 - \xi\right)^2$, i.e.

$$\varphi\left(\overline{x}, x_0\right) = \varphi\left(\overline{x}, \xi\right).$$

So $\varphi\left(\overline{x}\right) := \varphi\left(\overline{x}, x_0\right)$ does not depend on $x_0 > 0$.

e) In view of d), we define a bijection T of X' with $T \mid \Sigma = \tau$. Put $T\left(\infty\right) = \infty$ and

$$T\left(\overline{x} + x_0 t\right) := \varphi\left(\overline{x}\right) + x_0 t \tag{3.67}$$

for all real x_0 and all $\overline{x} \in H\left(t, 0\right)$. Since $T\left(t\right) = t$, we get $\varphi\left(0\right) = 0$. Hence $T\left(0\right) = 0$, by (3.67). Now

$$\delta\left(\overline{x} + t, \overline{y} + t\right) = \delta\left(\varphi\left(\overline{x}\right) + t, \varphi\left(\overline{y}\right) + t\right)$$

implies $\left(\overline{x} - \overline{y}\right)^2 = \left(\varphi\left(\overline{x}\right) - \varphi\left(\overline{y}\right)\right)^2$ and hence

$$\left(x - y\right)^2 = \left(T\left(x\right) - T\left(y\right)\right)^2$$

for all $x, y \in X$. Thus $T \in O\left(X\right)$, $T\left(t\right) = t$, i.e. the mapping $x \to T\left(x\right)$ is of type (γ). Hence $\psi = \pi^{-1}T \mid \Sigma$, i.e. the extension $\pi^{-1}T$ of ψ is a product of mappings $(\alpha), (\beta), (\gamma)$. □

Proposition 36. *Every $\mu \in G\left(\Sigma\right)$ can be written in the form β or in the form $\alpha\iota\beta$ with*

$$\alpha\left(x\right) \quad := \quad kx + a,\ k > 0,\ a \in t^{\perp},$$
$$\beta\left(x\right) \quad := \quad l\omega\left(x\right) + m,\ l > 0,\ m \in t^{\perp}, \omega \in O\left(X\right),\ \omega\left(t\right) = t.$$

Proof. According to step a) of the proof of Theorem 35, $G\left(\Sigma\right)$ is exactly the group of all h-distance preserving bijections of Σ. Let now ψ be an arbitrary h-distance preserving bijection of Σ, so an element of $G\left(\Sigma\right) = \Gamma\left(\Sigma\right)$. According to steps c), d), e) of the proof of Theorem 35, we get

$$hg\psi\left(x\right) = \omega\left(x\right),\ \omega \in O\left(X\right),\ \omega\left(t\right) = t,$$

with $h = \mathrm{id}$ for $c_0 = 2$ where $c := g\psi\left(2t\right)$, and $h = \sigma^{-1}\lambda\iota\sigma$ for $c_0 \neq 2$ such that

$$\sigma\left(x\right) = x - d,\ \lambda\left(x\right) = \left(1 + d^2\right)x, d := \frac{2\overline{c}}{2 - c_0},$$

$$g\left(x\right) = \frac{1}{b_0}x - \frac{1}{b_0}\overline{b},\ b := \psi\left(t\right).$$

In other words, if $c_0 = 2$, we obtain, by observing (3.66),

$$\psi(x) = g^{-1}\omega(x) = b_0\omega(x) + \bar{b}, \ b_0 > 0,$$

i.e. a mapping of the form β. If $c_0 \neq 2$, we get

$$\psi = g^{-1}h^{-1}\omega = g^{-1}\sigma^{-1}\iota\lambda^{-1}\sigma\omega,$$

i.e. $\psi = \alpha\iota\beta$ with

$$\alpha(x) = g^{-1}\sigma^{-1}(x) = b_0 x + (b_0 d + \bar{b}), \ b_0 > 0, \ b_0 d + \bar{b} \in t^{\perp},$$
$$\beta(x) = \lambda^{-1}\sigma\omega(x) = l\omega(x) + m, \ \omega(t) = t,$$

where $l = \frac{1}{1+d^2} > 0$ and $m = -\frac{d}{1+d^2}, \ m \in t^{\perp}$. $\square$

Theorem 37. *The geometries $(\Sigma, G(\Sigma))$ (Poincaré model) and $(X, M(X, \mathrm{hyp}))$ (Weierstrass model), both based on X, are isomorphic. Here $M(X, \mathrm{hyp})$ (see (2.60)) is the group of hyperbolic motions of X.*

Proof. a) The mapping $\sigma : \Sigma \rightarrow X$ with

$$\sigma(x) = \frac{\bar{x}}{x_0} + \frac{x^2 - 1}{2x_0}t \tag{3.68}$$

must be a bijection: if $y \in X$, the uniquely determined inverse image x is given by

$$\bar{x} = x_0\bar{y}, \ x_0 = \frac{y_0 + \sqrt{1+y^2}}{1+\bar{y}^2}.$$

Note that $x_0 > 0$, since $y^2 = \bar{y}^2 + y_0^2$. Important will be the relation

$$\delta(x, y) = \mathrm{hyp}\left(\sigma(x), \sigma(y)\right) \tag{3.69}$$

for all $x, y \in \Sigma$ between the distance notions δ, (3.63), and hyp,

$$\cosh \mathrm{hyp}(x, y) = \sqrt{1+x^2}\sqrt{1+y^2} - xy.$$

Equation (3.69) follows from

$$1 + \sigma^2(x) = \left(\frac{x^2+1}{2x_0}\right)^2,$$

i.e. from

$$\cosh \mathrm{hyp}\left(\sigma(x), \sigma(y)\right) = \frac{x^2 + y^2 - 2\bar{x}\,\bar{y}}{2x_0 y_0} = \cosh \delta(x, y)$$

for $x, y \in \Sigma$.

b) Define $\tau : G(\Sigma) \to M(X, \mathrm{hyp})$ by $\tau(\mu) = \sigma\mu\sigma^{-1}$ for all $\mu \in G(\Sigma)$, and observe, by (3.69),

$$
\begin{aligned}
\mathrm{hyp}\left(\sigma(x), \sigma(y)\right) &= \delta(x, y) = \delta\left(\mu(x), \mu(y)\right) \\
&= \mathrm{hyp}\left(\sigma\mu(x), \sigma\mu(y)\right) = \mathrm{hyp}\left(\tau\sigma(x), \tau\sigma(y)\right)
\end{aligned}
$$

for $x, y \in \Sigma$ and $\mu \in G(\Sigma)$. This implies that if μ is an h-distance preserving bijection of Σ, then $\tau(\mu)$ is a motion of (X, hyp). On the other hand, if f is a motion of (X, hyp), then $\mu := \sigma^{-1}f\sigma$ is in $G(\Sigma)$. Hence τ is an isomorphism between $G(\Sigma)$ and $M(X, \mathrm{hyp})$ satisfying

$$
\sigma\mu(x) = \tau(\mu)\sigma(x)
$$

for all $x, y \in \Sigma$ and $\mu \in G(\Sigma)$. $\square$

Proposition 38. *All hyperbolic lines l of Σ are given as follows:*

(α) *Take $m, r \in X$ with $m_0 = 0 = r_0$, $r \neq 0$. Then put*

$$
l = \{m + r\cos\varphi + \|r\| \cdot t\sin\varphi \mid 0 < \varphi < \pi\}.
$$

(β) *Take $p \in X$ with $p_0 = 0$ and put*

$$
l = \{p + \xi t \mid 0 < \xi \in \mathbb{R}\}.
$$

Proof. Let c be an M-circle orthogonal to $B = H(t, 0)$. Put $c \cap B := \{a, b\}$. If $b = \infty$, we get $c \cap \Sigma = \{a + \xi t \mid \xi > 0\}$ with $a_0 = 0$, since $a \in B$. On the other hand, given l from (β), then

$$
\{p + \xi t \mid \xi \in \mathbb{R}\} \cup \{\infty\}
$$

is an M-circle orthogonal to B. Assume now $\infty \notin \{a, b\} = c \cap B$ with $c \perp B$. Define

$$
\alpha(x) = x - \iota(a - b), \quad \beta(x) = x - b.
$$

Hence $\alpha\iota\beta(a) = 0$, $\alpha\iota\beta(b) = \infty$. Thus, by $l := c \cap \Sigma$,

$$
\alpha\iota\beta(l) = \{\xi t \mid \xi > 0\}, \text{ i.e.}
$$

$$
l = \beta^{-1}\iota\alpha^{-1}\left(\{\xi t \mid \xi > 0\}\right) = \left\{b + \frac{\iota(a - b) + \xi t}{\left(\iota(a - b) + \xi t\right)^2} \Big| \xi > 0\right\}.
$$

Because of $a \neq b$ and $a, b \in B = t^{\perp}$, we get $0 \neq a - b \in t^{\perp}$, i.e.

$$
\iota(a - b) = \frac{a - b}{(a - b)^2} \in t^{\perp}.
$$

Put $m := \frac{a+b}{2}$, $r := \frac{a-b}{2}$. Hence $\iota(a-b) = \frac{r}{2r^2}$ and

$$\left\{ b + \frac{\iota(a-b) + \xi t}{(\iota(a-b) + \xi t)^2} \,\middle|\, \xi > 0 \right\} \tag{3.70}$$

$$= \left\{ m + r\frac{1-\eta^2}{1+\eta^2} + \|r\|\frac{2\eta t}{1+\eta^2} \,\middle|\, \eta > 0 \right\},$$

by putting $\eta := 2\|r\|\xi$. Define

$$\sin\varphi := \frac{2\eta}{1+\eta^2}$$

with $\varphi \in {]}0, \pi[$ for $\eta > 0$. Hence l gets the form (α). Vice versa, if we are given l from (α), let c be the M-circle containing l, and put $m + r =: a$, $m - r =: b$. Hence $a \neq b$ and $a, b \in t^\perp$. Working again with (3.70), we obtain

$$\alpha\iota\beta\,(l) = \{\xi t \mid \xi > 0\},$$

i.e. $c \perp B$, since $\{\xi t \mid \xi \in \mathbb{R}\} \cup \{\infty\} \perp B$. □

3.9 Spears, Laguerre cycles, contact

The basis will be again a real inner product space X of (finite or infinite) dimension ≥ 2.

A *spear* of X is an ordered pair (A, E) where A is a euclidean hyperplane $H(a, \alpha)$ of X and where E is one of the sides of A. Two spears (A_i, E_i), $i = 1, 2$, are called *equal* if, and only if, $A_1 = A_2$ and $E_1 = E_2$.

Occasionally, it will be useful to apply the following notation for the two *sides* of $H(a, \alpha)$,

$$H^+(a, \alpha) := \{x \in X \mid ax > \alpha\},$$
$$H^-(a, \alpha) := \{x \in X \mid ax < \alpha\}.$$

Of course, this notation depends on a common real factor $\lambda \neq 0$ of a, α: so observe

$$H^+(a, \alpha) = H^-(\lambda a, \lambda\alpha)$$

in the case $\lambda < 0$, despite the fact that $H(a, \alpha) = H(\lambda a, \lambda\alpha)$. Obviously, there is no difference between the definition of the two sides of $H'(a, \alpha)$ of section 8 and that of the two sides $H^+(a, \alpha)$, $H^-(a, \alpha)$ of $H(a, \alpha)$. However, there will be a difference between the definition of the two sides of an M-ball $B(m, \varrho)$ and that of the two sides of the ball $B(m, \varrho)$, $\varrho > 0$, of X. We define these two *sides* as follows,

$$B^+(m, \varrho) := \{x \in X \mid (x - m)^2 > \varrho^2\},$$
$$B^-(m, \varrho) := \{x \in X \mid (x - m)^2 < \varrho^2\}.$$

A *Laguerre cycle* of X is a point of X or an ordered pair (C, D) where C is a ball $B(m, \varrho)$, $\varrho > 0$, of X, and D one of its sides. Let z_1, z_2 be Laguerre cycles. If $z_1 \in X$, we will say that z_1, z_2 are *equal* exactly in the case that z_2 is the same point. If z_1 is the Laguerre cycle (C, D), then the Laguerre cycles z_1, z_2 are called *equal* if, and only if, z_2 is the Laguerre cycle (C, D) as well.

In section 5 (see Proposition 12, c)) we defined the notion of a *tangent hyperplane* $H(a, \alpha)$ of a ball $B(m, \varrho)$, $\varrho > 0$, by means of

$$\#[B(m, \varrho) \cap H(a, \alpha)] = 1. \tag{3.71}$$

Proposition 39. (3.71) *holds true if, and only if,*

$$(\alpha - am)^2 = \varrho^2 a^2. \tag{3.72}$$

Proof. Assume (3.71), i.e. $B(m, \varrho) \cap H(a, \alpha) =: \{p\}$. Hence, by Proposition 12, c),we obtain $H(a, \alpha) = H(m - p, (m - p)p)$. Thus, by observing Proposition 12 of chapter 2, there exists a real $\lambda \neq 0$ with

$$a = \lambda(m - p), \ \alpha = \lambda(m - p)p.$$

This implies $(\alpha - am)^2 = \lambda^2(m - p)^2 \cdot (m - p)^2 = a^2 \cdot \varrho^2$. Assume, vice versa, (3.72). Define

$$p := m + \frac{\alpha - am}{a^2} a, \tag{3.73}$$

and observe $(p - m)^2 = \varrho^2$ and $ap = \alpha$, by (3.72). Note, moreover, $\alpha - am \neq 0$, because of (3.72), $\varrho \neq 0$, $a \neq 0$. Hence

$$p \in B(m, \varrho) \cap H(a, \alpha),$$
$$\lambda := \tfrac{am - \alpha}{a^2} \neq 0,$$

i.e., by (3.73) and $ap = \alpha$,

$$\lambda a = m - p, \ \lambda \alpha = \lambda ap = (m - p)p.$$

Now apply Proposition 12, c) on $H(a, \alpha) = H(m - p, (m - p)p)$. $\square$

If $a \neq 0$ is in X, then $X = a^{\perp} \oplus \mathbb{R}a$ holds true. Observe that $x = v + w$ with $x \in X$, $v \in a^{\perp}$, $w \in \mathbb{R}a$ implies

$$v = x - \frac{xa}{a^2} a, \ w = \frac{xa}{a^2} a. \tag{3.74}$$

Obviously, $H(a, \alpha) = x_0 + a^{\perp} := \{x_0 + y \mid y \in a^{\perp}\}$ for all $x_0 \in H$, and hence

$$H^+(a, \alpha) \quad = \quad \{p + \lambda a \mid p \in H \text{ and } \lambda > 0\}, \tag{3.75}$$

$$H^-(a, \alpha) \quad = \quad \{p + \lambda a \mid p \in H \text{ and } \lambda < 0\} : \tag{3.76}$$

if $x \in X$ satisfies $ax > \alpha$, then $x = p + \lambda a$ with

$$p := x + \frac{\alpha - ax}{a^2} a \in H(a, \alpha) \text{ and } \lambda := \frac{ax - \alpha}{a^2} > 0.$$

We already defined *spears* and *Laguerre cycles* of X. A third fundamental notion in our present context is that one of *contact* between a spear and a Laguerre cycle z. We say that the spear (A, E) *is in contact with* or *touches* z if, and only if,

    (1) $z \in A$ in the case that z is a point,

    (2) A is a tangent hyperplane of C and

$$E \subseteq D \text{ or } D \subseteq E \text{ in the case } z = (C, D).$$

If the spear s touches the Laguerre cycle z we will write $s - z$ or $z - s$, otherwise $s \not{-} z$ or $z \not{-} s$.

    Suppose that the spear $s = (A, E)$ is in contact with the Laguerre cycle z. The *point p of contact* of s and z is defined by z if $z \in X$, and by $\{p\} = A \cap C$ for $z = (C, D)$.

    The pair (m, ϱ) will be called the *(cycle) coordinates* of $(B(m, \varrho), B^+(m, \varrho))$ and $(m, -\varrho)$ those of (B, B^-). If $z \in X$, then its cycle coordinates are defined by $(z, 0)$.

    $(a, \sqrt{a^2}, \alpha)$ are called the *(spear) coordinates* of the spear

$$\left(H(a, \alpha), H^+(a, \alpha)\right)$$

and $(a, -\sqrt{a^2}, \alpha)$ those of $\left(H(a, \alpha), H^-(a, \alpha)\right)$.

    Since $H(a, \alpha) = H(\lambda a, \lambda \alpha)$ for a real $\lambda \neq 0$, we must determine the coordinates of

$$S_\lambda^+ := \left(H(\lambda a, \lambda \alpha), H^+(\lambda a, \lambda \alpha)\right),$$
$$S_\lambda^- := \left(H(\lambda a, \lambda \alpha), H^-(\lambda a, \lambda \alpha)\right),$$

and see how they depend on the coordinates of S_1^+, S_1^-.

Case 1. $\lambda > 0$.

The coordinates of S_λ^+, S_λ^- are

$$\begin{aligned}
(\lambda a, \sqrt{(\lambda a)^2}, \lambda \alpha) &= (\lambda a, \lambda \sqrt{a^2}, \lambda \alpha), \\
(\lambda a, -\sqrt{(\lambda a)^2}, \lambda \alpha) &= (\lambda a, -\lambda \sqrt{a^2}, \lambda \alpha),
\end{aligned}$$

respectively.

Case 2. $\lambda < 0$.

Similarly, we get $(\lambda a, -\lambda \sqrt{a^2}, \lambda \alpha)$, $(\lambda a, \lambda \sqrt{a^2}, \lambda \alpha)$, as coordinates for S_λ^+, S_λ^-, respectively.

Spear coordinates must be *homogeneous coordinates*: (a, ξ, α) with $0 \neq a^2 = \xi^2$ and (b, η, β) with $0 \neq b^2 = \eta^2$ determine the same spear if, and only if, there exists $\lambda \neq 0$ with $b = \lambda a$, $\eta = \lambda \xi$, $\beta = \lambda \alpha$. This follows from

$$S_\lambda^+ = S_1^+, \ S_\lambda^- = S_1^- \text{ for } \lambda > 0,$$

$$S_\lambda^+ = S_1^-, \ S_\lambda^- = S_1^+ \text{ for } \lambda < 0.$$

Proposition 40. *The spear (a, ξ, α) touches the Laguerre cycle (m, τ) if, and only if,*

$$am + \xi\tau = \alpha. \tag{3.77}$$

Proof. If $\tau = 0$, then (3.77) characterizes the fact that m is on the underlying hyperplane of (a, ξ, α). Suppose now that $\tau \neq 0$. Without loss of generality, we may assume $\xi = 1$, since otherwise we would work with

$$\left(\frac{a}{\xi}, 1, \frac{\alpha}{\xi} \right),$$

by observing that spear coordinates are homogeneous coordinates and that (3.77) does not depend on a common factor $\neq 0$ of a, ξ, α. Hence $a^2 = 1$ for $(a, 1, \alpha)$. Thus $(a, 1, \alpha)$ is given by $\left(H(a, \alpha), H^+ \right)$. Denote the Laguerre cycle (m, τ) by (B, B^*).

a) Assume that $(a, 1, \alpha)$ touches (m, τ). (3.72) implies

$$\alpha - am \in \{\tau, -\tau\}, \tag{3.78}$$

and $\{p\} = H \cap B$ is given, in view of (3.73), by

$$p = m + (\alpha - am)\, a. \tag{3.79}$$

Assume $\alpha - am = -\tau$. Hence, by (3.79), $p = m - \tau a$. For $\tau < 0$ we get $B^* = B^-$, i.e. $B^- \subseteq H^+$, by $(B, B^-) - (H, H^+)$, i.e. $m \in B^- \subseteq H^+$. Thus $m = p + \lambda a$, $\lambda > 0$, by (3.75), contradicting $\lambda = \tau < 0$. For $\tau > 0$ we get $B^* = B^+$, i.e. $H^+ \subseteq B^+$. Hence $m \in H^-$, since $m \in H \cup H^+$ would imply $m \in H$ or $m \in B^+$. Thus $m = p + \lambda a$, $\lambda < 0$, by (3.76), contradicting $\lambda = \tau > 0$.

We obtained that our assumption $\alpha - am = -\tau$ does not hold true. Hence $\alpha - am = \tau$, by (3.78). Thus (3.77) is satisfied.

b) Assume that (3.77) holds true, i.e. $am + \tau = \alpha$ for the spear $(a, 1, \alpha)$ and the cycle (m, τ), $\tau \neq 0$. The underlying hyperplane of $(a, 1, \alpha)$ is $H := H(a, \alpha)$, and the underlying ball of (m, τ) is $B := B(m, |\tau|)$. Observe that (3.77) (with $\xi = 1$) implies (3.72), in view of $a^2 = 1$, by $\xi = 1$, and $\varrho = |\tau|$. Hence H is a tangent hyperplane of B with

$$H \cap B = \{p\},$$

$p = m + (\alpha - am)\, a = m + \tau a$, by (3.73). Since $H = p + a^\perp$, we get, by (3.75),

$$H^+ = \{(p + v) + \lambda a \mid v \in a^\perp \text{ and } \lambda > 0\}.$$

Case $\tau > 0$.

Here $B^* = B^+$. We would like to show $H^+ \subseteq B^+$, because then $(B, B^+) - (H, H^+)$ holds true. But $x \in H^+$ implies

$$(m - x)^2 = \left(m - [(p + v) + \lambda a]\right)^2 = (\tau + \lambda)^2 + v^2 > \tau^2.$$

Case $\tau < 0$.

Here we get $B^- \subseteq H^+$. In fact! If $x \in B^-$, then

$$x - m \in X = a^\perp \oplus \mathbb{R}a,$$

i.e. $x - m =: v + \mu a$ with $v \in a^\perp$ and $\mu \in \mathbb{R}$, i.e.

$$\tau^2 > (x - m)^2 = v^2 + \mu^2 \geq \mu^2,$$

i.e. $-\tau > |\mu| \geq -\mu$, i.e. $\mu > \tau$. Moreover, by $p = m + \tau a$,

$$x = m + v + \mu a = (p + v) + (\mu - \tau)\, a \in H^+. \qquad \square$$

Sometimes it will be useful to identify a spear s with the set of all Laguerre cycles touching this spear s, and also a Laguerre cycle c with the set of all spears in contact with c.

Lemma 41. *Denote by Σ the set of all spears of X, and by Γ the set of all Laguerre cycles. If $s_1, s_2 \in \Sigma$ satisfy*

$$\{c \in \Gamma \mid c - s_1\} = \{c \in \Gamma \mid c - s_2\}, \tag{3.80}$$

then $s_1 = s_2$, and if $c_1, c_2 \in \Gamma$ satisfy

$$\{s \in \Sigma \mid s - c_1\} = \{s \in \Sigma \mid s - c_2\},$$

then $c_1 = c_2$.

Proof. a) Let $(a, 1, \alpha)$, $(b, 1, \beta)$ be coordinates of s_1, s_2, respectively. By (3.77) and (3.80) we know that

$$am + \tau = \alpha \text{ and } bm + \tau = \beta$$

have the same solutions (m, τ). If we look to the solutions $(x, 0)$, we obtain that the hyperplanes of equations $ax = \alpha$, $bx = \beta$ are identical. Hence, by Proposition 12, chapter 2, there exists a real $\lambda \neq 0$ with $b = \lambda a$ and $\beta = \lambda \alpha$. Furthermore, $a^2 = 1 = b^2$ implies $\lambda^2 = 1$. The case $\lambda = -1$ does not occur, since otherwise we would determine all solutions $(x, 1)$ with the consequence that the hyperplanes of equations $ax = \alpha - 1$ and $ax = \alpha + 1$ would coincide.

b) Let (m, τ), (n, σ) be coordinates of c_1, c_2, respectively. We hence know that

$$am + \tau = \alpha \text{ and } an + \sigma = \alpha$$

must have the same solutions $(a, 1, \alpha)$, $a^2 = 1$. Obviously, $(\varepsilon v, 1, \varepsilon vm + \tau)$ with $v \in X$, $v^2 = 1$, $\varepsilon \in \{1, -1\}$, solves $am + \tau = \alpha$. Hence

$$\varepsilon vn + \sigma = (\varepsilon vm + \tau), \ \varepsilon \in \{1, -1\},$$

i.e. $\sigma = \tau$, i.e. $v(n - m) = 0$ for all $v \in X$. Thus $m = n$. $\qquad \square$

Two euclidean hyperplanes $H_1 = H(a, \alpha)$, $H_2 = H(b, \beta)$ of X are called *parallel* (see section 7, chapter 2), $H_1 \parallel H_2$, provided $H_1 = H_2$ or $H_1 \cap H_2 = \emptyset$. Of course, $H_1 \parallel H_2$ is equivalent with $\mathbb{R}a = \mathbb{R}b$, since $\mathbb{R}a \neq \mathbb{R}b$ implies $\xi a + \eta b \in H_1 \cap H_2$ where

$$a(\xi a + \eta b) = \alpha,$$
$$b(\xi a + \eta b) = \beta,$$

by observing

$$\begin{vmatrix} a^2 & ab \\ ba & b^2 \end{vmatrix} = a^2 b^2 - (ab)^2 \neq 0$$

(see Lemma 1, chapter 1).

Two spears $s_1 \neq s_2$ are called *parallel*, $s_1 \parallel s_2$, provided there does not exist $c \in \Gamma$ with $s_1 - c - s_2$. Moreover, every spear is said to be *parallel* to itself.

Proposition 42. *Two spears (a, ξ, α) and (b, η, β) are parallel if, and only if, one of the following equivalent properties hold true.*

(i) $\xi b = \eta a$,

(ii) $H_1 \parallel H_2$ *and* $V_1 \subseteq V_2$ *or* $V_2 \subseteq V_1$, *where* (H_1, V_1), (H_2, V_2) *are the spears* (a, ξ, α), (b, η, β), *respectively.*

Proof. Without loss of generality we may assume $\xi = 1 = \eta$. If the two spears coincide, then $(a, 1, \alpha) = (b, 1, \beta)$, and hence (i) holds true. If they are distinct and parallel, then $H_1 \cap H_2 = \emptyset$, because $c \in H_1 \cap H_2$ would satisfy $c - (H_i, V_i)$ for $i = 1, 2$. Hence $H_1 \parallel H_2$, i.e. $b = \lambda a$, $\lambda^2 = 1$. If λ were -1, then, by (3.77),

$$\left(p - \frac{\alpha + \beta}{2} a, \frac{\alpha + \beta}{2} \right), p \in H_1,$$

would touch both spears.

(i) implies (ii). In fact! $a = b$ (observe $\xi = 1 = \eta$) yields $H_1 \parallel H_2$. Take $p_1 \in H$, i.e. $ap_1 = \alpha$, and put $p_2 := p_1 + (\beta - \alpha) a \in H_2$. If $\beta \geq \alpha$, then $H_2^+ \subseteq H_1^+$, by (3.75), if $\beta < \alpha$, then $H_1^+ \subset H_2^+$.

(ii) implies $(H_1, V_1) \parallel (H_2, V_2)$. If those spears are distinct, we have to show that there does not exist $c \in \Gamma$ touching both. Since $H_1 \parallel H_2$, we get $b = \lambda a$, i.e. $\lambda^2 = 1$. If $\lambda = -1$, $V_1 \subseteq V_2$ or $V_2 \subseteq V_1$ does not hold true. Hence $\lambda = 1$, i.e. the two distinct spears s_1, s_2 are $(a, 1, \alpha)$ and $(a, 1, \beta)$. Because of $\alpha \neq \beta$ and (3.77) there does not exist $c \in \Gamma$ with $s_1 - c - s_2$. $\qquad \square$

Remark. From Proposition 42, (i), follows that the parallel relation on Σ is an equivalence relation.

3.10 Separation, cyclographic projection

If (m_1, τ_1), (m_2, τ_2) are two Laguerre cycles of X, the real number

$$l(c_1, c_2) = (m_1 - m_2)^2 - (\tau_1 - \tau_2)^2$$

will be called their *separation*, or their *power*.

We say that $c_1 \in \Gamma$ *touches* $c_2 \in \Gamma$, designated by $c_1 - c_2$, if, and only if, $c_1 = c_2$ or

$$S(c_1, c_2) := \{s \in \Sigma \mid c_1 - s - c_2\}$$

consists of exactly one spear.

Proposition 43. *Let c_1, c_2 be Laguerre cycles. Then*

$$
\begin{array}{lll}
(\alpha) & c_1 - c_2 & \Leftrightarrow \quad l(c_1, c_2) = 0, \\
(\beta) & S(c_1, c_2) = \emptyset & \Leftrightarrow \quad l(c_1, c_2) < 0, \\
(\gamma) & S(c_1, c_2) \neq \emptyset & \Leftrightarrow \quad l(c_1, c_2) \geq 0
\end{array}
$$

hold true.

Proof. (β) follows from (γ). Let (m_i, τ_i) be the coordinates of c_1, c_2. Assume $l(c_1, c_2) \geq 0$. If $m_1 = m_2$, then

$$0 \leq l(c_1, c_2) = -(\tau_1 - \tau_2)^2,$$

i.e. $c_1 = c_2$, i.e. $S(c_1, c_2) \neq \emptyset$. If $m_1 \neq m_2$, take $b \in X$ with $b^2 = 1$ and $b \cdot (m_1 - m_2) = 0$. Define $\alpha := am_1 + \tau_1$ and $a \in X$ by

$$(m_1 - m_2)^2 \cdot a := b\sqrt{l(c_1, c_2) \cdot (m_1 - m_2)^2} - (m_1 - m_2)(\tau_1 - \tau_2). \qquad (3.81)$$

Hence $a^2 = 1$ and $a(m_1 - m_2) + (\tau_1 - \tau_2) = 0$. Thus, by Proposition 40, $(a, 1, \alpha)$ touches c_1 and c_2, i.e. $(S(c_1, c_2) \neq \emptyset$. In order to prove the other part of (γ), assume that $(a, 1, \alpha)$ touches c_1 and c_2. From

$$am_1 + \tau_1 = \alpha, \; am_2 + \tau_2 = \alpha, \; a^2 = 1 \qquad (3.82)$$

we get $(\tau_1 - \tau_2)^2 = [a(m_1 - m_2)]^2 \leq a^2(m_1 - m_2)^2$, i.e. $l(c_1, c_2) \geq 0$. Finally, we would like to prove (α). Assume $l(c_1, c_2) = 0$. There is nothing to show for $c_1 = c_2$. If $c_1 \neq c_2$, there exists, by (γ), $(a, 1, \alpha)$ touching c_1 and c_2. Hence (3.82) holds true and also, as before,

$$(\tau_1 - \tau_2)^2 = [a(m_1 - m_2)]^2 \leq a^2(m_1 - m_2)^2 = (m_1 - m_2)^2.$$

Since also $0 = l(c_1, c_2) = (m_1 - m_2)^2 - (\tau_1 - \tau_2)^2$ is satisfied, we even obtain

$$[a(m_1 - m_2)]^2 = a^2(m_1 - m_2)^2,$$

i.e., by Lemma 1, chapter 1, that $a, m_1 - m_2$ are linearly dependent. Hence $m_1 - m_2 = \lambda a$ with $\lambda \in \mathbb{R}$ and thus, by (3.82), $\lambda = \tau_2 - \tau_1$. Since $c_1 \neq c_2$, we get $\lambda \neq 0$. The spear $(a, 1, \alpha)$ is hence uniquely determined by

$$a = \frac{m_1 - m_2}{\tau_2 - \tau_1} \text{ and } \alpha = a m_1 + \tau_1. \tag{3.83}$$

Thus $c_1 - c_2$. Vice versa assume $c_1 - c_2$. We then must prove $l(c_1, c_2) = 0$. This is clear for $c_1 = c_2$. Hence suppose that $c_1 \neq c_2$. From (γ) we get $l(c_1, c_2) \geq 0$. Replacing b in (3.81) by $-b$, we also obtain a spear $(a', 1, \alpha')$ touching c_1 and c_2. Since $c_1 - c_2$, the spears $(a, 1, \alpha)$ and $(a', 1, \alpha')$ must be identical. This, by (3.81), implies $l(c_1, c_2) = 0$. □

Proposition 44. *Let* $s = (H, H^*)$ *be a spear and* c_1, c_2 *be Laguerre cycles with* $c_1 - s - c_2$. *If* p_i, $i = 1, 2$, *is the point of contact of* s *and* c_i, *then*

$$l(c_1, c_2) = (p_1 - p_2)^2. \tag{3.84}$$

Proof. Let $(a, 1, \alpha)$ be coordinates of s. Then, by (3.77), (3.73),

$$p_i = m_i + \tau_i a, \ i = 1, 2.$$

This and $a m_i + \tau_i = \alpha$, $i = 1, 2$, imply

$$\begin{aligned}
(p_1 - p_2)^2 &= [(m_1 - m_2) + (\tau_1 - \tau_2) a]^2 \\
&= (m_1 - m_2)^2 + 2(\tau_1 - \tau_2) a (m_1 - m_2) + (\tau_1 - \tau_2)^2 \\
&= l(c_1, c_2).
\end{aligned}$$

□

If $l(c_1, c_2) \geq 0$, i.e. by Proposition 43, (γ), that $S(c_1, c_2) \neq \emptyset$, the expression

$$\sqrt{l(c_1, c_2)} \tag{3.85}$$

is called the *tangential distance* of c_1, c_2. In other words, it exists exactly if there is at least one spear s touching c_1 and c_2. By (3.84), $\sqrt{l(c_1, c_2)}$ represents the euclidean distance between the two points p_i of contact of s and c_i, $i = 1, 2$.

If the separation $l(c_1, c_2)$ of c_1 and c_2 is negative, there is also a geometric interpretation of $l(c_1, c_2)$:

A. If $l(c_1, c_2) < 0$, there exists $c \in \Gamma$ with $c_1 - c - c_2$. Take $e \in X$ with $e^2 = 1$. Then

$$[e(m_1 - m_2)]^2 \leq e^2(m_1 - m_2)^2 = (m_1 - m_2)^2 < (\tau_1 - \tau_2)^2,$$

because of $l(c_1, c_2) < 0$. Hence $k := e(m_1 - m_2) - (\tau_1 - \tau_2) \neq 0$. Put

$$m := m_2 + \lambda e, \ \tau := \tau_2 + \lambda, \ 2k\lambda := l(c_1, c_2).$$

Now Proposition 43, (α), implies $c_1 - (m, \tau) - c_2$.

B. If there is no spear touching c_1 and c_2, i.e., by Proposition 43, (β), $l(c_1, c_2) < 0$, then, as we will prove,

$$l(c_1, c_2) = -4l(y, z) \tag{3.86}$$

holds true where

$$y := \left(\frac{m_1 + m_2}{2}, \frac{\tau_1 + \tau_2}{2} \right) \tag{3.87}$$

is the so-called *mid-cycle* of c_1 and c_2, and where z is a Laguerre cycle with $c_1 - z - c_2$ (see step A for the existence of such a z).

Since $l(c_1, c_2)$ is negative, we obtain $l(y, z) > 0$, and this latter separation can hence be interpreted by a tangential distance.

In order to prove (3.86), we again will work with the real vector space

$$Z := X \oplus \mathbb{R},$$

as in section 8, but now with the product

$$(m, \tau) \cdot (n, \sigma) := mn - \tau\sigma \tag{3.88}$$

instead of $(a, \alpha) \cdot (b, \beta) := ab + \alpha\beta$ as in (3.54). Observe the rules

$$xy = yx, \ (x + y) z = xz + yz, \ (\lambda x) \cdot y = \lambda \cdot (xy)$$

for all $x, y, z \in Z$ and $\lambda \in \mathbb{R}$. However, x^2 may be negative for $x \neq 0$: take $e \in X$ with $e^2 = 1$ and put $x = (e, 2)$. We will call Z the *vector space of Laguerre cycles of X*.

Equation (3.86) reads as $(c_1 - c_2)^2 = -4(y - z)^2$ in Z with

$$2y = c_1 + c_2 \text{ and } (c_1 - z)^2 = 0 = (c_2 - z)^2.$$

Note that we have to distinguish between the difference $x - y$ in Z and the contact $x - y$ of $x, y \in \Gamma$. A proof of (3.86) is now given by

$$-(c_1 - c_2)^2 = -\big((c_1 - z) - (c_2 - z)\big)^2 = 2(c_1 - z)(c_2 - z),$$
$$(2y - 2z)^2 = \big((c_1 - z) + (c_2 - z)\big)^2 = 2(c_1 - z)(c_2 - z). \qquad \square$$

The mapping which associates to a Laguerre cycle c of X its coordinates (m, τ) in the vector space Z is called *cyclographic projection*. We are now interested in the images of spears s,

$$s = \{c \in \Gamma \mid c - s\}, \tag{3.89}$$

under this mapping. Here we will work with the inner product (3.54). In this context we prefer to speak of the vector space Y (more precisely of the real inner product space Y), as we did in section 7. If s has coordinates (a, ξ, α), then (3.89) is given by all $(m, \tau) \in Y$ satisfying

$$(a, \xi) \cdot (m, \tau) = \alpha, \tag{3.90}$$

according to (3.77). This is a hyperplane of the real inner product space Y, but of course, a special hyperplane, since $a^2 = \xi^2$ holds true. Assuming $(a, \xi)^2 = 1$, i.e. $a^2 + \xi^2 = 1$, we obtain for the cosine of the angle between (a, ξ) and $(0, 1)$,

$$(a, \xi) \cdot (0, 1) = \xi \in \left\{ \frac{1}{\sqrt{2}}, -\frac{1}{\sqrt{2}} \right\}.$$

We call these hyperplanes of Y its *45°-hyperplanes*,

$$(a, \xi)(m, \tau) = \alpha, \ (a, \xi)^2 = 1, \ \sqrt{2}\xi \in \{1, -1\}.$$

The 45°-hyperplanes of Y are hence exactly the images of the spears under the cyclographic projection. Two spears are parallel if, and only if, the image hyperplanes are parallel (see Proposition 42, (i)).

In order to define the *cylinder model* for our structure of spears and Laguerre cycles take the cylinder

$$C := \{(y, \eta) \in Y \mid y^2 = 1\}$$

of Y. If $H\left((b, \beta), \gamma\right)$, $(b, \beta) \neq 0$, is a hyperplane of Y such that $(b, \beta) \not\perp (0, 1)$ holds true, then

$$C \cap H\left((b, \beta), \gamma\right) \tag{3.91}$$

is called a *hyperplane cut* of C. A line l of Y contained in C is said to be a *generator* of C. If $l \subset C$ with

$$l = \{(a, \alpha) + \lambda\left((b, \beta) - (a, \alpha)\right) \mid \lambda \in \mathbb{R}\},$$

$(a, \alpha), (b, \beta) \in C$, $(a, \alpha) \neq (b, \beta)$, then $[a + \lambda(b - a)]^2 = 1$ holds true for all $\lambda \in \mathbb{R}$, i.e. $ab = 1$, i.e. $a = b$, by Lemma 1, chapter 1. Hence l must be parallel to the *axis*

$$\mathbb{R} \cdot (0, 1)$$

of the cylinder C. Thus we get all generators g of C by taking $a \in X$ with $a^2 = 1$ and putting

$$g = \{(a, \lambda) \mid \lambda \in \mathbb{R}\}.$$

We will say that the hyperplane H of Y is *parallel* to the line h of Y provided there exists a line $h_0 \subset H$ parallel to h. This implies that $C \cap H\left((b, \beta), \gamma\right)$ is a hyperplane cut if, and only if, H is not parallel to the axis of C. If $(m, \tau) \neq (n, \sigma)$ are both in H such that

$$(m, \tau) - (n, \sigma) \in \mathbb{R}\,(0, 1),$$

then $m = n$, i.e. $bm + \beta\tau = \gamma = bm + \beta\sigma$, i.e. $(m, \tau) = (n, \sigma)$ for $\beta \neq 0$. Thus $(b, \beta) \perp (0, 1)$. If, vice versa, (3.91) is not a hyperplane cut, i.e. if $(b, \beta) \perp (0, 1)$, i.e. $\beta = 0$ holds true, then $b \neq 0$, by $(b, \beta) \neq 0$ and

$$\frac{\gamma}{b^2}\,(b, \beta) + \mathbb{R}\,(0, 1) \subset H,$$

i.e. H is parallel to the axis of C.

This is now the cylinder model. Associate to the spear of X with coordinates $(a, 1, \alpha)$ the point (a, α) of C. This is a bijection between C and the set Σ of all spears of X. The important thing is that the image of a Laguerre cycle

$$c = \{s \in \Sigma \mid s - c\} \tag{3.92}$$

is a hyperplane cut of C and that the inverse image of every hyperplane cut must be a Laguerre cycle of X. In fact, the image of (3.92) is the hyperplane cut

$$C \cap H\,\big((-m, 1), \tau\big) \tag{3.93}$$

where (m, τ) are the coordinates of c: this follows from

$$(-m, 1)(a, \alpha) - \tau = -ma + \alpha - \tau = 0$$

by observing (3.77) with $\xi = 1$, by noticing $a^2 = 1$, i.e. $(a, \alpha) \in C$, and $H \not\perp (0, 1)$. Let, vice versa, (3.91) be an arbitrary hyperplane cut of C. Hence $\beta \neq 0$. Put

$$(m, \tau) := \left(-\frac{b}{\beta}, \frac{\gamma}{\beta}\right)$$

and (3.93) becomes (3.91).

Lemma 45. *If both sides of $C \cap H\,\big((b, \beta), \gamma\big) = C \cap H\,\big((b', \beta'), \gamma'\big)$ are hyperplane cuts, then the hyperplanes involved coincide.*

Proof. We may assume $\beta = 1 = \beta'$ without loss of generality. Take arbitrarily $a \in X$ with $a^2 = 1$. Then $(a, \gamma - ba) \in C \cap H\,\big((b, 1), \gamma\big)$ and hence

$$b'a + \gamma - ba = \gamma'.$$

Take $a \perp b' - b$. Hence $\gamma = \gamma'$. Thus $(b' - b)\,a = 0$ for all $a \in X$ with $a^2 = 1$. Hence $b' = b$. $\qquad\square$

Note that two spears are parallel if, and only if, their images on C are on the same generator of C (see Proposition 42, (i)).

Remark. We would like to emphasize that we designated by Y the real vector space $X \oplus \mathbb{R}$ equipped with the inner product (3.54), and that Z denotes the real vector space $X \oplus \mathbb{R}$ furnished with the product (3.88).

3.11   Pencils and bundles

In this section we shall work with the vector space Z, so with $X \oplus \mathbb{R}$ and the product (3.88). Of course, X is a subspace of Z in the sense that x of X is identified with the element $(x, 0)$ of Z. As far as elements a, b of X are concerned, we get for these

$$ab = (a, 0)(b, 0) = ab - 0 \cdot 0,$$

i.e. we get their inner product in X. Hence for elements a, b of Z which already belong to X, we may apply, for instance, Lemma 1 of chapter 1.

Note

$$l\left(c_1, c_2\right) = \left(c_1 - c_2\right)^2$$

for elements c_1, c_2 of Z.

Let $c_1 \neq c_2$ be Laguerre cycles satisfying $c_1 - c_2$. Then

$$B_p(c_1, c_2) = \{c \in \Gamma \mid c_1 - c - c_2\} \tag{3.94}$$

is called a *parabolic pencil* of Laguerre cycles.

Proposition 46. *Let $c_1 \neq c_2$ be elements of Γ with $c_1 - c_2$. Then the following statements $(\alpha), (\beta), (\gamma)$ hold true.*

(α) *If $v \neq w$ are in $B_p(c_1, c_2)$, then $v - w$ and $B_p(v, w) = B_p(c_1, c_2)$.*

(β) *The spear s touching c_1, c_2 touches every $v \in B_p(c_1, c_2)$.*

(γ) $B_p(c_1, c_2) = \{c_1 + \varrho\left(c_2 - c_1\right) \mid \varrho \in \mathbb{R}\}$.

Proof. If $c \in B_p(c_1, c_2)$, then, by (3.94) and Proposition 43, $l\left(c, c_i\right) = 0$, $i = 1, 2$ holds true. Put

$$x_i := c - c_i = (m_i, \tau_i), \ i = 1, 2.$$

Then $x_i^2 = 0$ for $i = 1, 2$, and, by $l\left(c_1, c_2\right) = 0$, also $(x_1 - x_2)^2 = 0$, i.e. $x_1 x_2 = 0$, i.e. $m_1 m_2 = \tau_1 \tau_2$. Hence, by $x_i^2 = 0$,

$$(m_1 m_2)^2 = \tau_1^2 \tau_2^2 = m_1^2 m_2^2.$$

Thus, by Lemma 1, chapter 1, m_1, m_2 are linearly dependent. Even x_1, x_2 are linearly dependent. This is clear for $x_1 = 0$ or $x_2 = 0$. Assume $x_1 \neq 0 \neq x_2$. If $m_j = \lambda m_i$, $\lambda \in \mathbb{R}$, for suitable $\{i, j\} = \{1, 2\}$, then $m_1 m_2 = \tau_1 \tau_2$ implies

$$\lambda m_i^2 = \tau_i \tau_j, \tag{3.95}$$

and $m_j^2 = \tau_j^2$, by $x_j^2 = 0$, yields $(\lambda m_i)^2 = m_j^2 = \tau_j^2$, i.e. $\lambda^2 \tau_i^2 = \lambda^2 m_i^2 = \tau_j^2$. Hence $\tau_j \in \{\lambda \tau_i, -\lambda \tau_i\}$. If $\tau_j = \lambda \tau_i$, then $x_j = \lambda x_i$. If $\tau_j \neq \lambda \tau_i$, then $\tau_j = -\lambda \tau_i$ and $\lambda \neq 0$. Hence $0 \leq m_i^2 = -\tau_i^2 \leq 0$, by (3.95), i.e. $m_i = 0 = \tau_i$. But $x_i = 0$ contradicts $x_1 \neq 0 \neq x_2$. Thus $x_j = \lambda x_i$ and x_1, x_2 are linearly dependent.

Now $x_j = \lambda x_i$ yields $c - c_j = \lambda (c - c_i)$. Since $c_1 \neq c_2$, we obtain $\lambda \neq 1$, and hence

$$c = c_i + \gamma (c_j - c_i) = c_j + \delta (c_i - c_j) =: c_1 + \varrho (c_2 - c_1)$$

with $\gamma \cdot (1 - \lambda) := 1$ and $\delta := 1 - \gamma$. Thus

$$B_p(c_1, c_2) \subseteq \{c_1 + \varrho (c_2 - c_1) \mid \varrho \in \mathbb{R}\}.$$

Moreover, we have to show $d := c_1 + \varrho (c_2 - c_1) \in B_p(c_1, c_2)$ for every $\varrho \in \mathbb{R}$: this follows from

$$(d - c_1)^2 = \varrho^2 (c_2 - c_1)^2 = 0, \ (d - c_2)^2 = (\varrho - 1)^2 (c_2 - c_1)^2 = 0,$$

i.e. $c_1 - d - c_2$. This proves (γ). Statement (α) follows form (γ). In order to show (β), put

$$s = (a, 1, \alpha), \ c_i =: (m_i, \tau_i) \text{ for } i = 1, 2.$$

Now $am_1 + \tau_1 = \alpha$ and $am_2 + \tau_2 = \alpha$ imply

$$am + \tau = \alpha$$

for $(m, \tau) := c_1 + \varrho (c_2 - c_1)$. $\qquad \square$

Exactly the lines

$$B_p(c_1, c_2) = \{c_1 + \varrho (c_2 - c_1) \mid \varrho \in \mathbb{R}\}, \ c_1 \neq c_2, \ l(c_1, c_2) = 0,$$

of $Z = X \oplus \mathbb{R}$ represent the parabolic pencils.

Proposition 47. *Let $c_1 \neq c_2$ be Laguerre cycles with $c_1 - c_2$. The only spear s touching c_1 and c_2 is given as subset of Γ by $N \cup B_p(c_1, c_2)$ with*

$$N := \{c \in \Gamma \mid \text{ there is no } w \in B_p(c_1, c_2) \text{ touching } c\}.$$

Proof. If $c_i = (m_i, \tau_i)$, $i = 1, 2$, then, by (3.83),

$$s = (a, 1, m_1 a + \tau_1)$$

with $(\tau_2 - \tau_1) a := m_1 - m_2$. All $z = (m, \tau) \in \Gamma$ touching s are given by the equation

$$am + \tau = am_1 + \tau_1,$$

i.e. by $(m_1 - m_2)(m - m_1) - (\tau_1 - \tau_2)(\tau - \tau_1) = 0$, i.e. by

$$P := \{z \in \Gamma \mid (c_1 - c_2)(z - c_1) = 0\}. \tag{3.96}$$

We have to show $P = N \cup B_p(c_1, c_2) =: Q$. Obviously,

$$l(z, c_1 + \varrho (c_2 - c_1)) = l(z, c_1) + 2\varrho (c_1 - c_2)(z - c_1) \tag{3.97}$$

for all $z \in \Gamma$, in view of $l(c_1, c_2) = 0$.

If $z \in P$, then (3.97) implies $l(z, c_1 + \varrho(c_2 - c_1)) = l(z, c_1)$ for all $\varrho \in \mathbb{R}$. Hence, if $l(z, c_1) = 0$, so $l(z, c_2) = 0$ (case $\varrho = 1$), and we get $c_1 - z - c_2$, i.e. $z \in B_p(c_1, c_2) \subseteq Q$. If $l(z, c_1) \neq 0$, we obtain $l(z, c_1 + \varrho(c_2 - c_1)) \neq 0$ for all $\varrho \in \mathbb{R}$, i.e. $z \in N \subseteq Q$. Also $Q \subseteq P$ holds true. In fact! Obviously, $B_p(c_1, c_2) \subseteq P$. If $z \in N$, then $l(z, c_1 + \varrho(c_2 - c_1)) \neq 0$ for all $\varrho \in \mathbb{R}$. If $(c_1 - c_2)(z - c_1)$ were $\neq 0$, there would exist a real number ϱ_0 such that the right-hand side of (3.97) were 0, i.e. $l(z, c_1 + \varrho_0(c_2 - c_1)) = 0$ would hold true.   □

Let c_1, c_2 be distinct Laguerre cycles such that there exist at least two distinct spears touching both c_1 and c_2. Then

$$B_e(c_1, c_2) := \{c \in \Gamma \mid c_1 - s - c_2 \Rightarrow s - c \text{ for all } s \in \Sigma\}$$

is called an *elliptic pencil*.

Proposition 48. *Exactly the lines*

$$B_e(c_1, c_2) = \{c_1 + \varrho(c_2 - c_1) \mid \varrho \in \mathbb{R}\}, \ l(c_1, c_2) > 0,$$

of $Z = X \oplus \mathbb{R}$ represent the elliptic pencils.

Proof. Let (m_i, τ_i), $i = 1, 2$, be the coordinates of $c_i \in \Gamma$. We assume $c_1 \neq c_2$ and that there are at least two distinct spears touching c_1 and c_2. Hence $l(c_1, c_2) > 0$, according to Proposition 43, and thus $m_1 \neq m_2$. If $b \in X$ satisfies $b^2 = 1$ and $b \cdot (m_1 - m_2) = 0$, then, compare (3.81), $(a, 1, am_1 + \tau_1)$ where

$$(m_1 - m_2)^2 \cdot a := b\sqrt{l(c_1, c_2)(m_1 - m_2)^2} - (m_1 - m_2)(\tau_1 - \tau_2) \qquad (3.98)$$

touches c_1 and c_2. This spear must also touch any $c \in B_e(c_1, c_2)$. Put $c := (m, \tau)$. Hence $a(m - m_1) = \tau_1 - \tau$ and thus, by (3.98),

$$w^2(\tau_1 - \tau) = b(m - m_1)\sqrt{l(c_1, c_2)w^2} - w(m - m_1)(\tau_1 - \tau_2) \qquad (3.99)$$

where $w := m_1 - m_2$. This equation which only depends on c_1, c_2 and on a suitable chosen b must also hold true if we replace b by $-b$, whence $b \cdot (m - m_1) = 0$ for all $b \in X$ satisfying $b^2 = 1$ and $bw = 0$. Since $X = w^\perp \oplus \mathbb{R}w$, the element $t := m - m_1$ of X must be of the form $t = \alpha b_0 + \beta w$ with $b_0^2 = 1$ and $b_0 w = 0$ where b_0 is a suitable chosen element of $w^\perp$. Applying this b_0 for (3.98) we hence have $b_0 \cdot (m - m_1) = 0$, i.e.

$$0 = b_0 t = \alpha,$$

i.e. $t = \beta w$. This together with (3.99) yields

$$w^2(\tau_1 - \tau) = -\beta w^2(\tau_1 - \tau_2),$$

i.e. $\tau = \tau_1 + \beta(\tau_1 - \tau_2)$. Hence $c = c_1 + (-\beta)(c_2 - c_1)$ with $t = \beta w$, by

$$(m, \tau) = (m_1, \tau_1) + \beta(m_1 - m_2, \tau_1 - \tau_2).$$

Let vice versa $s = (d, 1, \delta)$ be a spear with $c_1 - s - c_2$, i.e. with $dm_i + \tau_i = \delta$. This implies

$$d\left(m_1 + \varrho\left(m_2 - m_1\right)\right) + \tau_1 + \varrho\left(\tau_2 - \tau_1\right) = \delta,$$

i.e. $c \in B_e(c_1, c_2)$ for $c = c_1 + \varrho\left(c_2 - c_1\right)$. $\qquad\square$

Lemma 49. *If s is a spear and c a Laguerre cycle, there exists exactly one spear t satisfying $t - c$ and $t \parallel s$.*

Proof. If $(a, 1, \alpha)$ are the coordinates of s, and (m, τ) those of c, all spears t with $s \parallel t$ and $t - c$ are given by $am + \tau = \beta$, since $t \parallel s$ must have coordinates $(a, 1, \beta)$, by Proposition 42. Hence there is exactly one such spear t, namely

$$(a, 1, am + \tau). \qquad\square$$

As a consequence of Lemma 49 we obtain that parallel spears s, t touching the same Laguerre cycle must coincide. However, this statement is already part of the definition of parallelism of spears.

Suppose that c_1, c_2 are Laguerre cycles such that there is no spear touching c_1 and c_2. If λ is a real number, we define the Laguerre cycle (c_1, c_2, λ) as follows. If the spear s touches c_1, let A_s be the point of contact of s and c_1, and B_s be the point of contact of t and c_2 where the spear t is defined by $s \parallel t$ and $t - c_2$. If $t(s, \lambda)$ denotes the spear which is parallel to s and which touches the Laguerre cycle with coordinates

$$\left(A_s + \lambda\left(B_s - A_s\right), 0\right),$$

then

$$(c_1, c_2, \lambda) := \{t(s, \lambda) \mid s - c_1\} \tag{3.100}$$

is a Laguerre cycle concerning the interpretation of Lemma 41, and

$$B_h(c_1, c_2) := \{(c_1, c_2, \lambda) \mid \lambda \in \mathbb{R}\} \tag{3.101}$$

is called a *hyperbolic pencil.*

In fact! If $\{t(s, \lambda) \mid s - c_1\}$ is the set of all spears touching a Laguerre cycle $c = (m, \tau)$, if $c_i =: (m_i, \tau_i)$ for $i = 1, 2$, and if $(a, 1, am_2 + \tau_1)$ is an arbitrary spear s with $s - c_1$, then $t = (a, 1, am_2 + \tau_2)$ and

$$t(s, \lambda) = \left(a, 1, a\left[A_s + \lambda\left(B_s - A_s\right)\right]\right),$$

in view of $s \parallel t(s, \lambda)$ and $t(s, \lambda) - z$, $z = \left(A_s + \lambda\left(B_s - A_s\right), 0\right)$. Since (see (3.79))

$$A_s = m_1 + \tau_1 a, \; B_s = m_2 + \tau_2 a,$$

we obtain from $t(s, \lambda) - z$,

$$a\left(m - m_1 - \lambda\left(m_2 - m_1\right)\right) = -\left(\tau - \tau_1 - \lambda\left(\tau_2 - \tau_1\right)\right). \tag{3.102}$$

If we choose $a_0 = a \in X$, $a^2 = 1$, such that the left-hand side of this equation is 0, then

$$\tau = \tau_1 + \lambda\,(\tau_2 - \tau_1). \tag{3.103}$$

Hence the left-hand side of (3.102) must be 0 for all $a \in X$, $a^2 = 1$. This implies

$$m = m_1 + \lambda\,(m_2 - m_1). \tag{3.104}$$

The cycle of c is thus, if it exists, uniquely determined. We now would like to verify that $c := (m, \tau)$ with (3.103), (3.104) touches all $t\,(s, \lambda)$, $s - c_1$, but no other spear. Of course, every

$$t\,(s, \lambda) = \big(a, 1, a\,[m_1 + \lambda\,(m_2 - m_1)] + [\tau_1 + \lambda\,(\tau_2 - \tau_1)]\big)$$

with arbitrary $a \in X$, $a^2 = 1$, touches c. On the other hand, if the spear r touches c, take $s - c_1$ parallel to r, whence $r \parallel s \parallel t\,(s, \lambda)$ and $t\,(s, \lambda) - c - r$, i.e. $r = t\,(s, \lambda)$. □

Formulas (3.103), (3.104) also prove

Proposition 50. *Exactly the lines*

$$B_h(c_1, c_2) = \{c_1 + \varrho\,(c_2 - c_1) \mid \varrho \in \mathbb{R}\},\ l\,(c_1, c_2) < 0,$$

of $Z = X \oplus \mathbb{R}$ represent the hyperbolic pencils.

If $B_p(c_1, c_2)$ is a parabolic pencil, we already know that there is exactly one spear touching all c in $B_p(c_1, c_2)$. We will call this spear the *underlying spear* of the pencil. Two parabolic pencils B^1, B^2 are said to be *parallel*, designated by $B^1 \parallel B^2$, provided their underlying spears are parallel.

Let B^1, B^2 be parabolic pencils having exactly one Laguerre cycle in common. Then the union of all parabolic pencils B satisfying

$$B \cap B^1 \neq \emptyset \text{ and } B \parallel B^2 \tag{3.105}$$

will be called a *bundle*, designated by $\mathfrak{B} = \mathfrak{B}\,(B^1, B^2)$.

Proposition 51. (α) *Two parabolic pencils are parallel if, and only if, their associated lines in $Z = X \oplus \mathbb{R}$ (see Proposition 46, (γ)) are parallel.*

(β) *Let $B^1 = B_p(c, c_1)$, $B^2 = B_p(c, c_2)$ be parabolic pencils having exactly c in common. The union of all parabolic pencils B satisfying (3.105) is then given by*

$$\mathfrak{B}\,(B^1, B^2) = \{c + \alpha\,(c_1 - c) + \beta\,(c_2 - c) \mid \alpha, \beta \in \mathbb{R}\}, \tag{3.106}$$

where the elements $c_1 - c$, $c_2 - c$ of Z are linearly independent in Z.

(γ) *If a, b, p are Laguerre cycles such that $a, b \in Z$ are linearly independent, then*

$$\mathfrak{B} = \{p + \alpha a + \beta b \mid \alpha, \beta \in \mathbb{R}\} \tag{3.107}$$

is a bundle if, and only if,

$$(ab)^2 > a^2 \cdot b^2. \tag{3.108}$$

(δ) *If B_h is a hyperbolic pencil, then there exist bundles $\mathfrak{B}^1, \mathfrak{B}^2$ with*

$$B_h = \mathfrak{B}^1 \cap \mathfrak{B}^2. \tag{3.109}$$

Proof. (α) If $B_p(c_1, c_2)$ with $c_i = (m_i, \tau_i)$, $i = 1, 2$, is a parabolic pencil, then, by (3.83), $s = (a, 1, m_1 a + \tau_1)$ is the underlying spear satisfying $(\tau_2 - \tau_1) a := m_1 - m_2$. Observe $\tau_1 \neq \tau_2$, since otherwise $c_1 = c_2$. Suppose now that $B_p(c_1, c_2)$, $B_p(c_3, c_4)$ are parabolic pencils. They are hence parallel, by Proposition 42, if, and only if,

$$\frac{m_1 - m_2}{\tau_2 - \tau_1} = \frac{m_3 - m_4}{\tau_4 - \tau_3},$$

i.e. if, and only if,

$$\frac{c_1 - c_2}{\tau_2 - \tau_1} = \frac{c_3 - c_4}{\tau_4 - \tau_3}$$

holds true.

(β) Since $B^i = \{c + \varrho_i(c_i - c) \mid \varrho_i \in \mathbb{R}\}$, $i = 1, 2$, have exactly c in common, $c_1 - c$, $c_2 - c$ must be linearly independent. Hence $\mathfrak{B}(B^1, B^2)$ is given by the union of all lines

$$l_\varrho := \{c + \varrho(c_1 - c) + \lambda(c_2 - c) \mid \lambda \in \mathbb{R}\}, \varrho \in \mathbb{R},$$

in view of (α).

(γ) Assume that (3.108) holds true. Put $c := p$ and

$$u := a^2, v := b^2, w := ab.$$

Hence $w^2 > uv$. Define $\xi := \sqrt{w^2 - uv}$ and

$$c_1 - c := (\xi - w) a + ub, \quad c_2 - c := -(\xi + w) a + ub$$

for $u \neq 0$, and

$$c_1 - c := a, \quad c_2 - c := va - 2wb$$

for $u = 0$. Then $l(c, c_i) = 0$, $i = 1, 2$, and $c_1 - c$, $c_2 - c$ are linearly independent in Z. Hence (3.107) has the form (3.106). Vice versa, we are now assuming that (3.107) is a bundle, say the bundle (3.106). Since

$$p + 2a, \, p + a \in \mathfrak{B} = \mathfrak{B}(B^1, B^2),$$

we get $p + 2a = c + \alpha_1(c_1 - c) + \beta_1(c_2 - c)$, i.e.

$$a = (p + 2a) - (p + a) =: \alpha(c_1 - c) + \beta(c_2 - c),$$

and similarly,

$$b = \gamma(c_1 - c) + \delta(c_2 - c)$$

with $\alpha\delta - \beta\gamma \neq 0$, since a, b are linearly independent. Hence

$$a^2 = \alpha^2(c_1 - c)^2 + 2\alpha\beta\sigma + \beta^2(c_2 - c)^2 = 2\alpha\beta\sigma$$

with $\sigma := (c_1 - c)(c_2 - c)$. Moreover,

$$b^2 = 2\gamma\delta\sigma, \ ab = (\alpha\delta + \beta\gamma)\sigma.$$

Hence $(ab)^2 - a^2b^2 = (\alpha\delta - \beta\gamma)^2\sigma^2$. We, finally, have to show that $\sigma \neq 0$. But $\sigma = 0$ would imply

$$(c_1 - c_2)^2 = [(c_1 - c) + (c - c_2)]^2 = -2\sigma = 0,$$

i.e. c_1 would touch c_2, i.e. $c_1 - c_2 - c$, i.e. $c_2 \in B^1$, i.e. $B^1 = B^2$, by Proposition 46, (α), a contradiction.

(δ) Let $B_h = \{p + \alpha a \mid \alpha \in \mathbb{R}\}$ be a hyperbolic pencil. Hence $l(a, 0) < 0$, by Proposition 50, i.e. $a^2 < 0$. Thus $\tau \neq 0$ for

$$a =: (m, \tau) \in Z = X \oplus \mathbb{R}.$$

If $m \neq 0$, take $e \in X$ with $e^2 = 1$ and $e \in m^\perp$, and put

$$b := (e, 1), \ c := (e, -1).$$

Then a, b, c are linearly independent, and, by (γ),

$$\{p + \alpha a + \beta b \mid \alpha, \beta \in \mathbb{R}\}, \ \{p + \alpha a + \gamma c \mid \alpha, \gamma \in \mathbb{R}\}$$

are bundles with B_h as their intersection.— If $m = 0$, take $e_1, e_2 \in X$ with $e_1^2 = e_2^2 = 1$, $e_1 e_2 = 0$, and put $b := (e_1, 1)$, $c := (e_2, 1)$. Then proceed as above. $\square$

3.12   Lie cycles, groups Lie (X), Lag (X)

A *Lie cycle* of X is a Laguerre cycle of X or a spear of X, or a new object which will be designated by the symbol ∞. The set of all Lie cycles of X will be denoted by $\Delta = \Delta(X)$. Hence $\Delta = \Gamma \cup \Sigma \cup \{\infty\}$. Another fundamental notion in the context of Lie cycles is again the *contact relation*. This is defined to be a reflexive and symmetric relation "–" on Δ satisfying

  (i) ∞ touches every spear, but no Laguerre cycle,

(ii) the spear s_1 touches the spear s_2 if, and only if, $s_1 \parallel s_2$,

(iii) the spear or Laguerre cycle g touches the Laguerre cycle c if, and only if, $g - c$ holds true.

A *Lie transformation* of X is a bijection

$$\lambda : \Delta\,(X) \to \Delta\,(X)$$

such that for all $x, y \in \Delta$ the statements $x - y$ and $\lambda\,(x) - \lambda\,(y)$ are equivalent.

By Lie (X) we denote the group of all Lie transformations of X where the multiplication is defined to be the usual product of bijections.

A Laguerre transformation of X is a Lie transformation λ satisfying $\lambda\,(\infty) = \infty$. If λ is such a Laguerre transformation, then $\lambda\,(\Sigma) \subseteq \Sigma$ and $\lambda\,(\Gamma) \subseteq \Gamma$, because of (i) and

$$\infty - s \Leftrightarrow \infty - \lambda\,(s),$$

$$\infty \neq c \Leftrightarrow \infty \neq \lambda\,(c)$$

for all $s \in \Sigma$ and $c \in \Gamma$.

We shall identify Laguerre transformations with their restrictions on

$$\Sigma \cup \Gamma = \Delta \backslash \{\infty\}.$$

The group of all Laguerre transformations of X, Lag (X), is a subgroup of Lie (X). An important subgroup of Lag (X) is $\mathrm{Lag}_*(X)$ consisting of all separation preserving Laguerre transformations of X.

The geometry $\big(\Delta\,(X), \mathrm{Lie}\,(X)\big)$ is called *Lie sphere geometry* over X, and $\big(\Sigma \cup \Gamma, \mathrm{Lag}\,(X)\big)$ *Laguerre sphere geometry* over X. Since Lag (X) is the stabilizer of Lie (X) in ∞, obviously, Laguerre geometry concerns Lie geometry with respect to a fixed Lie cycle, namely the cycle ∞. The geometry $\big(\Gamma, \mathrm{Lag}_*(X)\big)$ is called *proper Laguerre sphere geometry* over X. Of course, Σ defines an invariant notion (Σ, φ) of $\big(\Gamma, \mathrm{Lag}_*(X)\big)$ as well as of the geometry $\big(\Gamma, \mathrm{Lag}\,(X)\big)$, by means of

$$\varphi\,(f, s) = \{f\,(c) \mid c - s\}.$$

Proposition 52. *A bijection* $\lambda : \Sigma \cup \Gamma \to \Sigma \cup \Gamma$ *with* $\lambda\,(\Sigma) \subseteq \Sigma$ *and* $\lambda\,(\Gamma) \subseteq \Gamma$ *is a Laguerre transformation of X if, and only if,*

$$s - c \Leftrightarrow \lambda\,(s) - \lambda\,(c) \tag{3.110}$$

holds true for all $s \in \Sigma$ *and* $c \in \Gamma$.

Proof. Obviously, $\lambda\,(\Sigma) = \Sigma$ and $\lambda\,(\Gamma) = \Gamma$ are satisfied, since λ is a bijection. If λ is a Laguerre transformation, then (3.110) holds true. If, vice versa, λ satisfies (3.110), we extend it by $\lambda\,(\infty) = \infty$, and, of course, we extend the relation "−" on

Δ by (i), (ii), (iii) and in such a way that it is reflexive and symmetric. Now we must prove

$$x - y \Leftrightarrow \lambda(x) - \lambda(y) \tag{3.111}$$

for all $x, y \in \Delta$. Instead of (3.111) we will show

$$x - y \Rightarrow \lambda(x) - \lambda(y) \text{ and } x \not{-} y \Rightarrow \lambda(x) \not{-} \lambda(y) \tag{3.112}$$

for all $x, y \in \Delta$. If $\infty \in \{x, y\}$, (3.112) holds true, because of $\lambda(\Sigma) = \Sigma$, $\lambda(\Gamma) = \Gamma$. If one of the elements x, y of Δ is a spear and the other one a Laguerre cycle, we know (3.112) from (3.110). So it remains to consider the two cases

$$1) \{x, y\} \subset \Sigma, \quad 2) \{x, y\} \subset \Gamma$$

where we even may assume $x \neq y$. In the first case, $x - y$ implies that there is no Laguerre cycle c with $x - c - y$, and $x \not{-} y$ yields that there is such a Laguerre cycle having this property. Hence (3.112) follows from (3.110) in case 1). Also in the second case (3.112) follows from (3.110). $\qquad\square$

On the basis of Lemma 41, there exist two other possibilities to define a Laguerre transformation.

Proposition 53. *Identify $s \in \Sigma$ with $\{c \in \Gamma \mid s - c\}$. A Laguerre transformation can now be defined as a bijection λ of Γ such that images and pre-images of spears are spears.*

Proof. We must extend $\lambda : \Gamma \to \Gamma$ to $\lambda : \Sigma \cup \Gamma \to \Sigma \cup \Gamma$ with $\lambda(\Sigma) \subseteq \Sigma$ and show that (3.110) holds true. The image of the spear $s = \{c \in \Gamma \mid s - c\}$ must be a spear,

$$\lambda(s) := \{\lambda(c) \mid c \in \Gamma \text{ and } s - c\},$$

since images of spears are supposed to be spears. Observe

$$s - c \Rightarrow \lambda(c) \in \lambda(s) \Rightarrow \lambda(s) - \lambda(c).$$

The pre-image of $s = \{c \in \Gamma \mid s - c\}$, namely

$$\lambda^{-1}(s) := \{\lambda^{-1}(c) \mid c \in \Gamma \text{ and } s - c\},$$

is also a spear. Observe $\lambda(\lambda^{-1}s)) = s$. Hence

$$s - c \Rightarrow \lambda^{-1}(c) \in \lambda^{-1}(s) \Rightarrow \lambda^{-1}(s) - \lambda^{-1}(c),$$

i.e. $\lambda(s) - \lambda(c) \Rightarrow \lambda^{-1}[\lambda(s)] - \lambda^{-1}[\lambda c] \Rightarrow s - c$. $\qquad\square$

Proposition 54. *Identify $c \in \Gamma$ with $\{s \in \Sigma \mid c - s\}$. A Laguerre transformation can also be defined as a bijection λ of Σ such that images and pre-images of Laguerre cycles are Laguerre cycles.*

Proof. We now must extend $\lambda : \Sigma \to \Sigma$ to $\lambda : \Sigma \cup \Gamma \to \Sigma \cup \Gamma$ with $\lambda(\Gamma) \subseteq \Gamma$ and show that (3.110) holds true. Mutatis mutandis, we may follow the procedure of the proof of Proposition 53 in order to complete the present proof. $\square$

Examples of Laguerre transformations of X are the (Laguerre) *dilatations*. Let δ be a real number and define

$$\lambda_\delta(m, \tau) := (m, \tau + \delta), \lambda_\delta(a, 1, \alpha) := (a, 1, \alpha + \delta)$$

for the Laguerre cycle with coordinates (m, τ) and the spear $(a, 1, \alpha)$. Since

$$am + \tau = \alpha \text{ and } am + (\tau + \delta) = \alpha + \delta$$

are equivalent, (3.110) is satisfied by Proposition 40.

If, for instance, $\delta = 1$, then the Laguerre cycle $(0, 0)$, which is actually the point $0 \in X$, has as image under λ_1 the Laguerre cycle $(0, 1)$ which is not a point of X. So Laguerre transformations need not be induced by point transformations of X. The set $D(X)$ of dilatations of X is a subgroup of Lag (X).

If $\omega \in O(X)$, then, obviously,

$$\lambda_\omega(m, \tau) := \big(\omega(m), \tau\big), \lambda_\omega(a, 1, \alpha) := \big(\omega(a), 1, \alpha\big)$$

defines a Laguerre transformation. If d is a fixed element of X, and if $\sigma \neq 0$ is a fixed real number, then also

$$\lambda(m, \tau) := (\sigma \cdot m + d, \sigma \cdot \tau), \lambda(a, 1, \alpha) := (a, 1, \sigma \cdot \alpha + a \cdot d)$$

defines a Laguerre transformation designated by $\lambda_{d,\sigma}$. If $z_1 = (m_1, \tau_1)$, $z_2 = (m_2, d_2)$ are Laguerre cycles, then, obviously,

$$\lambda_{\tau_2 - \tau_1} \cdot \lambda_{m_2 - m_1, 1}(z_1) = z_2,$$

so that Lag (X) operates transitively on Γ. The same group acts transitively on Σ as well. This will be shown as follows. If $a, b \in X$ satisfy $a^2 = 1 = b^2$, there exists $\omega \in O(X)$ with $\omega(a) = b$ (see step A of the proof of Theorem 7, chapter 1). Suppose now that $(a, 1, \alpha)$ and $(b, 1, \beta)$ are spears. Take $\omega \in O(X)$ with $\omega(a) = b$. Then

$$\lambda_{\beta - \alpha} \lambda_\omega(a, 1, \alpha) = (b, 1, \beta).$$

Reversing the orientation,

$$\varphi(m, \tau) := (m, -\tau), \varphi(a, 1, \alpha) := (a, -1, \alpha),$$

also represents a Laguerre transformation.

Proposition 55. *Suppose that λ is a Laguerre transformation. Then images and pre-images of parabolic, elliptic, hyperbolic pencils are parabolic, elliptic, hyperbolic pencils, respectively. Images and pre-images of bundles are bundles.*

Proof. This statement follows immediately from the definitions, with the exception of the assertion that images and pre-images of hyperbolic pencils are hyperbolic pencils. So let B_h be a hyperbolic pencil. We are interested in

$$\lambda(B_h) := \{\lambda(c) \mid c \in B_h\}.$$

Suppose that $\mathfrak{B}^1, \mathfrak{B}^2$ are bundles satisfying (3.109). Hence

$$\lambda(B_h) = \lambda(\mathfrak{B}^1) \cap \lambda(\mathfrak{B}^2)$$

is a line g of $Z = X \oplus \mathbb{R}$, say the line $c_1 + \mathbb{R}(c_2 - c_1)$. Here $l(c_1, c_2)$ must be negative, since otherwise g (and then also $B_h = \lambda^{-1}(g)$) would represent a parabolic or an elliptic pencil.   □

3.13 Lie cycle coordinates, Lie quadric

Put $\mathbb{R}^* := \mathbb{R} \setminus \{0\}$. If $c = (m, \tau)$ is in Γ, define

$$\left(m, -\tau, \frac{1 + l(0, c)}{2}, \frac{1 - l(0, c)}{2} \right)$$

to be *coordinates* of the Lie cycle c. Moreover, *coordinates* of the spear $(a, 1, \alpha)$ are $(a, 1, \alpha, -\alpha)$, and those of ∞ are $(0, 0, 1, -1)$. We would like to have coordinates of Lie cycles as *homogenous coordinates*. Therefore we say that every quadruple contained in

$$\mathbb{R}^* \left(m, -\tau, \frac{1 + l(0,c)}{2}, \frac{1 - l(0,c)}{2} \right)$$
$$= \left\{ \lambda m, -\lambda \tau, \frac{\lambda + \lambda l(0,c)}{2}, \frac{\lambda - \lambda l(0,c)}{2} \,\middle|\, \lambda \in \mathbb{R}^* \right\}$$

are *coordinates* of (m, τ). Similarly we consider

$$\mathbb{R}^*(a, 1, \alpha, -\alpha) \quad \text{and} \quad \mathbb{R}^*(0, 0, 1, -1)$$

for spears, for ∞, respectively. Observe that $\mathbb{R}^*(v, \xi_1, \xi_2, \xi_3)$ is a subset of $V := X \oplus \mathbb{R}^3$ and that

$$v^2 - \xi_1^2 - \xi_2^2 + \xi_3^2 = 0 \tag{3.113}$$

holds true for the coordinates of a Lie cycle. Given

$$\mathbb{R}^*(v, \xi_1, \xi_2, \xi_3) \in V$$

with (3.113) and $(v, \xi_1, \xi_2, \xi_3) \neq 0$, we distinguish the following cases.

Case: $\xi_2 + \xi_3 \neq 0$.

Then $\mathbb{R}^*(\alpha v, \alpha\xi_1, \alpha\xi_2, \alpha\xi_3)$ with $\alpha \cdot (\xi_2 + \xi_3) := 1$ represents the Laguerre cycle $c = (\alpha v, -\alpha\xi_1)$, by observing

$$\frac{1+c^2}{2} = \frac{1+\alpha^2(v^2-\xi_1^2)}{2} = \frac{1}{2}\left(1 + \alpha^2(\xi_2^2 - \xi_3^2)\right) = \alpha\xi_2,$$

$$\frac{1-c^2}{2} = 1 - \frac{1+c^2}{2} = 1 - \alpha\xi_2 = \alpha\xi_3.$$

Here and in the remaining sections of chapter 3 we are still working with the vector space Z of Laguerre cycles of X and with the product (3.88).

Case: $\xi_2 + \xi_3 = 0$.

Hence (3.113) implies $v^2 = \xi_1^2$. If $\xi_1 = 0$, we get coordinates of ∞, and if $\xi_1 \neq 0$, we obtain the spear (v, ξ_1, ξ_2).

The set $LQ(X)$ of all subsets $\mathbb{R}^*(v, \xi_1, \xi_2, \xi_3)$ of $X \oplus \mathbb{R}^3$ such that $(v, \xi_1, \xi_2, \xi_3) \neq (0,0,0,0)$ and (3.113) hold true is called the *Lie quadric* of the Lie geometry over X.

What we proved before, is the

Proposition 56. *Associate to every Lie cycle of X of coordinates (v, ξ_1, ξ_2, ξ_3) the element $\mathbb{R}^*(v, \xi_1, \xi_2, \xi_3)$ of $LQ(X)$. This mapping*

$$\psi : \Delta(X) \to LQ(X)$$

is a bijection between the set $\Delta(X)$ of all Lie cycles of X and $LQ(X)$.

Also important in connection with Proposition 56 is the following

Proposition 57. *Let x, y be Lie cycles and let*

$$(v, \xi_1, \xi_2, \xi_3), \ (w, \eta_1, \eta_2, \eta_3)$$

be coordinates of x, y, respectively. Then $x - y$ holds true if, and only if,

$$vw - \xi_1\eta_1 - \xi_2\eta_2 + \xi_3\eta_3 = 0 \tag{3.114}$$

is satisfied.

Proof. Assume $x - y$. If $x = y$, (3.114) follows from $\psi(x) \in LQ(X)$. If $\infty \in \{x, y\}$, say $x = \infty \neq y$, (3.114) holds true for all spears y. In the case $x = (a, 1, \alpha)$, $y = (b, 1, \beta)$, we get $a = b$ from $x - y$, i.e.

$$ab - 1 \cdot 1 - \alpha \cdot \beta + (-\alpha)(-\beta) = 0.$$

If x is a spear and y a Laguerre cycle, $x - y$ implies

$$am + \tau = \alpha$$

with $x = (a, 1, \alpha)$, $y = (m, \tau)$, i.e. $am + \tau - \alpha \frac{1+y^2}{2} - \alpha \frac{1-y^2}{2} = 0$. In the case that x, y are both Laguerre cycles, $(m, \tau), (n, \sigma)$, respectively, $x - y$ yields $0 = l\,(x, y) = (x - y)^2$. Hence

$$mn - \tau\sigma - \frac{1}{4}(1 + x^2)(1 + y^2) + \frac{1}{4}(1 - x^2)(1 - y^2) = -\frac{1}{2}(x - y)^2 = 0,$$

by $mn - \tau\sigma = xy$.

Vice versa, assume (3.114) for the Lie cycles x, y. If $x = \infty$, then we get $\eta_2 + \eta_3 = 0$ from (3.114). Hence y must be a spear or ∞. If $x = (a, 1, \alpha)$, then

$$aw - \eta_1 - \alpha\eta_2 - \alpha\eta_3 = 0. \tag{3.115}$$

Of course, $y = \infty$ solves this equation. If $y = (b, 1, \beta)$ is a solution, then $ab - 1 = 0$, i.e. $a = b$, in view of Lemma 1, chapter 1. Hence $y \parallel x$, i.e. $x - y$. For $y = (m, \tau)$, we obtain

$$am + \tau - \frac{\alpha}{2}(1 + y^2) - \frac{\alpha}{2}(1 - y^2) = 0$$

from (3.115), i.e. $am + \tau = \alpha$, i.e. $x - y$. We, finally, consider the case $x = (m, \tau)$. Hence, by (3.114),

$$mw + \tau\eta_1 - \frac{1}{2}(1 + x^2)\,\eta_2 + \frac{1}{2}(1 - x^2)\,\eta_3 = 0. \tag{3.116}$$

Since the contact relation is symmetric and also (3.114) in x, y, we only need to check the case $y = (n, \sigma)$ in (3.116). Here we get

$$mn - \tau\sigma - \frac{1}{4}(1 + x^2)(1 + y^2) + \frac{1}{4}(1 - x^2)(1 - y^2) = 0,$$

i.e. $(x - y)^2 = 0$, i.e. $x - y$. $\qquad\square$

Proposition 58. *If x, y are Lie cycles, then there exists $\lambda \in$ Lie (X) with $\lambda\,(x) = y$.*

Proof. 1) We only need to show that to $z \in \Delta$ there exists $\alpha \in$ Lie (X) with $\alpha\,(\infty) = z$ because $\alpha\,(\infty) = x$ and $\beta\,(\infty) = y$ imply $(\beta\alpha^{-1})(x) = y$.

2) Define $\varepsilon : LQ\,(X) \to LQ\,(X)$ by

$$\varepsilon\left(\mathbb{R}^*(v, \xi_1, \xi_2, \xi_3)\right) = \mathbb{R}^*(v, \xi_1, \xi_2, -\xi_3).$$

Since $\varepsilon\varepsilon\,(x) = x$ for all $x \in LQ\,(X)$, ε must be a bijection of $LQ\,(X)$: $\varepsilon\,(x)$ is also the inverse image of x, and $\varepsilon\,(x) = \varepsilon\,(y)$ implies $x = \varepsilon\varepsilon\,(x) = \varepsilon\varepsilon\,(y) = y$ (compare the consideration in section 7, chapter 1, before the definition of a translation group). Identifying in the following a Lie cycle z with its image $\psi\,(z)$ on $LQ\,(X)$, we can say that ε is a Lie transformation, since (3.114) implies

$$vw - \xi_1\eta_1 - \xi_2\eta_2 + (-\xi_3)(-\eta_3) = 0.$$

3) Note that $\varepsilon(\infty)$ is the Laguerre cycle $(0,0)$ with $m = 0$ and $\tau = 0$. Since Lag (X) operates transitively on Γ (see section 12), to $z \in \Gamma$ there exists $\alpha \in$ Lie (X) with $\alpha(0,0) = z$. Hence $(\alpha\varepsilon)(\infty) = z$.

4) Take $a \in X$ with $a^2 = 1$. Then $\mathbb{R}^*(a, 1, 1, -1)$ is a spear s and $\varepsilon(s)$ a Laguerre cycle x. Since Lag (X) operates transitively on Σ (see section 12), to $z \in \Sigma$ exists $\alpha \in$ Lie (X) with $\alpha(s) = z$. If $\beta(\infty) = x$ for a suitable $\beta \in$ Lie (X) (see step 3)), then $\alpha\varepsilon\beta(\infty) = z$.

5) If $z = \infty$, then id $(\infty) = \infty$. $\qquad\qquad\qquad\qquad\qquad\qquad\qquad\square$

Let $\mathbb{P} = \mathbb{P}(X \oplus \mathbb{R}^3)$ be the set of all 1-dimensional subspaces of $X \oplus \mathbb{R}^3$ and define $\mathbb{L} = \mathbb{L}(X \oplus \mathbb{R}^3)$ to be the set of all 2-dimensional subspaces of the vector space $X \oplus \mathbb{R}^3$. The elements of $\mathbb{P}$ are called *points* and the elements of $\mathbb{L}$ *lines*. Let p be a point and l be a line. We say that p is on l or that l goes through p provided $p \subset l$. In this case we also say that p is *incident* with l or that l is *incident* with p. Lines may be considered as sets of points by identifying a line l with the set of points on l,

$$\{p \in \mathbb{P} \mid p \subset l\}.$$

Projective transformations are defined as bijections of $\mathbb{P}$ such that images and pre-images of lines are lines. We thus may speak of the group $G = G(X \oplus \mathbb{R}^3)$ of projective transformations. The geometry

$$(\mathbb{P}, G)$$

is called the *projective geometry* $\Pi = \Pi(X \oplus \mathbb{R}^3)$ *over* $X \oplus \mathbb{R}^3$ and G is called its *projective group*.

If p_1, p_2 are distinct points, there is exactly one line l through p_1, p_2. A subset S of $\mathbb{P}$ is called *collinear* provided there exists a line containing all points of S.

It is important that the Lie quadric $LQ(X)$ is a subset of the set of points of the projective geometry Π over $X \oplus \mathbb{R}^3$.

Proposition 59. *Again we do not distinguish between a Lie cycle of X and its image on $LQ(X)$ in $\Pi(X \oplus \mathbb{R}^3)$. Let c_1, c_2 be distinct Lie cycles of X. Then the following properties are equivalent.*

(α) $c_1 - c_2$,

(β) *the line of Π through c_1, c_2 is contained in $LQ(X)$,*

(γ) *there exists $c \in LQ(X) \backslash \{c_1, c_2\}$ such that c, c_1, c_2 are collinear in Π.*

Moreover, the following holds true.

If c_1, c_2, c_3 are pairwise distinct Lie cycles which are pairwise in contact, then $\{c_1, c_2, c_3\}$ is collinear in Π.

Proof. Assume (α). Let c_i, $i = 1, 2$, be given by

$$\mathbb{R}^*(v_i, \xi_{i1}, \xi_{i2}, \xi_{i3}).$$

Then $\mathbb{R}^*(\alpha v_1 + \beta v_2, \alpha \xi_{11} + \beta \xi_{21}, \dots)$ is the line through c_1, c_2 where (α, β) runs over $\mathbb{R}^2 \backslash \{(0,0)\}$. That all these points are in $LQ(X)$ follows immediately from

$$v_i^2 - \xi_{i1}^2 - \xi_{i2}^2 + \xi_{i3}^2 = 0, \ i = 1, 2, \tag{3.117}$$

$$v_1 v_2 - \xi_{11}\xi_{21} - \xi_{12}\xi_{22} + \xi_{13}\xi_{23} = 0 \tag{3.118}$$

via $(\alpha v_1 + \beta v_2)^2 - (\alpha \xi_{11} + \beta \xi_{21})^2 - (\alpha \xi_{12} + \beta \xi_{22})^2 + (\alpha \xi_{13} + \beta \xi_{23})^2 = \cdots = 0$.
That (β) implies (γ) is trivial.

Assume (γ). There hence exists $\alpha \neq 0$ and $\beta \neq 0$ such that c is given by

$$\mathbb{R}^*(\alpha v_1 + \beta v_2, \dots).$$

Observing $c_1, c_2, c \in LQ(X)$ and $\alpha \neq 0 \neq \beta$, we get from

$$0 = (\alpha v_1 + \beta v_2)^2 - (\alpha \xi_{11} + \beta \xi_{21})^2 - \cdots$$

and (3.117), obviously, (3.118), i.e. $c_1 - c_2$. In order to prove the last part of Proposition 59, we distinguish several cases.

Case 1: $\infty \in \{c_1, c_2, c_3\}$. Hence $\{c_1, c_2, c_3\} = \{\infty, x, y\}$ where $x \neq y$ are parallel spears. Observe

$$\mathbb{R}^*(0,0,1,-1) = \mathbb{R}^*\big(1 \cdot (a,1,\alpha,-\alpha) + (-1)(a,1,\beta,-\beta)\big)$$

with $x =: (a,1,\alpha)$, $y =: (a,1,\beta)$, $\alpha \neq \beta$.

For the remaining cases we assume $\infty \notin \{c_1, c_2, c_3\}$.

Case 2: c_1, c_2, c_3 are three spears. Hence $c_i = (a,1,\alpha_i)$, $i = 1,2,3$, where $\alpha_1, \alpha_2, \alpha_3$ are pairwise distinct. Observe

$$(a,1,\alpha_3,-\alpha_3) = \lambda(a,1,\alpha_1,-\alpha_1) + (1-\lambda)(a,1,\alpha_2,-\alpha_2)$$

with $\lambda(\alpha_2 - \alpha_1) := \alpha_2 - \alpha_3$.

Case 3: $c_1 = (a,1,\alpha)$, $c_2 = (a,1,\beta)$, $c_3 = (m,\tau)$, $\alpha \neq \beta$. Hence, by the definition of parallelism of spears, this case does not occur.

Case 4: $c_1 = (a,1,\alpha)$, $c_2 = (m,\tau) \neq (n,\sigma) = c_3$. Hence

$$ma + \tau = \alpha = na + \sigma, \ (m-n)^2 = (\tau - \sigma)^2.$$

Thus $(m-n)a = \sigma - \tau$, i.e. $(\sigma - \tau)^2 = [(m-n)a]^2 \leq (m-n)^2 = (\tau - \sigma)^2$. This implies, by Lemma 1, chapter 1,

$$m - n =: \lambda a, \ \text{i.e.} \ \lambda = (\lambda a)a = (m-n)a = \sigma - \tau.$$

Now observe, by $c_3^2 = n^2 - \sigma^2$,

$$\left(m, -\tau, \frac{1 + m^2 - \tau^2}{2}, \frac{1 - m^2 + \tau^2}{2}\right)$$

$$= \lambda(a,1,\alpha,-\alpha) + \left(n, -\sigma, \frac{1 + c_3^2}{2}, \frac{1 - c_3^2}{2}\right).$$

Case 5: c_1, c_2, c_3 are all Laguerre cycles,

$$c_1 =: (a, \alpha), \ c_2 =: (b, \beta), \ c_3 =: (c, \gamma).$$

Hence $(a - b)^2 = (\alpha - \beta)^2$, $(b - c)^2 = (\beta - \gamma)^2$, $(c - a)^2 = (\gamma - \alpha)^2$. Observe that $a = b$ implies $\alpha = \beta$, and that $\alpha = \beta$ yields $a = b$. Thus a, b, c must be pairwise distinct, and also α, β, γ. Note that

$$(\gamma - \alpha)^2 = \big((a - b) + (b - c)\big)^2 = (\alpha - \beta)^2 + 2(a - b)(b - c) + (\beta - \gamma)^2$$

implies $(a - b)(b - c) = (\alpha - \beta)(\beta - \gamma)$, i.e. $(\alpha - \beta)^2(\beta - \gamma)^2 = [(a - b)(b - c)]^2 \le (a - b)^2(b - c)^2 = (\alpha - \beta)^2(\beta - \gamma)^2$. Hence

$$b - c = \mu\,(a - b), \ \mu \in \mathbb{R},$$

by $a \ne b$. Together with $(a - b)(b - c) = (\alpha - \beta)(\beta - \gamma)$, we obtain

$$\mu = \frac{\beta - \gamma}{\alpha - \beta},$$

by noticing that α, β, γ are pairwise distinct. We now verify, by

$$c_1^2 = a^2 - \alpha^2, \ c_2^2 = b^2 - \beta^2, \ c_3^2 = c^2 - \gamma^2,$$

that

$$\left(c, -\gamma, \frac{1 + c_3^2}{2}, \frac{1 - c_3^2}{2}\right) = -\mu\left(a, -\alpha, \frac{1 + c_1^2}{2}, \frac{1 - c_1^2}{2}\right)$$

$$+ (1 + \mu)\left(b, -\beta, \frac{1 + c_2^2}{2}, \frac{1 - c_2^2}{2}\right)$$

holds true: an essential step for this purpose is to show

$$(a^2 - \alpha^2)(\beta - \gamma) + (b^2 - \beta^2)(\gamma - \alpha) + (c^2 - \gamma^2)(\alpha - \beta) = 0.$$

However, this equation follows from elementing

$$c = (1 + \mu)\,b - \mu a \ \text{and} \ \gamma = (1 + \mu)\,\beta - \mu\alpha,$$

and by applying $a^2 - \alpha^2 + b^2 - \beta^2 = 2(ab - \alpha\beta)$. $\qquad\square$

3.14 Lorentz boosts

Lemma 60. *(α) If $w \in Z = X \oplus \mathbb{R}$, there exist reals α, β and linearly independent $a, b \in Z$ satisfying*

$$a^2 = 0, \ b^2 = 0, \ w = \alpha a + \beta b. \tag{3.119}$$

(β) *Let the linearly independent* $a, b \in Z$ *satisfy* $a^2 = 0 = b^2$. *Then* $ab \neq 0$. *Moreover,* $(\alpha a + \beta b)^2 = 0$ *for* $\alpha, \beta \in \mathbb{R}$ *implies* $\alpha = 0$ *or* $\beta = 0$.

Proof. (α) In the case $w = (0, \tau)$ take $e \in X$ with $e^2 = 1$. Then

$$w = \frac{\tau}{2}(e, 1) - \frac{\tau}{2}(e, -1).$$

If $w = (m, \tau)$, $m \neq 0$, note

$$w = \left(\frac{1}{2} + \frac{\tau}{2\sqrt{m^2}}\right)(m, \sqrt{m^2}) + \left(\frac{1}{2} - \frac{\tau}{2\sqrt{m^2}}\right)(m, -\sqrt{m^2}).$$

(β) Put $a =: (m_1, \tau_1)$, $b =: (m_2, \tau_2)$. Assume $ab = 0$. Now $m_1^2 = \tau_1^2$, $m_2^2 = \tau_2^2$ and $m_1 m_2 = \tau_1 \tau_2$. Hence

$$(m_1 m_2)^2 = m_1^2 m_2^2,$$

i.e. there exist $(\xi, \eta) \in \mathbb{R}^2 \setminus \{(0, 0)\}$ such that $\xi m_1 + \eta m_2 = 0$, by Lemma 1, chapter 1. Thus

$$\begin{aligned}(\xi a + \eta b)^2 &= \xi^2 a^2 + 2\xi \eta ab + \eta^2 b^2 = 0, \\ 0 = (\xi a + \eta b)^2 &= (\xi m_1 + \eta m_2)^2 - (\xi \tau_1 + \eta \tau_2)^2,\end{aligned}$$

i.e $\xi \tau_1 + \eta \tau_2 = 0$, by $\xi m_1 + \eta m_2 = 0$, whence $\xi a + \eta b = 0$. But a, b are linearly independent.—

The last part of statement (β) follows from $ab \neq 0$. $\square$

If $x = (m, \tau) \in Z$, we shall write $\bar{x} := m$ and $x_0 := \tau$. Hence $x = (\bar{x}, x_0)$ will be the typical element of $Z = X \oplus \mathbb{R}$.

Let p be an element of X with $p^2 < 1$ and let $k \neq -1$ be a real number satisfying $k^2 \cdot (1 - p^2) = 1$. Define

$$A_p(x) := (x_0 p, \bar{x} p)$$

for $x \in Z$. Obviously, $A_p : Z \to Z$ is a linear mapping of Z. Define

$$B_{p,k} := E + k A_p + \frac{k^2}{k+1} A_p^2,$$

where E designates the identity mapping of Z. Moreover, put

$$B_{0,-1}(x) := (\bar{x}, -x_0).$$

The mappings $B_{p,k}$ are linear and they are bijections of Z, since

$$B_{p,k} \cdot B_{-p,k} = E \tag{3.120}$$

holds true (compare the first Remark of section 7, chapter 1, and also the following Remark). $B_{p,k}$ is called a *proper Lorentz boost* for $k > 0$ and an *improper* one otherwise.

Remark. Obviously, $B_{p,k} \cdot B_{-p,k} = E$ holds true for $k = -1$. In the case $k \neq -1$, we get

$$A_{-p}(x) = -A_p(x), \; A^2_{-p}(x) = A^2_p(x).$$

Hence

$$B_{p,k} \cdot B_{-p,k} = E - k^2 \frac{k-1}{k+1} A^2_p + \frac{k^4}{(k+1)^2} A^4_p,$$

i.e., by $q := kp$, and $q^2 = k^2 - 1$ from $k^2(1 - p^2) = 1$,

$$B_{p,k} \cdot B_{-p,k}(x) = x + \left((\overline{x}q) \, q, x_0 q^2 \right) \left(-\frac{k-1}{k+1} + \frac{k^2 - 1}{(k+1)^2} \right) = x.$$

This proves (3.120). □

Theorem 61. *Suppose that* $\lambda : Z \to Z$ *is a mapping from* $Z = X \oplus \mathbb{R}$ *into itself satisfying*

$$l(x, y) = l\left(\lambda(x), \lambda(y) \right)$$

for all $x, y \in Z$ *where* $l(x, y) = (\overline{x} - \overline{y})^2 - (x_0 - y_0)^2$ *designates the separation of* x, y. *Then there exist a uniquely determined Lorentz boost* $B_{p,k}$ *and a uniquely determined orthogonal mapping* ω *of* X *such that*

$$\lambda(x) = B_{p,k}\omega(x) + \lambda(0) \tag{3.121}$$

holds true for all $x \in Z$ *where we put* $\omega(\overline{x}, x_0) := \left(\omega(\overline{x}), x_0 \right)$ *for all* $\overline{x} \in X$, $x_0 \in \mathbb{R}$.

Proof. 1) Define $\lambda_1(x) := \lambda(x) - \lambda(0)$ for $x \in Z$. Also λ_1 preserves separations, i.e.

$$l(x, y) = l\left(\lambda_1(x), \lambda_1(y) \right)$$

holds true for all $x, y \in Z$. Put $\tau := \lambda_1(t)$ with $t := (0, 1) \in Z$. Hence

$$-1 = l(0, t) = l\left(\lambda_1(0), \lambda_1(t) \right) = \overline{\tau}^2 - \tau_0^2,$$

i.e. $\tau_0^2 = 1 + \overline{\tau}^2 > 0$. Define $p \cdot \tau_0 := \overline{\tau}$, $k := \tau_0$ and observe $p \in X$ and

$$k^2(1 - p^2) = \tau_0^2 \left(1 - \frac{\overline{\tau}^2}{\tau_0^2} \right) = 1.$$

This implies $p = 0$ in the case $k = \tau_0 = -1$ and hence $\tau = B_{0,-1}(t)$. In the case $k = \tau_0 \neq -1$ we get $\tau = B_{p,k}(t)$ on account of $A_p(t) = p$ and $A^2_p(t) = p^2 t$. Define $\lambda_2 = B_{-p,k} \cdot \lambda_1$. Every Lorentz boost B satisfies (see step 2) below)

$$l(x, y) = l\left(B(x), B(y) \right) \tag{3.122}$$

for all $x, y \in Z$. Hence also λ_2 preserves separations.

2) Equation (3.122) is clear for $B_{0,-1}$. So assume $k \neq -1$. Observe

$$l\left(B(x), B(y) \right) = \left(B(x) - B(y) \right)^2 = [B(x - y)]^2,$$

since B is linear. So we have to prove $L := \left(B\left(z\right)\right)^2 = z^2$ for all $z \in Z$. With $q := kp$ we obtain

$$L = \left(\overline{z} + z_0 q + \frac{(\overline{z}q)\,q}{k+1}\right)^2 - \left(z_0 + \overline{z}q + \frac{z_0 q^2}{k+1}\right)^2,$$

i.e. $L = z^2 + \left((z_0 q)^2 - (\overline{z}q)^2\right)\left(1 - \frac{2}{k+1} - \frac{q^2}{(k+1)^2}\right) = z^2$, in view of

$$q^2 = k^2 - 1 \text{ from } k^2(1 - p^2) = 1.$$

3) Observe $\lambda_2(0) = 0$ and $\lambda_2(t) = t$, because of (3.120). Suppose that $x \in X$ and that $\lambda_2(x) =: y$. Then

$$l\left(0, x\right) = l\left(0, y\right) \text{ and } l\left(t, x\right) = l\left(t, y\right)$$

imply $x^2 = \overline{y}^2 - y_0^2$ and $x^2 - 1 = \overline{y}^2 - (y_0 - 1)^2$, i.e. $y_0 = 0$. The restriction η of λ_2 on X is hence a mapping of X into itself. Suppose that x, z are elements of X. Then

$$l\left(x, z\right) = l\left(\lambda_2(x), \lambda_2(z)\right)$$

implies $(x - z)^2 = \left(\eta\left(x\right) - \eta\left(z\right)\right)^2$ and η must hence be an orthogonal mapping $\omega : X \to X$ of X. We now would like to show that

$$\lambda_2(\overline{x} + x_0 t) = \omega\left(\overline{x}\right) + x_0 t \tag{3.123}$$

holds true for all $x \in Z$. Put $\lambda_2(\overline{x} + x_0 t) =: \overline{y} + y_0 t$. Then $l\left(0, x\right) = l\left(0, y\right)$ and $l\left(t, x\right) = l\left(t, y\right)$ imply $x_0 = y_0$. Hence

$$\lambda_2(\overline{x} + x_0 t) = \overline{y} + x_0 t,$$

i.e. $l\left(\overline{x}, x\right) = l\left(\omega\left(\overline{x}\right), \overline{y} + x_0 t\right)$. Thus

$$-x_0^2 = \left(\overline{y} - \omega\left(\overline{x}\right)\right)^2 - x_0^2,$$

i.e. $\overline{y} = \omega\left(\overline{x}\right)$. This proves (3.123). Finally, we obtain

$$\lambda\left(\overline{x}, x_0\right) = \lambda_1\left(\overline{x}, x_0\right) + \lambda\left(0\right) = B_{p,k}\lambda_2\left(\overline{x}, x_0\right) + \lambda\left(0\right),$$

i.e., by (3.123),

$$\lambda\left(\overline{x}, x_0\right) = B_{p,k}\omega\left(\overline{x}, x_0\right) + \lambda\left(0\right)$$

for all $(\overline{x}, x_0) \in Z = X \oplus \mathbb{R}$.

4) Suppose now that

$$B_{p,k}\omega\left(x\right) + \lambda\left(0\right) = B_{p',k'}\omega'(x) + \lambda\left(0\right)$$

holds true for all $x \in Z$. In the case $x = t = (0, 1)$ we get

$$B_{p,k}(t) = B_{p',k'}(t),$$

i.e. $kp + kt = k'p' + k't$, i.e. $k = k'$ and $p = p'$. Hence, $\omega\left(x\right) = \omega'(x)$, i.e. $\omega\left(\overline{x}\right) = \omega'(\overline{x})$ for all $\overline{x} \in X$. Thus $\omega = \omega'$. $\square$

Theorem 62. *All Laguerre transformations λ of X are given as follows. Let σ be a positive real number and ω be an element of $O(X)$. Moreover, let $B_{p,k}$ be a Lorentz boost and d be an element of $X \oplus \mathbb{R}$. Then*

$$\lambda(\overline{x}, x_0) = \sigma \cdot B_{p,k}\omega(\overline{x}, x_0) + d \qquad (3.124)$$

for all Laguerre cycles $(\overline{x}, x_0)$.

Proof. 1) We already introduced the Laguerre transformations

$$
\begin{aligned}
\lambda_\delta(x) &= B_{0,1}(x) + (0, \delta),\\
\lambda_\omega(x) &= B_{0,1}\omega(\overline{x}, x_0),\\
\lambda_{d,\sigma}(x) &= \sigma \cdot B_{0,1}(x) + (d,0) \text{ for } d \in X,\\
\varphi(x) &= B_{0,-1}(x).
\end{aligned}
$$

In order to be sure that $\lambda_{d,\sigma}$, $\sigma < 0$, is also of the form (3.124), observe

$$\lambda_{d,\sigma}(x) = (-\sigma) \cdot B_{0,-1}\big(\omega(\overline{x}), x_0\big) + (d,0)$$

with $\omega(y) = -y$ for $y \in X$. We also must show that

$$\lambda(x) = B_{p,k}(x), \qquad (3.125)$$

$k \neq -1$, is a Laguerre transformation. We will do this via Proposition 53. The mapping $\lambda : \Gamma \to \Gamma$ is a bijection. Let $(a, 1, \alpha)$ be a spear and observe $ap \neq 1$, since otherwise

$$1 = (ap)^2 \leq a^2 p^2 = p^2 < 1,$$

by $k^2(1 - p^2) = 1$, would be the consequence. The image of $(a, 1, \alpha)$ under λ is $(b, 1, \beta)$ with

$$b := \frac{a}{k(1 - ap)} - \left(k + \frac{1}{1 - ap}\right)\frac{p}{k+1}, \qquad (3.126)$$

$$\beta := \frac{\alpha}{k(1 - ap)}. \qquad (3.127)$$

A simple calculation yields $b^2 = 1$. Moreover, $a\overline{x} + x_0 = \alpha$ is equivalent with $b\overline{y} + y_0 = \beta$ for all $x \in X \oplus \mathbb{R}$ with

$$y := B_{p,k}(x) = \left(\overline{x} + kx_0 p + \frac{k^2(\overline{x}p)p}{k+1},\, kx_0 + k\overline{x}p\right):$$

this follows from the identity

$$b\overline{y} + y_0 - \beta = \frac{a\overline{x} + x_0 - \alpha}{k(1 - ap)}.$$

The pre-image of the spear $(b', 1, \beta')$ is given by $(a', 1, \alpha') = B_{-p,k}(b', 1, \beta')$ with

$$a' \quad := \quad \frac{b'}{k\,(1 + b'p)} \quad + \left(k + \frac{1}{1 + b'p} \right) \frac{p}{k+1},$$

$$\alpha' \quad := \quad \frac{\beta'}{k\,(1 + b'p)} \; :$$

this is a consequence of

$$B_{p,k}[B_{-p,k}(b', 1, \beta')] = (b', 1, \beta')$$

(see (3.120)) and of (3.126), (3.127) where we replace a, α, p by $b', \beta', -p$, respectively.

2) Suppose that λ is an arbitrary Laguerre transformation of X. Hence λ is a bijection of $X \oplus \mathbb{R}$ such that lines of $X \oplus \mathbb{R}$ are mapped onto lines, in view of Proposition 55. Applying now step b) of Theorem 3 where we replace X by $X \oplus \mathbb{R}$, and the euclidean lines of X by

$$\{(\bar{p}, p_0) + \xi\,(\bar{v}, v_0) \mid \xi \in \mathbb{R}\},$$

$(\bar{v}, v_0) \neq 0$, there exists a bijective linear mapping μ of the vector space $Z = X \oplus \mathbb{R}$ satisfying

$$\lambda\,(x) = \mu\,(x) + \lambda\,(0) \qquad\qquad (3.128)$$

for all $x \in Z$. Put $d := -\lambda\,(0)$. Then also

$$x \to \mu\,(x) = \lambda\,(x) + d$$

must be a Laguerre transformation of X. What we would like to prove is that μ has a representation

$$\mu\,(x) = \sigma \cdot B_{p,k}\omega\,(\bar{x}, x_0)$$

with a suitable real $\sigma > 0$, a suitable $\omega \in O\,(X)$ and suitable $p \in X, k \in \mathbb{R}$ satisfying $k^2(1 - p^2) = 1$. We know that $c_1 - c_2$ is equivalent with $\mu\,(c_1) - \mu\,(c_2)$ for all $c_1, c_2 \in \Gamma = Z$, since $\mu \in \mathrm{Lag}\,(X)$. According to (α), Proposition 43, $l\,(c_1, c_2) = 0$ is hence the same as $l\,(\mu\,(c_1), \mu\,(c_2)) = 0$. If $c^2 = 0$ holds true for $c \in Z$, then, obviously, also $l\,(c, 0) = 0$, i.e. $l\,(\mu\,(c), \mu\,(0)) = 0$, i.e. $(\mu\,(c))^2 = 0$.

3) Suppose that $a, b \in Z$ are linearly independent and that they satisfy $a^2 = 0$, $b^2 = 0$. If $w = \alpha a + \beta b$ with $\alpha, \beta \in \mathbb{R}$ and $\alpha \cdot \beta \neq 0$, then $ab \neq 0$ and

$$(\mu\,(w))^2 = \frac{\mu\,(a)\,\mu\,(b)}{ab}\,w^2 \neq 0 \qquad\qquad (3.129)$$

hold true.

In fact! Because of Lemma 60, (β), we get $ab \neq 0$. Now $a^2 = 0$, $b^2 = 0$ imply, by step 2), $(\mu(a))^2 = 0$, $(\mu(b))^2 = 0$. Since $\mu : Z \to Z$ is linear,

$$\mu(w) = \alpha\mu(a) + \beta\mu(b),$$

i.e. $(\mu(w))^2 = 2\alpha\beta\mu(a)\mu(b)$. Now $w^2 = 2\alpha\beta ab$ and $\alpha\beta \neq 0$ imply (3.129), by noticing that $\mu(a)$, $\mu(b)$ are linearly independent, since μ is a linear bijection, and that hence $\mu(a)\mu(b) \neq 0$, by (β), Lemma 60.

4) If $a, b \in Z$ are linearly independent satisfying $a^2 = 0$ and $b^2 = 0$, we will say that a, b are *strongly independent*. The elements $(e, 1)$, $(e, -1)$ of Z with $e^2 = 1$, for instance, are strongly independent. If $a, b \in Z$ are strongly independent, we define

$$\gamma(a, b) := \frac{\mu(a)\mu(b)}{ab}.$$

Observe $\gamma(b, a) = \gamma(a, b) = \gamma(\varrho a, b)$ for every real $\varrho \neq 0$.

If a, b are strongly independent, and also b, c, then $\gamma(a, b) = \gamma(b, c)$.

In fact! We may assume that the second components of $a, b, c \in Z$ are all equal to 1, since, for instance, $a \neq 0 = a^2$ and $a =: (m, \tau)$ yield $\tau \neq 0$, whence a may be replaced by $\frac{1}{\tau}a$ leading to $\gamma(a, b) = \gamma\left(\frac{1}{\tau}a, b\right)$.

Case 1: a, b, c are linearly dependent. Then $c = \alpha a + \beta b$ with suitable real α, β. The second part of (β), Lemma 60, implies $\alpha\beta = 0$, i.e. $\beta = 0$, since b, c are linearly independent. Hence $\alpha \neq 0$ and $\gamma(a, b) = \gamma(c, b)$.

Case 2: a, b, c are linearly independent. Take the two distinct elements $p, q \in Z$ on the line through

$$w_1 = \frac{a + b}{2}, \quad w_2 = \frac{b + c}{2} \tag{3.130}$$

satisfying $p^2 = 0 = q^2$: writing $a =: (A, 1)$, $b =: (B, 1)$, $c =: (c_1)$ we have to solve in $\varrho \in \mathbb{R}$,

$$\left(w_1 + \varrho(w_2 - w_1)\right)^2 = 0,$$

i.e.

$$\left(\frac{A + B}{2} + \varrho\frac{C - A}{2}\right)^2 = 1. \tag{3.131}$$

$a^2 = b^2 = c^2 = 0$ imply $A^2 = B^2 = C^2 = 1$. We get $B \neq A \neq C$, because $A = B$, for instance, would lead to

$$ab = AB - 1 = A^2 - 1 = 0,$$

contradicting $ab \neq 0$, in view of Lemma 60, (β). Hence $A \neq B$ are points of $B(0, 1)$ of X with $\frac{A+B}{2} \in B^-(0, 1)$. There are thus exactly two solutions ϱ of

(3.131), because a euclidean line of X through a point of $B^-(0,1)$ cuts $B(0,1)$ in exactly two points.

Having the distinct elements $p, q \in Z$ on the line of Z through w_1, w_2 of (3.130), we would like to verify that p, q are linearly independent, so strongly independent, since $p^2 = 0 = q^2$: but since w_1, w_2 are linearly independent, so must be

$$p = w_1 + \varrho_1(w_2 - w_1), \ q = w_1 + \varrho_2(w_2 - w_1).$$

We now obtain from (3.12),

$$\gamma(a,b)\, w_1^2 = [\mu(w_1)]^2 = \gamma(p,q)\, w_1^2,$$
$$\gamma(b,c)\, w_2^2 = [\mu(w_2)]^2 = \gamma(p,q)\, w_2^2,$$

i.e. $\gamma(a,b) = \gamma(p,q) = \gamma(b,c)$.

5) *If a, b are strongly independent, and also c, d, then $\gamma(a,b) = \gamma(c,d)$.*

If b, c are linearly dependent, we get $b = \beta c$, $\beta \neq 0$, i.e. $\gamma(c,d) = \gamma(b,d)$. Hence

$$\gamma(c,d) = \gamma(b,d) = \gamma(a,b),$$

from 4). If b, c are linearly independent, then

$$\gamma(a,b) = \gamma(b,c) \text{ and } \gamma(b,c) = \gamma(c,d),$$

also from 4).

6) *There exists a real constant $\varrho > 0$ with $[\mu(w)]^2 = \varrho \cdot w^2$ for all $w \in Z$.*

Take strongly independent $a, b \in Z$, for instance, $(e, 1), (e, -1)$ with $e^2 = 1$. Put $\varrho := \gamma(a,b)$. If $w^2 = 0$, then $[\mu(w)]^2 = 0$, i.e. $[\mu(w)]^2 = \varrho w^2$ holds true. If $w^2 \neq 0$, then, by (α), Lemma 60, there exist strongly independent c, d satisfying

$$w = \alpha c + \beta d.$$

$w^2 \neq 0$ implies $\alpha\beta \neq 0$. Hence, by (3.129) and 5),

$$[\mu(w)]^2 = \gamma(c,d)\, w^2 = \varrho \cdot w^2,$$

i.e. $l(\mu(w), 0) = \varrho l(w, 0)$, i.e. $\varrho > 0$, in view of Proposition 43 and Proposition 55.

7) If $v, w \in Z$, then

$$l(\mu(v), \mu(w)) = [\mu(v-w)]^2 = \varrho \cdot l(v, w),$$

in view of 6). Hence

$$l(\delta^{-1}\mu(v), \delta^{-1}\mu(w)) = l(v, w)$$

for all $v, w \in Z$ for the Laguerre transformation

$$\delta(x) := \sqrt{\varrho} \cdot x.$$

The Laguerre transformation $\delta^{-1}\mu$ is hence of the form, by Theorem 61,

$$\delta^{-1}\mu(x) = B_{p,k}\omega(x) + \delta^{-1}\mu(0),$$

i.e., we obtain

$$\mu(x) = \sqrt{\varrho} \cdot B_{p,k}\omega(x)$$

for all $x \in Z$. $\qquad\qquad\qquad\qquad\qquad\qquad\qquad\qquad\qquad\qquad\quad\square$

Remark. Lorentz boosts were discovered in 1911 by G. Herglotz [1] and in 1912 by A. von Brill [1] in the form of special matrices, the so-called Herglotz–Brill matrices (see W. Benz [2]).

3.15 $\mathbb{M}(X)$ as part of Lie (X)

Every M-transformation μ of X leads in a natural way to a Lie transformation λ. We will show this for similitudes and for the inversion ι in order to be sure, by Theorem 3, that it holds true for all M-transformations. In both cases, similitude or ι, we define $\lambda(c) := \mu(c)$ for the Lie cycle $c \in X \cup \{\infty\}$. If μ is a similitude, put

$$\lambda\big(H(a,\alpha), H^*(a,\alpha)\big) := \big(\mu(H), \mu(H^*)\big), \qquad (3.132)$$

by observing Proposition 31, for $H^* \in \{H^+, H^-\}$ (see section 9), and, for $\varrho > 0$,

$$\lambda\big(B(m,\varrho), B^*\big) := \big(\mu(B), [\mu(B)]^*\big) \qquad (3.133)$$

(see section 9 for the definition of B^+ and B^-). We must notice here that, because of $\mu(\infty) = \infty$,

$$\mu(B) =: B(n,\sigma) \text{ implies } \mu(B^*) = B^*(n,\sigma)$$

for $* \in \{+,-\}$. In fact! Because of Proposition 31 we know that

$$\{\mu(B^+ \cup \{\infty\}), \mu(B^-)\} = \{B^+(n,\sigma) \cup \{\infty\}, B^-(n,\sigma)\}.$$

If now $\mu(B^+ \cup \{\infty\} = B^-(n,\sigma)$ and $\mu(B^-) = B^+(n,\sigma) \cup \{\infty\}$ would hold true, $\mu\{\infty\}$ could not be equal to ∞.

It is not difficult to verify that the induced mapping λ must be a Lie transformation, by noticing that if a side Σ_1 is a subset of another side Σ_2, of course, $\mu(\Sigma_1) \subseteq \mu(\Sigma_2)$ must be the consequence.

If μ is the *inversion* ι, we put, by applying the cases (1), (2), (3), (4) (between formulas (3.7) and (3.8)) of section 1,

$$\lambda\big(B(c,\varrho), B^*\big) \quad := \quad \big(\iota(B), [\iota(B)]^*\big) \text{ for } |c| > \varrho,$$
$$\lambda\big(B(c,\varrho), B^*\big) \quad := \quad \big(\iota(B), [\iota(B)]^\circ\big) \text{ for } |c| < \varrho,$$

with

$$\iota\left(B\right) = B\left(\frac{c}{c^2 - \varrho^2}, \frac{\varrho}{|c^2 - \varrho^2|}\right),$$

for $* \in \{+, -\}$ and $\{*, \circ\} = \{+, -\}$.

Moreover, we put $\lambda^2 = \mathrm{id}$ and

$$\begin{aligned}
\lambda\left(B\left(c, \varrho\right), B^*\right) &:= \left(H\left(2c, 1\right), H^\circ\right) \text{ for } |c| = \varrho, \\
\lambda\left(H\left(a, 0\right), H^*\right) &:= \left(H\left(a, 0\right), H^*\right),
\end{aligned}$$

again for $* \in \{+, -\}$ and $\{*, 0\} = \{+, -\}$.

Also here it is easy to verify that the induced mapping λ must be a Lie transformation. In step 2) of the proof of Proposition 58 we defined the Lie transformation

$$\varepsilon : LQ\left(X\right) \to LQ\left(X\right)$$

by means of

$$\varepsilon\left[\mathbb{R}^*(v, \xi_1, \xi_2, \xi_3)\right] = \mathbb{R}^*(v, \xi_1, \xi_2, -\xi_3).$$

The mapping ε is exactly the Lie transformation induced by the inversion ι.

There exist infinitely many Lie transformations which are not induced by M-transformations, for instance all Laguerre dilatations

$$\lambda_\xi(m, \tau) = (m, \tau + \xi), \quad \lambda_\xi(a, 1, \alpha) = (a, 1, \alpha + \xi)$$

with $0 \neq \xi \in \mathbb{R}$, since the image of the Laguerre cycle $(m, -\xi)$, $\xi \neq 0$, is the point m, and since there is no M-transformation transforming an M-ball $(m, |-\xi|)$ into the point m. So the question arises how to characterize the Möbius group $\mathbb{M}\left(X\right)$ within Lie $\left(X\right)$.

In section 12 we defined a Laguerre transformation,

$$(m, \tau) \to (m, -\tau) \text{ and } (a, 1, \alpha) \to (a, -1, \alpha),$$

which reverses the orientation of every Laguerre cycle (m, τ), $\tau \neq 0$, and of every spear. This Lie transformation can be written in coordinates as

$$\delta : \left[\mathbb{R}^*(v, \xi_1, \xi_2, \xi_3)\right] = \mathbb{R}^*(v, -\xi_1, \xi_2, \xi_3).$$

We are also interested in the *centralizer* of δ,

$$C\left(\delta\right) := \{\lambda \in \text{Lie}\left(X\right) \mid \delta\lambda = \lambda\delta\},$$

within Lie $\left(X\right)$.

Theorem 63. $\mathbb{M}\left(X\right) \cong C\left(\delta\right)/\{\mathrm{id}, \delta\}$.

Proof. M-transformations are Lie transformations which are permutations on the set $X \cup \{\infty\}$ of points, and which transform balls or hyperplanes apart from their chosen side. We would like to show that a Lie transformation λ has these properties if, and only if, $\lambda\delta = \delta\lambda$ holds true. So assume $\lambda\delta = \delta\lambda$ and that c is a Lie cycle which is a point. Hence $\delta(c) = c$, i.e.

$$\lambda(c) = \lambda\delta(c) = \delta\lambda(c).$$

If $\lambda(c)$ were not a point, then it would be equal to the Lie cycle based on the same ball or hyperplane as $\lambda(c)$, but with the opposite orientation, a contradiction. Let now d be a ball or a hyperplane and s_1, s_2 its sides. Put

$$\lambda(d, s_1) =: (e, \sigma_1), \tag{3.134}$$

by observing that $\lambda(d, s_1)$ cannot be a point c, since otherwise

$$(d, s_1) = \lambda^{-1}(c) = \lambda^{-1}\delta(c) = \delta\lambda^{-1}(c) = \delta(d, s_1) = (d, s_2)$$

would hold true. Hence

$$\lambda(d, s_2) = \lambda\delta(d, s_1) = \delta\lambda(d, s_1) = (e, \sigma_2),$$

i.e. λ transforms d into e apart from its chosen side.

Assume now that λ is a Lie transformation which is a permutation on $X \cup \{\infty\}$ and which satisfies $\lambda(d, s_2) = (e, \sigma_2)$, whenever $\lambda(d, s_1) = (e, \sigma_1)$. For every Lie cycle c which is a point we get

$$\lambda\delta(c) = \lambda(c) = \delta\lambda(c),$$

since $\lambda(c)$ is also a point. For any other cycle (d, s_1) we get, by (3.134),

$$\lambda\delta(d, s_1) = \lambda(d, s_2) = (e, \sigma_2) = \delta(e, \sigma_1) = \delta\lambda(d, s_1).$$

Hence $\lambda\delta = \delta\lambda$.

Our result is that exactly the Lie transformations λ in the centralizer $C(\delta)$ of δ lead to M-transformations. Obviously, this correspondence is a homomorphism of $C(\delta)$ onto $\mathbb{M}(X)$. In order to determine the kernel of this homomorphism, we ask for all $\lambda \in C(\delta)$ inducing the identity of $\mathbb{M}(X)$. If $\lambda \neq \mathrm{id}$ in $C(\delta)$ induces the identity, then $\lambda(\infty) = \infty$, i.e. λ must be a Laguerre transformation, i.e., by Theorem 62,

$$\lambda(\overline{x}, x_0) = \sigma B_{p,k}(\omega(\overline{x}), x_0) + d. \tag{3.135}$$

Because of $\lambda(\overline{x}, 0) = (\overline{x}, 0)$ for all $\overline{x} \in X$, and because of

$$\lambda(0, 1) \in \{(0, 1), (0, -1)\}, \tag{3.136}$$

we obtain

$$d = 0 \text{ from } \lambda(0,0) = (0,0),$$

$$(p,k) \in \{(0,1),\ (0,-1)\} \text{ and } \sigma = 1 \text{ from } (3.135),\ (3.136),$$

$$\omega = \text{id from } \lambda(\overline{x},0) = (\overline{x},0) \text{ for all } \overline{x} = X.$$

The only mapping $\lambda \neq \text{id in } C(\delta)$ which induces id $\in \mathbb{M}(X)$, is hence $\lambda(\overline{x},x_0) = B_{0,-1}(\overline{x},x_0) = (\overline{x},-x_0)$, i.e. $\lambda = \delta$. $\qquad\square$

3.16   A characterization of Lag (X)

The following result characterizes the elements of Lag (X) under mild hypotheses.

Theorem 64. *Let λ be a bijection of the set Γ of all Laguerre cycles of X such that*

$$\forall_{c_1,c_2 \in \Gamma}\ c_1 - c_2 \Rightarrow \lambda(c_1) - \lambda(c_2) \tag{3.137}$$

holds true. Then there exists a uniquely determined Laguerre transformation Δ of X with $\Delta \mid \Gamma = \lambda$ and, moreover, λ has the form (3.124).

Proof. 1) Let λ be a bijection of $\Gamma = Z = X \oplus \mathbb{R}$ satisfying (3.137). Instead of $\lambda(x)$, $x \in Z$, we will write x'. Assume that there exist $a, b \in Z$ with $a' - b'$ and $a \neq b$. Then $a \neq b$ and hence $a' \neq b'$. Designate the parabolic pencil $B_p(a',b')$ by B. Also now it will be important to distinguish between subtraction $c_1 - c_2$ and contactness $c_1 - c_2$ of two elements $c_1, c_2 \in Z$. In view of (α), Lemma 60, there exist linearly independent v, w in $X \oplus \mathbb{R}$ and $\alpha, \beta \in \mathbb{R}$ with

$$b - a = \alpha v + \beta w, \tag{3.138}$$

$v^2 = 0 = w^2$. Since $(b - a)^2 \neq 0$ by (α), Proposition 43, we get $\alpha\beta \cdot vw \neq 0$ from (3.138). Hence

$$x := a + \alpha v = b + (-\beta)w \notin \{a,b\}$$

and $y := a + \beta w = b + (-\alpha)v \notin \{a,b\}$. Thus $a - x - b$ and $a - y - b$. This implies $x', y' \in B$ by (3.137) and hence

$$(a + \xi v)',\ (y + \xi v)' \in B$$

for every real ξ since $l(a, a + \xi v) = 0 = l(a + \xi v, x)$ and

$$l(y, y + \xi v) = 0 = l(y + \xi v, b).$$

Thus $z' \in B$ for all $z = a + \xi v + \eta w$ with reals ξ, η because of

$$l(a + \xi v, z) = 0 = l(y + \xi v, z).$$

Finally, we would like to show $c' \in B$ also for those $c \in \Gamma$ which do not belong to the bundle
$$\mathfrak{B} := \{a + \xi v + \eta w \mid \xi, \eta \in \mathbb{R}\}.$$
Observe here (γ), Proposition 51, and $vw \neq 0$. There exist $q_1 \neq q_2$,
$$q_i = a + \xi_i v + \eta_i w, \ i = 1, 2,$$
in $\mathfrak{B}$ with $(q_i - c)^2 = 0$, $i = 1, 2$, since this latter equation has the form
$$(\xi_i P + A)(\eta_i P + B) = C, \ i = 1, 2, \tag{3.139}$$
with $P := vw \neq 0$, $A := (a - c) w$, $B := (a - c) v$ and
$$C := AB - \frac{1}{2} P \cdot (a - c)^2 :$$
take reals $\xi_1 \neq \xi_2$, both unequal to $-P^{-1}A$, and calculate η_i, $i = 1, 2$, according to (3.139). Now $l(c, q_i) = 0$ and $q'_i \in B$ for $i = 1, 2$ imply $c' \in B$.

Hence $\lambda(\Gamma) \subseteq B$ which contradicts the fact that $\lambda : \Gamma \to \Gamma$ is a bijection. We thus proved that $a' - b'$ and $a \neq b$ for $a, b \in \Gamma$ is not possible. This implies
$$\forall_{c_1, c_2 \in \Gamma} \ \lambda(c_1) - \lambda(c_2) \Rightarrow c_1 - c_2. \tag{3.140}$$

2) A Laguerre transformation of X can be defined, by Proposition 53, as a bijection of Γ such that images and pre-images of spears are spears by identifying a spear $s \in \Sigma$ with $\{c \in \Gamma \mid s - c\}$. Let now s be an arbitrary spear and let c_1, c_2 be elements of Γ satisfying
$$c_1 \neq c_2, \ c_1 - c_2, \ c_1 - s - c_2.$$
Then, in view of Proposition 47, s is given by $N \cup B_p(c_1, c_2)$ with
$$N = \{c \in \Gamma \mid \text{there is no } w \in B_p(c_1, c_2) \text{ touching } c\}.$$
If we look to $\lambda(N \cup B_p(c_1, c_2))$ for a mapping as considered in Theorem 64, we obtain by (3.137), (3.140) and $z' := \lambda(z)$, $z \in \Gamma$,
$$\lambda(B_p(c_1, c_2)) = \{c' \in \Gamma \mid c_1 - c - c_2\} = \{c' \in \Gamma \mid c'_1 - c' - c'_2\} = B_p(c'_1, c'_2).$$
Since $N = N(c_1, c_2)$ can be written as
$$\{c \in \Gamma \mid w \neq c \text{ for all } w \in B_p(c_1, c_2)\},$$
we get by (3.137), (3.140),
$$\lambda(N) = \{c' \in \Gamma \mid w' \neq c' \text{ for all } w' \in B_p(c'_1, c'_2)\},$$
and hence $\lambda(N) = N(c'_1, c'_2)$. In view of Proposition 47, the set
$$N(c'_1, c'_2) \cup B_p(c'_1, c'_2)$$
is the spear touching c'_1, c'_2. The image of a spear under λ must hence be a spear. The same holds true for the pre-image since also λ^{-1} satisfies (3.137), (3.140). If we now apply Theorem 62, then Theorem 64 is proved. $\qquad \square$

3.17   Characterization of the Lorentz group

The basic structure of *Lorentz–Minkowski geometry* is $Z = X \oplus \mathbb{R}$. The *Lorentz–Minkowski distance* of $(\overline{x}, x_0)$, $(\overline{y}, y_0) \in Z$ is defined by

$$l\,(x, y) := (\overline{x} - \overline{y})^2 - (x_0 - y_0)^2. \tag{3.141}$$

In section 10 this expression was called the separation of the two Laguerre cycles $x, y \in Z$. A connection between classical Lorentz–Minkowski geometry and classical Laguerre geometry was dicovered in its first steps by H. Bateman [1] (1910) and H.E. Timerding [1] (1912). W. Blaschke [2] (1929) then realized that this connection was indeed very close. A mapping $\lambda : X \oplus \mathbb{R} \to X \oplus \mathbb{R}$ is called a *Lorentz transformation* of $Z = X \oplus \mathbb{R}$ provided

$$l\,(x, y) = l\,\big(\lambda\,(x),\, \lambda\,(y)\big) \tag{3.142}$$

holds true for all $x, y \in Z$. The *Lorentz group* $\mathbb{L}\,(Z)$ of Z is the set of all bijective Lorentz transformations of Z equipped with the usual product of permutations. The geometry $\big(Z, \mathbb{L}\,(Z)\big)$ is called *Lorentz–Minkowski geometry* over Z. The case

$$\big(Z = \mathbb{R}^3 \oplus \mathbb{R},\ \mathbb{L}\,(Z)\big)$$

is called *classical Lorentz–Minkowski geometry*, as well as *classical proper Laguerre geometry* in the earlier context (see section 3.12).

Theorem 61 then determines all Lorentz transformations $\lambda : Z \to Z$ in the context of Laguerre geometry.

Here we would like to present an important and immediate consequence of Theorems 64 and 62.

Theorem 65. *If the bijection* $\lambda : X \oplus \mathbb{R} \to X \oplus \mathbb{R}$ *satisfies*

$$\forall_{c_1, c_2 \in X \oplus \mathbb{R}}\ l\,(c_1, c_2) = 0 \Rightarrow l\,\big(\lambda\,(c_1),\, \lambda\,(c_2)\big) = 0, \tag{3.143}$$

then

$$\lambda\,(\overline{x}, x_0) = \sigma \cdot B_{p,k}\big(\omega\,(\overline{x}),\, x_0\big) + d \tag{3.144}$$

holds true for all $x = (\overline{x}, x_0)$ *in* Z *where* $d \in Z$, $p \in X$, $k \in \mathbb{R}$ *with* $k^2(1 - p^2) = 1$, $\sigma \in \mathbb{R}$, $\omega \in O\,(X)$ *are suitable elements.*

Proof. Because of (α), Proposition 43, the properties (3.137), (3.143) coincide. Moreover, the mapping (3.144) is bijective if, and only if, $\omega : X \to X$ is bijective, i.e. $\omega \in O\,(X)$; this follows from the fact that Lorentz boosts are bijective. $\square$

Remark. Concerning generalizations of Lorentz transformations and generalizations of these see also J. Lester [2], [9], W.F. Pfeffer [1], H. Schaeffer [1], E.M. Schröder [3], [6]. Configurations in Lorentz–Minkowski structures were studied by H.J. Samaga [1].

3.18 Another fundamental theorem

Theorem 66. (α) *A bijection* $\lambda : \Delta(X) \to \Delta(X)$ *satisfying*

$$\forall_{x,y \in \Delta} \; x - y \Rightarrow \lambda(x) - \lambda(y) \tag{3.145}$$

is already a Lie transformation of X.

(β) *Lie* (X) *consists exactly of all finite products of elements of* Lag $X \cup \{\varepsilon\}$. *More precisely, every* $\lambda \in$ Lie (X) *is of the form*

$$\alpha \text{ or } \beta_1 \varepsilon \beta_2 \text{ or } \gamma_1 \varepsilon \gamma_2 \varepsilon \gamma_3$$

with $\alpha, \beta_1, \beta_2, \gamma_1, \gamma_2, \gamma_3$ *in* Lag (X).

Proof. 1) Let the bijection $\lambda : \Delta \to \Delta$ satisfy (3.145). By Proposition 58 there exists $\mu \in$ Lie (X) with $\mu(\lambda(\infty)) = \infty$. The bijection $\lambda' := \mu\lambda$ of Δ also satisfies (3.145). If s is a spear, then $s - \infty$ implies $\lambda'(s) - \infty$, i.e. $\lambda'(s) \in \Sigma$. Hence $\lambda'(\Sigma) \subseteq \Sigma$, $\lambda'(\infty) = \infty$. We also would like to show $\lambda'(\Gamma) \subseteq \Gamma$. Assume that $c \in \Gamma$ satisfies $c' := \lambda'(c) \subseteq \Sigma$. If $s \in Z$ is given arbitrarily, by Lemma 49, there exists $s_1 \in \Sigma$ with $c - s_1$ and $s_1 \parallel s$. Hence $c' - s_1'$ and $s_1' \parallel s'$, i.e. $c' \parallel s'$, since c' is a spear: all spears in $\lambda'(\Sigma)$ are thus parallel to the spear c'. Take now a spear $s \nparallel c'$. The pre-image c_1 of s must hence be in Γ. So we get, as before, that every spear in $\lambda'(\Sigma)$ must be parallel to c_1'. This implies

$$s = c_1' \parallel t \parallel c'$$

where t is taken arbitrarily from $\lambda'(\Sigma)$, i.e. $s \parallel c'$, contradicting $s \nparallel c'$. Hence our assumption that there exists $c \in \Gamma$ with $c' \in \Sigma$ was wrong. This implies $\lambda'(\Gamma) \subseteq \Gamma$. Since $\lambda' : \Delta \to \Delta$ is bijective,

$$\lambda'(\infty) = \infty, \; \lambda'(\Sigma) \subseteq \Sigma, \; \lambda'(\Gamma) \subseteq \Gamma$$

leads to $\lambda'(\Sigma) = \Sigma$ and $\lambda'(\Gamma) = \Gamma$. From Theorem 64 we then get (α) by observing $\lambda = \mu^{-1}\lambda'$.

2) Let λ be a Lie transformation. If $\lambda(\infty) = \infty$, then $\lambda \in$ Lag (X). If $\lambda(\infty) =: z$ is in Γ, then, with $\alpha\varepsilon(\infty) = z$ (see step 3 of the proof of Proposition 58), $\alpha \in$ Lag (X), we get

$$\lambda^{-1}\alpha\varepsilon(\infty) = \infty,$$

i.e. $\lambda^{-1}\alpha\varepsilon \in$ Lag (X). In the case $\lambda(\infty) =: z \in \Sigma$, we refer to the spear s and the Laguerre cycle $x = \varepsilon(s)$ as defined in step 4 of the proof of Proposition 58. Take α, β in Lag (X) with $\alpha(0) = x$ and $\beta(s) = z$. Since ∞ remains unaltered under $\lambda^{-1}\beta\varepsilon\alpha\varepsilon$, we get that $\lambda^{-1}\beta\varepsilon\alpha\varepsilon$ is in Lag (X). $\square$

With Proposition 59, part (α) of Theorem 66 can be presented equivalently in the following form.

Theorem 67. *A bijection λ of $LQ(X)$ is a Lie transformation of X if $\lambda(l)$ is collinear for every line l of $\Pi(X \oplus \mathbb{R}^3)$ contained in $LQ(X)$.*

This theorem already appears in U. Pinkall (Math. Ann. 270 (1985) 427–440) under the additional assumptions that $\dim X < \infty$ and that λ is a line preserving diffeomorphism.

Remark. Concerning finite-dimensional sphere geometries of Möbius, Laguerre and Lie see, for instance, R. Artzy [1], W. Benz [2], W. Blaschke [2], H. Schaeffer [2], E.M. Schröder [2], H. Schwerdtfeger [1], concerning generalizations in this context A. Blunck [1], H. Havlicek [1], [2], A. Herzer [1]. The arbitrary-dimensional case was developed by W. Benz [9], [12].

Chapter 4

Lorentz Transformations

As in the chapters before, X denotes a real inner product space of arbitrary (finite or infinite) dimension ≥ 2.

4.1 Two characterization theorems

Define the so-called Lorentz–Minkowski spacetime $Z := X \oplus \mathbb{R}$ with the product

$$(\overline{x}, x_0) \cdot (\overline{y}, y_0) := \overline{x}\,\overline{y} - x_0 y_0 \tag{4.1}$$

as in (3.88) where $\overline{x}, \overline{y} \in X$ and $x_0, y_0 \in \mathbb{R}$, or, in other words where $(\overline{x}, x_0)$, $(\overline{y}, y_0)$ are elements of Z. In the present context, the elements of Z are called *events*, and

$$\mathbb{R}(0,1) = \{(0, \xi) \in Z \mid \xi \in \mathbb{R}\}$$

is said to be the *time axis* of Z. Instead of events we also will speak of the *points* of Z. The Lorentz–Minkowski distance of $x = (\overline{x}, x_0)$, $y = (\overline{y}, y_0) \in Z$ is defined by (3.141),

$$l(x, y) := (\overline{x} - \overline{y})^2 - (x_0 - y_0)^2 = (x - y)^2. \tag{4.2}$$

Recall that this expression was defined to be the separation of the two Laguerre cycles x, y (see section 10 of chapter 3). A mapping $\lambda : Z \to Z$ is said to be a Lorentz transformation of Z (see section 17, chapter 3) provided

$$l(x, y) = l(\lambda(x), \lambda(y)) \tag{4.3}$$

holds true for all $x, y \in Z$.

Theorem 61 of chapter 3 immediately yields

Theorem 1. *Let* $\lambda : Z \to Z$ *be a Lorentz transformation of* Z. *Then there exist a uniquely determined Lorentz boost* $B_{p,k}$ *and a uniquely determined orthogonal mapping* ω *of* X *such that*

$$\lambda (x) = B_{p,k}\, \omega (x) + \lambda (0) \tag{4.4}$$

holds true for all $x \in Z$ *where we put* $\omega (\overline{x}, x_0) := \big(\omega (\overline{x}),\, x_0\big)$ *for all* $\overline{x} \in X$, $x_0 \in \mathbb{R}$.

On the other hand, all mappings (4.4) must be Lorentz transformations provided $B_{p,k}$ is a boost and ω an orthogonal mapping of X. Moreover, all mappings (4.4) are injective, since ω must be injective (see section 5 of chapter 1) and $B_{p,k}$ bijective (see (3.120)). There exist real inner product spaces X and orthogonal mappings $\omega : X \to X$ which are not bijective (see section 5 of chapter 1). However, if X is finite-dimensional, every orthogonal mapping ω of X must be bijective. In fact, if $a_1, \ldots, a_n$ is a basis of X, then $\omega (a_1), \ldots, \omega (a_n)$ is a basis as well.

Since $B_{p,k}$ in (4.4) is bijective and ω injective, a mapping

$$\lambda (x) = B_{p,k}\, \omega (x) + \lambda (0)$$

must be bijective if, and only if, ω is bijective, i.e. $\omega \in O\, (X)$. The Lorentz group $\mathbb{L}\, (Z)$ of Z is the set of all bijective Lorentz transformations with the permutation product as multiplication. Hence, the Lorentz group of Z consists of all mappings (4.4) satisfying $\omega \in O\, (X)$. The geometry $\big(Z, \mathbb{L}\, (Z)\big)$ in the sense of section 9, chapter 1, is called *Lorentz–Minkowski geometry* over $Z = X \oplus \mathbb{R}$. The elements of $\mathbb{L}\, (Z)$ are also called *motions* of this geometry.

The following structure theorem was already proved in the context of Laguerre geometry (Theorem 65, chapter 3).

Theorem 2. *If the bijection* $\lambda : X \oplus \mathbb{R} \to X \oplus \mathbb{R}$ *satisfies*

$$\forall_{x,y \in X \oplus \mathbb{R}} \ (x - y)^2 = 0 \Rightarrow \big(\lambda (x) - \lambda (y)\big)^2 = 0, \tag{4.5}$$

then

$$\lambda (x) = \sigma \cdot B_{p,k}\, \omega (x) + d \tag{4.6}$$

holds true for all $x \in Z$ *where* $d \in Z$, $p \in X$, $k \in \mathbb{R}$ *with* $k^2(1 - p^2) = 1$, $0 \neq \sigma \in \mathbb{R}$, $\omega \in O\, (X)$ *are suitable elements and where* $\omega (x)$ *is defined by* $\big(\omega (\overline{x}), x_0\big)$ *for all* $x = (\overline{x}, x_0) \in Z$.

Remark. Observe $\dim(X \oplus \mathbb{R}) \geq 3$, because of $\dim X \geq 2$. Theorem 2 was proved for $\dim(X \oplus \mathbb{R}) < \infty$ and under the stronger assumption

$$\forall_{x,y \in X \oplus \mathbb{R}} \ (x - y)^2 = 0 \Leftrightarrow \big(\lambda (x) - \lambda (y)\big)^2 = 0, \tag{4.7}$$

by A.D. Alexandrov [1, 2, 3], however not precisely in the form (4.6), but in the form $\lambda = \sigma \lambda'$ with

$$\forall_{x,y \in X \oplus \mathbb{R}} \ (x - y)^2 = \big(\lambda' (x) - \lambda' (y)\big)^2. \tag{4.8}$$

June Lester [1] and E.M. Schröder [5] solved the general field case, E.M. Schröder even in the infinite-dimensional case. The first mathematician who weakened (4.7) into (4.5) was F. Cacciafesta [1], however, under the assumption $\dim X \oplus \mathbb{R} < \infty$. The general form was proved by W. Benz [9].

4.2 Causal automorphisms

Let $x = (\overline{x}, x_0)$, $y = (\overline{y}, y_0)$ be elements of Z. We put

$$x \le y$$

if, and only if,

$$l(x, y) = (x - y)^2 \le 0 \text{ and } x_0 \le y_0$$

hold true. A bijection $\sigma : Z \to Z$ is called a *causal automorphism* of Z if, and only if,

$$x \le y \Leftrightarrow \sigma(x) \le \sigma(y)$$

for all $x, y \in Z$.

Of course, $x < y$ stands for $x \le y$ and $x \ne y$, $x \ge y$ for $y \le x$, and $x > y$ for $y < x$.

Proposition 3. *Let x, y, z be elements of Z and let k be a real number. Then the following statements hold true.*

(1) $x \le x$,

(2) $x \le y$ *and* $y \le x$ *imply* $x = y$,

(3) $x \le y$ *and* $y \le z$ *imply* $x \le z$,

(4) $x \le y$ *implies* $x + z \le y + z$,

(5) $x \le y$ *implies* $kx \le ky$ *for* $k \ge 0$,

(6) $x \le y$ *implies* $kx \ge ky$ *for* $k < 0$.

Proof. (1) follows from $(x - x)^2 = 0 \le 0$ and $x_0 \le x_0$. Obviously, $x \le y \le x$ implies

$$0 \ge (x - y)^2 = (\overline{x} - \overline{y})^2 - (x_0 - y_0)^2$$

and $x_0 \le y_0 \le x_0$, i.e. $x_0 = y_0$ and hence $0 \ge (\overline{x} - \overline{y})^2 \ge 0$. From $x \le y \le z$ we obtain

$$(\overline{x} - \overline{y})^2 \le (x_0 - y_0)^2, \ (\overline{y} - \overline{z})^2 \le (y_0 - z_0)^2, \ x_0 \le y_0 \le z_0.$$

Hence, by the triangle inequality,

$$\|\overline{x} - \overline{z}\| \le \|\overline{x} - \overline{y}\| + \|\overline{y} - \overline{z}\| \le (y_0 - x_0) + (z_0 - y_0),$$

i.e. $(\overline{x} - \overline{z})^2 \le (x_0 - z_0)^2$. Together with $x_0 \le z_0$ we get (3). The proof of (4), (5), (6) is trivial. $\square$

If $x, y \in Z$ satisfy $x < y$,

$$[x, y] := \{z \in X \mid x \leq z \leq y\}$$

is called *ordered* if, and only if, $u \leq v$ or $v \leq u$ holds true for all $u, v \in [x, y]$.

Proposition 4. *If $x, y \in Z$ satisfy $x < y$, then $[x, y]$ is ordered if, and only if, $(x - y)^2 = 0$.*

Proof. a) Assume $l(x, y) = (x - y)^2 = 0$ and $u \in [x, y]$, i.e.

$$x_0 \leq u_0 \leq y_0, \|\bar{u} - \bar{x}\| \leq u_0 - x_0, \ \|\bar{y} - \bar{u}\| \leq y_0 - u_0.$$

$l(x, y) = 0$ implies $\|\bar{y} - \bar{x}\| = y_0 - x_0$. Hence

$$y_0 - x_0 = \|\bar{y} - \bar{x}\| \leq \|\bar{y} - \bar{u}\| + \|\bar{u} - \bar{x}\| \leq y_0 - x_0, \qquad (4.9)$$

and thus $\|\bar{y} - \bar{x}\| = \|\bar{y} - \bar{u}\| + \|\bar{u} - \bar{x}\|$. Because of Lemma 2, chapter 2, $\bar{y} - \bar{u}$, $\bar{u} - \bar{x}$ must be linearly dependent. Hence there exists $\alpha \in \mathbb{R}$ with

$$\bar{u} = \bar{x} + \alpha(\bar{y} - \bar{x}), \qquad (4.10)$$

in view of $\bar{x} \neq \bar{y}$, since $\bar{x} = \bar{y}$ and $\|\bar{y} - \bar{x}\| = y_0 - x_0$ would lead to $x = y$. Now (4.9), (4.10) yield

$$\|\bar{y} - \bar{x}\| = \|\bar{y} - \bar{u}\| + \|\bar{u} - \bar{x}\| = |1 - \alpha| \, \|\bar{y} - \bar{x}\| + |\alpha| \|\bar{y} - \bar{x}\|,$$

i.e. $1 = |1 - \alpha| + |\alpha|$, i.e. $0 \leq \alpha \leq 1$. Hence, with $\xi := y_0 - x_0$,

$$\xi = (1 - \alpha)\xi + \alpha\xi = \|\bar{y} - \bar{u}\| + \|\bar{u} - \bar{x}\| \leq (y_0 - u_0) + (u_0 - x_0) = \xi,$$

i.e. $\|\bar{y} - \bar{u}\| = y_0 - u_0$, $\|\bar{u} - \bar{x}\| = u_0 - x_0$, i.e., by (4.10),

$$u = x + \alpha(y - x).$$

Similarly, $v \in [x, y]$ implies

$$v = x + \beta(y - x), \ 0 \leq \beta \leq 1.$$

Hence $u \leq v$ for $\alpha \leq \beta$, and $v \leq u$ for $\beta \leq \alpha$.

b) Assume that $[x, y]$ is ordered and that $l(x, y) \neq 0$. Hence, by $x < y$, we obtain $l(x, y) < 0$ and $x_0 \leq y_0$, i.e.

$$(\bar{y} - \bar{x})^2 < (y_0 - x_0)^2 \text{ and } x_0 < y_0.$$

Choose $e = (\bar{e}, e_0) \in Z$ with $\bar{e}^2 = 1$, $e_0 = 0$, and $\varepsilon \in \mathbb{R}$ with

$$0 < 2\varepsilon < (y_0 - x_0) - \|\bar{y} - \bar{x}\|, \qquad (4.11)$$

and put

$$u := \frac{x+y}{2}, \ v := \frac{x+y}{2} + \varepsilon e.$$

Observe $u_0 = v_0$ and $l(u,v) = \varepsilon^2 > 0$, i.e.

$$u \not\leq v \text{ and } v \not\leq u. \tag{4.12}$$

Moreover,

$$u, v \in [x, y]. \tag{4.13}$$

In order to prove (4.13), we observe

$$x_0 \leq u_0 \leq y_0 \text{ and } x_0 \leq v_0 \leq y_0,$$

by $u_0 = v_0 = \frac{1}{2}(x_0 + y_0)$, and, moreover,

$$l(x, u) = \frac{1}{4} l(x, y) = l(u, y),$$

i.e. $l(x, u) = l(u, y) < 0$. The triangle inequality yields

$$\|(\overline{y} - \overline{x}) \pm 2\varepsilon e\| \leq \|\overline{y} - \overline{x}\| + 2\varepsilon,$$

i.e., by (4.11),

$$\|(\overline{y} - \overline{x}) \pm 2\varepsilon e\| < y_0 - x_0.$$

Hence $[(\overline{y} - \overline{x}) \pm 2\varepsilon e]^2 < (y_0 - x_0)^2$, i.e. $l(x, v)$ and $l(v, y)$ are negative. Because of (4.12), (4.13), $[x, y]$ is not ordered, a contradiction. Hence $l(x, y) = 0$. □

A Lorentz transformation of Z is called *orthochronous* if, and only if, it is also a causal automorphism.

Theorem 5. *The orthochronous Lorentz transformations λ of Z are exactly given by all mappings*

$$\lambda(x) = B_{p,k} \, \omega(x) + d \tag{4.14}$$

with $\omega \in O(X)$, $d \in X$, $1 \leq k \in \mathbb{R}$, $p \in X$, $k^2(1 - p^2) = 1$.

Proof. a) Let λ be an arbitrary orthochronous Lorentz transformation, say (4.4). Since λ is bijective, so must be $\omega : X \to X$. Moreover, $0 \leq t := (0, 1)$ implies $\lambda(0) \leq \lambda(t)$, i.e., by (4) of Proposition 3,

$$0 \leq B_{p,k} \, \omega(\overline{t}, t_0) = B_{p,k}(t) = kp + kt.$$

Hence $(0,0) \leq (kp, k)$, i.e. $0 \leq k$, i.e. $1 \leq k$, in view of $k^2 \geq 1$.

b) Let λ be a mapping (4.14) with proper $B_{p,k}$ and $\omega \in O(X)$. We have to prove

$$a \leq b \Leftrightarrow \lambda(a) \leq \lambda(b)$$

for all $a, b \in Z$. This is clear for $\lambda(x) = x + d$, by (4), Proposition 3. It is also clear for $\lambda(x) = \omega(x)$ because of

$$l(a, b) = l\big(\lambda(a), \lambda(b)\big)$$

and $\lambda(x) = \omega(\overline{x}, x_0) = \big(\omega(\overline{x}), x_0\big)$, i.e. $[\lambda(x)]_0 = x_0$, and because of $\omega^{-1} \in O(X)$.

Finally, we consider the case $\lambda(x) = B_{p,k}(x)$ with $k \geq 1$. Since we have $B_{p,k}^{-1} = B_{-p,k}$ (see (3.120)), we only need to prove

$$a \leq b \Rightarrow \lambda(a) \leq \lambda(b)$$

for all $a, b \in Z$, i.e. $0 \leq b - a \Rightarrow 0 \leq \lambda(b) - \lambda(a)$. But this will be a consequence of

$$0 \leq x \Rightarrow 0 \leq \lambda(x), \tag{4.15}$$

because of the linearity of $B_{p,k}$, as soon as (4.15) is proved. Since $l(0, x) = l\big(\lambda(0), \lambda(x)\big) = l\big(0, \lambda(x)\big)$, it remains to show

$$0 \leq x_0 \Rightarrow [B_{p,k}(x)]_0 \geq 0$$

under the assumption $l(0, x) \leq 0$. Obviously, $[B_{p,k}(x)]_0 = k\overline{x}p + kx_0 =: R$. If $\overline{x}p \geq 0$, then $R \geq 0$, since $x_0 \geq 0$ and $k \geq 1$. If $\overline{x}p < 0$, then

$$(\overline{x}p)^2 \leq x_0^2 p^2 \leq x_0^2,$$

by $\overline{x}^2 - x_0^2 \leq 0$ and $p^2 < 1$. Hence $-\overline{x}p = |\overline{x}p| \leq x_0$. □

Theorem 6. *All causal automorphisms of Z are exactly given by all mappings*

$$\lambda(x) = \gamma \cdot B_{p,k}\,\omega(x) + d \tag{4.16}$$

where $\gamma > 0$ is a real number, $B_{p,k}$ a proper Lorentz boost, $d \in Z$, $\omega \in O(X)$ with $\omega(\overline{x}, x_0) = \big(\omega(\overline{x}), x_0\big)$ for $x = (\overline{x}, x_0) \in Z$.

Proof. Observe that $\mu(x) := \gamma x$ for $x \in Z$ defines a causal automorphism for a real constant $\gamma > 0$. Hence, by Theorem 5, (4.16) must be a causal automorphism.

Suppose now that $\lambda : Z \to Z$ is an arbitrary causal automorphism. If $x \neq y$ are elements of Z with $l(x, y) = 0$, we may assume $x_0 \leq y_0$, because otherwise, $x_0 > y_0$, we would interchange x and y. Hence $x < y$. Thus, by Proposition 4, $[x, y]$ is ordered. Since λ is a causal automorphism, also $[\lambda(x), \lambda(y)]$ must be ordered and $\lambda(x) < \lambda(y)$ holds true. Hence, by Proposition 4, $l\big(\lambda(x), \lambda(y)\big) = 0$. Now Theorem 2 implies that

$$\lambda(x) = m \cdot \lambda_1(x) \tag{4.17}$$

for all $x \in Z$ where λ_1 is a Lorentz transformation and $m \neq 0$ a real constant. We may assume $m > 0$ without loss of generality, since otherwise we would work with

$$\lambda(x) = (-m) \cdot \big(-\lambda_1(x)\big),$$

by considering that also $x \to -\lambda_1(x)$ is a Lorentz transformation. Hence

$$x \to \frac{1}{m}\lambda(x)$$

is a causal automorphism, and thus, by (4.17), λ_1 must be an orthochronous Lorentz transformation. In view of Theorem 5, we hence obtain for λ the form (4.16) with the properties described in Theorem 6. □

Remark. If X is finite-dimensional, Theorem 6 is a well-known theorem of Alexandrov–Ovchinnikova [1], Zeeman [1]. In the general form it is contained in W. Benz [11].

4.3 Relativistic addition

If $\lambda_i : Z \to Z$, $i = 1, 2$, are Lorentz transformations, then, of course, also $\lambda_1\lambda_2$ is such a transformation, because of

$$l(x, y) = l(\lambda_2(x), \lambda_2(y)) = l(\lambda_1[\lambda_2(x)], \lambda_1[\lambda_2(y)])$$

for all $x, y \in Z$. From Theorem 1 we get, say,

$$\lambda_1(x) = B_{a,\alpha}\,\omega_1(x) + d_1,$$
$$\lambda_2(x) = B_{b,\beta}\,\omega_2(x) + d_2,$$

i.e., by the linearity of ω_1 and $B_{a,\alpha}$,

$$\lambda_1\lambda_2(x) = B_{a,\alpha}\,\omega_1 B_{b,\beta}\,\omega_2(x) + d_3, \qquad (4.18)$$

$$d_3 := B_{a,\alpha}\,\omega_1(d_2) + d_1.$$

The problem now is to find a boost $B_{c,\gamma}$ and an orthogonal mapping $\omega : X \to X$ satisfying

$$B_{c,\gamma}\,\omega = B_{a,\alpha}\,\omega_1 B_{b,\beta}\,\omega_2.$$

Theorem 1 guarantees that these objects $B_{c,\gamma}$ and ω exist and that they are uniquely determined. It is easy to verify

$$\omega B_{p,k} = B_{\omega(p),k}\omega. \qquad (4.19)$$

Hence from (4.19),

$$B_{a,\alpha}B_{\omega_1(b),\beta} = B_{c,\gamma}\omega'$$

with the orthogonal mapping $\omega' := \omega\omega_2^{-1}\omega_1^{-1}$.

Theorem 7. *If $a, b \in X$ and $\alpha, \beta \in \mathbb{R}$ are given with*

$$\alpha^2(1 - a^2) = 1, \quad \beta^2(1 - b^2) = 1,$$

then $\delta := 1 + ab > 0$ holds true, and $B_{a,\alpha} B_{b,\beta} = B_{c,\gamma} \omega$ has the uniquely determined solution $\omega = B_{-c,\gamma} B_{a,\alpha} B_{b,\beta}$ and

1) $\quad c = \dfrac{a(1 + a\delta)}{(1 + \alpha)\delta} + \dfrac{b}{\alpha\delta} \quad$ *and $\gamma = \alpha\beta\delta \quad$ for $-1 \notin \{\alpha, \beta\}$,*

2) $\quad c = -b \qquad\qquad\qquad$ *and $\gamma = -\beta \quad$ for $\alpha = -1 \neq \beta$,*

3) $\quad c = a \qquad\qquad\qquad\quad$ *and $\gamma = -\alpha \quad$ for $\alpha \neq -1 = \beta$,*

4) $\quad c = 0 \qquad\qquad\qquad\quad$ *and $\gamma = 1 \qquad$ for $\alpha = -1 = \beta$.*

Proof. Observe $1 - a^2 > 0$, $1 - b^2 > 0$, i.e.

$$-ab \leq |ab| \leq \sqrt{a^2}\,\sqrt{b^2} < 1,$$

i.e. $\delta > 0$. Obviously, $\alpha \neq 0$ and $\beta \neq 0$. In all four cases we will examine the equation

$$L := B_{a,\alpha} B_{b,\beta}(t) = B_{c,\gamma}\, \omega\,(t) = B_{c,\gamma}(t). \tag{4.20}$$

Notice here $t = (0, 1)$ and $\omega\,(x) = \omega\,(\overline{x}, x_0) := \big(\omega\,(\overline{x}), x_0\big)$. In the case that α, β are both unequal to -1, we obtain

$$L = B_{a,\alpha}(\beta b + \beta t) = \beta B_{a,\alpha}(b) + \beta\,(\alpha a + \alpha t) = \gamma_1 c_1 + \gamma_1 t$$

with

$$\gamma_1 := \alpha\beta\delta, \quad \gamma_1 c_1 := \beta b + \frac{\alpha\beta a\,(1 + a\delta)}{1 + \alpha}.$$

We verify $\gamma_1^2(1 - c_1^2) = 1$. Hence B_{c_1, γ_1} is a boost and

$$L = \gamma_1 c_1 + \gamma_1 t = B_{c_1, \gamma_1}(t)$$

holds true. Let now $B_{c,\gamma}$ be a boost also satisfying $L = B_{c,\gamma}(t)$, i.e.

$$\gamma_1 c_1 + \gamma_1 t = \gamma c + \gamma t \text{ for } \gamma \neq -1,$$

and $\gamma_1 c_1 + \gamma_1 t = -t$ for $\gamma = -1$ (and hence $c = 0$ from $\gamma^2(1 - c^2) = 1$). In the first case we get, by $c_1, c \in X$, i.e. by $c_1 = (c_1, 0) \in X \oplus \mathbb{R}$, $c = (c, 0)$ and $t = (0, 1)$,

$$\gamma = \gamma_1 \text{ and } c = c_1,$$

in the second $\gamma_1 = -1 = \gamma$ and $c_1 = 0 = c$ with $\gamma_1^2(1 - c_1^2) = 1$.

If $\alpha = -1 \neq \beta$, (4.20) implies

$$L = B_{0,-1}(\beta b + \beta t) = \beta b - \beta t = \gamma c + \gamma t \text{ for } \gamma \neq -1,$$

and $L = \beta b - \beta t = -t$ for $\gamma = -1$, i.e. $\gamma = -\beta$, $c = -b$ for $\gamma \neq -1$, and $\beta = 1$, i.e. $b = 0 = c$ for $\gamma = -1$.

If $\alpha \neq -1 = \beta$, (4.20) implies

$$L = B_{a,\alpha}(-t) = -\alpha a - \alpha t = \gamma c + \gamma t \text{ for } \gamma \neq -1,$$

and $L = -\alpha a - \alpha t = -t$ for $\gamma = -1$, i.e. $\gamma = -\alpha$, $c = a$ for $\gamma \neq -1$, and $\alpha = 1 = -\gamma$, i.e. $c = 0 = a$, for $\gamma = -1$.

If $\alpha = -1 = \beta$, (4.20) implies

$$L = t = \gamma c + \gamma t \text{ for } \gamma \neq -1,$$

and $L = t = -t$ for $\gamma = -1$. Since the second case does not occur, we obtain $\gamma = 1$, $c = 0$ $\qquad\qquad\square$

We are now interested in the case

$$B_{a,\alpha} B_{b,\beta} = B_{c,\gamma} \omega$$

where the factors on the left-hand side of this equation are proper boosts, hence where $\alpha \geq 1$ and $\beta \geq 1$. This implies

$$\alpha = \frac{1}{\sqrt{1 - a^2}}, \quad \beta = \frac{1}{\sqrt{1 - b^2}},$$

by $\alpha^2(1 - a^2) = 1 = \beta^2(1 - b^2)$. Thus Theorem 7, case 1), yields

$$\gamma = \alpha\beta\delta > 0,$$

i.e. $B_{c,\gamma}$ is also a proper boost. Moreover, by Theorem 7, case 1), we obtain

$$c = \frac{a}{\delta} \frac{1 + \alpha\delta}{1 + \alpha} + \frac{b}{\alpha\delta} = \frac{a}{\delta} \frac{1 + \alpha(1 + ab)}{1 + \alpha} + \frac{b}{\alpha\delta},$$

i.e.

$$c = \frac{a}{\delta}\left(1 + \frac{\alpha}{1 + \alpha} ab\right) + \frac{b}{\alpha\delta}, \qquad (4.21)$$

i.e., by

$$\frac{\alpha}{1 + \alpha} = \frac{\alpha - 1}{\alpha a^2}$$

from $\alpha^2(1 - a^2) = 1$, we get with $\delta = 1 + ab$,

$$c = \frac{1}{1 + ab}\left[\frac{b}{\alpha} + \frac{\alpha - 1}{\alpha a^2}(ab) a + a\right].$$

This is in the case $X = \mathbb{R}^3$ the relativistic sum $c =: a \oplus b$ of the velocities $a, b \in X$ (see, for instance, R.U. Sexl and H.K. Urbantke, Relativität, Gruppen, Teilchen. Springer-Verlag. Wien–New York, 1976, page 34).

By observing $\alpha \geq 1$ and $\alpha^2(1 - a^2) = 1$, we get

$$\frac{1}{\alpha} = \frac{1 + \alpha}{\alpha(1 + \alpha)} = \frac{\alpha^2(1 - a^2) + \alpha}{\alpha(1 + \alpha)} = 1 - \frac{\alpha a^2}{1 + \alpha},$$

i.e., by (4.21),

$$\begin{aligned}
c &= \frac{a}{\delta}\left(1 + \frac{\alpha}{1 + \alpha}\, ab\right) + \frac{b}{\delta}\left(1 - \frac{\alpha}{1 + \alpha}\, a^2\right) \\
&= \frac{a + b}{\delta} + \frac{1}{1 + \frac{1}{\alpha}}\, \frac{(ab)\, a - a^2 b}{\delta}.
\end{aligned}$$

Hence, by $\alpha^2(1 - a^2) = 1$, $\alpha > 0$,

$$a \oplus b := c = \frac{a + b}{1 + ab} + \frac{1}{1 + \sqrt{1 - a^2}}\, \frac{(ab)\, a - a^2 b}{1 + ab}. \tag{4.22}$$

Remark. Let $g(x, y)$ (see (2.40) in section 12, chapter 2) denote the hyperbolic distance of $x, y \in P := \{x \in X \mid x^2 < 1\}$ in the Cayley–Klein model. Then

$$g(a, b) = g(x \oplus a,\, x \oplus b)$$

holds true for all $a, b, x \in P$. On the basis of this functional equation we characterized the function $f(x, y) = x \oplus y$ in W. Benz [10].

4.4 Lightlike, timelike, spacelike lines

Already in connection with pencils of Laguerre geometry we discussed the notion of a line of Z. The sets

$$l = \{(\bar{p}, p_0) + \lambda(\bar{v}, v_0) \mid \lambda \in \mathbb{R}\} = p + \mathbb{R}v \tag{4.23}$$

where $p = (\bar{p}, p_0)$, $v = (\bar{v}, v_0) \neq 0$ are elements of Z, are called the *lines* of Z. Let us denote for a moment the set of lines of Z by N. Then

$$\varphi : \mathbb{L}(Z) \times N \to N$$

with $f(l) := \{f(p) + \lambda B_{a,\alpha}\, \omega(v) \mid \lambda \in \mathbb{R}\}$ for

$$f(x) = B_{a,\alpha}\, \omega(x) + d,\ \omega \in O(X), \tag{4.24}$$

defines an action. Hence (N, φ) is an invariant notion of $(Z, \mathbb{L}(Z))$.

An element z in Z is called *timelike* provided $z^2 < 0$ holds true. $z \neq 0$ is said to be *lightlike, spacelike* if, and only if, $z^2 = 0$, $z^2 > 0$, respectively, is satisfied. The line (4.23) is defined to be timelike, lightlike, spacelike provided v has the corresponding property. Observe that

$$p + \mathbb{R}v = q + \mathbb{R}w$$

implies $q = p + \alpha v$ for a suitable real α, i.e.

$$p + \mathbb{R}v = p + \mathbb{R}w.$$

Hence $w = \mu v$ for a $\mu \neq 0$ in $\mathbb{R}$. Thus the character of l to be timelike,..., does not depend on the special chosen v.

Remark. If $x, y \in \mathbb{R}^3 \oplus \mathbb{R}$ are classical events, i.e. points $\overline{x}, \overline{y}$ of $\mathbb{R}^3$ at certain fixed moments x_0, y_0, the event x has the chance to influence the event y provided there exists a signal from x to y travelling from $\overline{x}$ to $\overline{y}$ along a line of $\mathbb{R}^3$ with a constant velocity $\mu \geq 0$ less than or equal to the speed (which we designate as 1) of light, starting at time x_0 and ending at time $y_0 \geq x_0$. This means

$$\sqrt{(y_1 - x_1)^2 + (y_2 - x_2)^2 + (y_3 - x_3)^2} = \mu \cdot (y_0 - x_0) \leq y_0 - x_0$$

on the basis of

distance travelled = velocity $\mu \cdot$ time taken,

i.e. $l(x, y) \leq 0$ and $x_0 \leq y_0$, i.e. $x \leq y$, where we applied the classical inner product of $\mathbb{R}^3$. Timelike or lightlike lines $x + \mathbb{R}(y - x)$, $x \neq y$, $(y - x)^2 \leq 0$, represent possible signals and lightlike lines signals with speed 1, i.e. the speed of light. In fact! A motion

$$\big(x_1(\tau), x_2(\tau), x_3(\tau)\big),$$

τ the time, $\tau \in [\alpha, \beta]$, leads to the set of events

$$\big\{ \big((x_1(\tau), x_2(\tau), x_3(\tau), \tau \big) \mid \tau \in [\alpha, \beta] \big\}$$

of $\mathbb{R}^3 \oplus \mathbb{R}$, called the *world-line* of the motion. The world-line

$$\{ (p_1 + \mu\tau a_1, p_2 + \mu\tau a_2, p_3 + \mu\tau a_3, p_0 + \tau \mid \tau \in [\alpha, \beta] \},$$

$a_1^2 + a_2^2 + a_3^2 = 1$, of a signal travelling with constant velocity $\mu \in [0, 1]$ along a line of $\mathbb{R}^3$ can be written in $\mathbb{R}^3 \oplus \mathbb{R}$ as

$$\{ p + \tau v \mid \tau \in [\alpha, \beta] \}$$

with $p := (p_1, p_2, p_3, p_0)$, $v := (\mu a_1, \mu a_2, \mu a_3, 1)$ such that

$$v^2 = \mu^2 - 1 \leq 0.$$

In this sense possible signals are represented by timelike or lightlike lines of $\mathbb{R}^3 \oplus \mathbb{R}$. Spacelike lines do not occur as signals.

The image line $f(l)$ of (4.23) under (4.24) is

$$f(p) + \mathbb{R} \cdot B_{a,\alpha}\, \omega\,(v),$$

as we already know. Because of

$$l(v,0) = l\,(B_{a,\alpha}\,\omega\,(v),\ B_{a,\alpha}\,\omega\,(0)),$$

i.e.

$$v^2 = [B_{a,\alpha}\,\omega\,(v)]^2,$$

$f(l)$ has the same character, namely to be timelike,..., as l. We hence get the invariant notions of timelike lines,....

From Propositions 46 and 43, chapter 3, we know that the parabolic pencils correspond to the lightlike lines, the elliptic pencils to the spacelike lines (see Proposition 48), the hyperbolic pencils to the timelike lines (see Proposition 50).

If $l := p + \mathbb{R}v = (\bar{p}, p_0) + \mathbb{R}\,(\bar{v}, v_0)$ is a line of Z, we will call

$$\pi\,(l) := \bar{p} + \mathbb{R}\bar{v}$$

its *projection* into X. This projection is a point if, and only if, $v \in \mathbb{R}t$. We define the angle measure

$$\sphericalangle\bigl(l,\, \pi\,(l)\bigr)$$

to be 90° provided $\pi\,(l)$ is a point, and otherwise by $\varphi \in [0°, 90°[$ satisfying

$$\cos^2\varphi = \frac{\bar{v}^2}{\bar{v}^2 + v_0^2}.$$

For $0 = v^2 = \bar{v}^2 - v_0^2$ (of course, with $v \neq 0$) we get $\varphi = 45°$. This angle characterizes the lightlike lines of Z. Moreover, l is

$$\text{timelike} \iff 45° < \varphi \le 90°,$$

and spacelike $\iff 0 \le \varphi < 45°$: this follows from

$$0 > v^2 = \bar{v}^2 - v_0^2 \iff \frac{1}{2} > \frac{\bar{v}^2}{\bar{v}^2 + v_0^2} \ge 0,$$

$$0 < v^2 = \bar{v}^2 - v_0^2 \iff \frac{1}{2} < \frac{\bar{v}^2}{\bar{v}^2 + v_0^2} \le 1.$$

4.5   Light cones, lightlike hyperplanes

If $p \in Z$,

$$C\,(p) := \{x \in Z \mid l\,(p, x) = 0\}$$

is called the *light cone* with *vertex p*. This leads also to an invariant notion of $(Z, \mathbb{L}(Z))$ with

$$f\left(C\left(p\right)\right) = C\left(f\left(p\right)\right)$$

for all $f \in \mathbb{L}(Z)$. The *future* of $p \in Z$ is defined by

$$F\left(p\right) := \{x \in Z \mid p \leq x\},$$

the *past* by

$$P\left(p\right) := \{x \in Z \mid x \leq p\}.$$

Take $\lambda \in \mathbb{L}(Z)$ with $\lambda\left(x\right) = 2p - x$ for all $x \in Z$. Then

$$\lambda\left(F\left(p\right)\right) = P\left(p\right).$$

Hence $N := \{F\left(p\right) \mid p \in Z\}$ with

$$\varphi\left(f\; F\left(p\right)\right) := f\left(F\left(p\right)\right) \tag{4.25}$$

for $f \in \mathbb{L}(Z)$ does not lead to an invariant notion. Here we consider the group $\mathbb{L}^+(Z)$ of all orthochronous Lorentz transformations of Z. Observe that the index of $\mathbb{L}^+(Z)$ in $\mathbb{L}(Z)$ is 2. Of course, (4.25) yields an invariant notion for $(Z, \mathbb{L}^+(Z))$ under the assumption that the mappings f of (4.25) are in $\mathbb{L}^+(Z)$.

Proposition 8. a) *A line l is lightlike if, and only if, there exists a light cone containing l.*

  b) *A line l is timelike or lightlike if, and only if, there exist $x, y \in l$ with $x < y$.*

  c) *The light cone $C\left(p\right)$ is the union of all lightlike lines through p.*

  d) *Let S be the set of all timelike or lightlike lines through p. Then*

$$F\left(p\right) = \bigcup_{l \in S} l^+(p), \quad P\left(p\right) = \bigcup_{l \in S} l^-(p)$$

  where $l^+(p) := \{x \in l \mid p \leq x\}$ and $l^-(p) := \{x \in l \mid x \leq p\}$.

Proof. a) If $p + \mathbb{R}v$ is lightlike, it is contained in $C\left(p\right)$, since

$$l\left(p, p + \lambda v\right) = \left(\left(p + \lambda v\right) - p\right)^2 = \lambda^2 v^2 = 0.$$

If the line $p + \mathbb{R}v$ is contained in $C\left(r\right)$, then

$$\left(\left(p + \lambda v\right) - r\right)^2 = \left(p - r\right)^2 + 2\lambda\left(p - r\right)v + \lambda^2 v^2$$

must be 0 for all real λ. Hence $v^2 = 0$.

b) Let $l = p + \mathbb{R}v$ be timelike or lightlike, i.e. assume $v^2 \leq 0$. Then $p < p + v$ for $v_0 \geq 0$, and $p < p - v$ for $v_0 \leq 0$. Vice versa, let l be a line containing x, y

with $x < y$. Hence $x - y = \varrho v$ with a suitable real $\varrho \neq 0$. Now $l(x,y) \leq 0$ implies $(x-y)^2 \leq 0$, i.e. $v^2 \leq 0$.

c) If $x \neq p$ is in $C(p)$, then $l(p,x) = 0$, i.e.

$$x \in p + \mathbb{R}(x-p)$$

with $(x-p)^2 = 0$ holds true. If $p + \mathbb{R}v$ is a lightlike line, then $v^2 = 0$ implies

$$l(p, p + \lambda v) = 0$$

for all $\lambda \in \mathbb{R}$.

d) If $x \neq p$ is in $F(p)$, then $p < x$ holds true. Hence

$$x \in l := p + \mathbb{R}(x-p), \ (x-p)^2 \leq 0,$$

and even $x \in l^+$ since $p \leq x$. Vice versa, if l is a line in S and x a point in l^+, then

$$x \in l = p + \mathbb{R}v, \ v^2 \leq 0, \ p \leq x,$$

i.e. $x \in F(p)$. $\qquad\qquad\qquad\qquad\qquad\qquad\qquad\qquad\qquad\qquad\qquad\square$

Another important invariant notion of $(Z, \mathbb{L}(Z))$ is based on what we call a *lightlike hyperplane* of Z. First of all, however, we would like to define the notion of a hyperplane of Z. If $a \neq 0$ is an element of Z and α a real number, the set of points,

$$\{x \in Z \mid ax = \alpha\}$$

will be called a *hyperplane* of Z. This set can also be written in the form $\{x \in Z \mid \overline{a}\,\overline{x} - a_0 x_0 = \alpha\}$.

Remark. Of course, the hyperplanes of the real inner product space $Y = X \oplus \mathbb{R}$ equipped with the inner product (3.54) (see the last Remark of section 10, chapter 3) coincide as sets of points with the hyperplanes of Z, since ax in Z is $(\overline{a}, -a_0)\,x$ in Y.

A special type of hyperplanes now will be of interest for the final discussions of this section 4.

If $v \in Z$ satisfies $v \neq 0 = v^2$, and if $\alpha \in \mathbb{R}$, then

$$\{x \in Z \mid vx = \alpha\} \qquad\qquad\qquad\qquad (4.26)$$

is said to be a *lightlike hyperplane* of Z.

For $\lambda \in \mathbb{L}(Z)$ define the image of (4.26) under λ by

$$\{\lambda(x) \mid x \in Z \text{ and } vx = \alpha\}.$$

We will prove that this image is also a lightlike hyperplane. This is clear for $\lambda(x) = x + d$, since then the image is given by

$$\{y \in Z \mid vy = vd + \alpha\}.$$

If $\lambda(x) = B_{p,k}\,\omega(x)$, we observe $l(0, a) = l\left(0, \lambda(a)\right)$ for all $a \in Z$, i.e.

$$a^2 = [\lambda(a)]^2. \tag{4.27}$$

Moreover, $l(a, b) = l\left(\lambda(a), \lambda(b)\right)$ for all $a, b \in Z$, i.e.

$$(a - b)^2 = [\lambda(a) - \lambda(b)]^2,$$

i.e., by (4.27),

$$ab = \lambda(a)\,\lambda(b) \tag{4.28}$$

for all $a, b \in Z$. *We would like to emphasize that (4.28) holds true for all Lorentz transformations λ of Z satisfying $\lambda(0) = 0$.*

Applying (4.28) on (4.26), we obtain with

$$y = B_{p,k}\,\omega(x) =: \lambda(x),$$

obviously,

$$\lambda\left(\{x \in Z \mid vx = \alpha\}\right) = \{y \in Z \mid \alpha = vx = \lambda(v)\,\lambda(x) = \lambda(v)\,y\}$$

with $[\lambda(v)]^2 = v^2 = 0$ and $\lambda(v) \neq \lambda(0) = 0$, since λ is injective.

Hence we proved

Proposition 9. *Define $LH(Z)$ to be the set of all lightlike hyperplanes of Z and $\lambda(E)$ for $\lambda \in \mathbb{L}(Z)$ and $E \in LH(Z)$ to be $\{\lambda(x) \mid x \in E\}$, then $\left(LH(Z), \varphi\right)$ with $\varphi(\lambda, E) := \lambda(E)$ is an invariant notion of $\left(Z, \mathbb{L}(Z)\right)$.*

Remark. Let $H(Z)$ be the set of all hyperplanes of Z and define

$$\lambda(E) = \{\lambda(x) \mid x \in E\}$$

for $\lambda \in \mathbb{L}(Z)$ and $E \in H(Z)$, then, with almost the same arguments, we obtain that $\left(H(Z), \varphi_1\right)$ with $\varphi_1(\lambda, E) := \lambda(E)$ is an invariant notion of $\left(Z, \mathbb{L}(Z)\right)$.

If $a \neq p$ is a point of the light cone $C(p)$,

$$\{x \in Z \mid (x - p)(a - p) = 0\}$$

will be called the *tangential hyperplane* of $C(p)$ in a.

Proposition 10. *If $a \neq p$ is in $C(p)$, then*

$$C(p) \cap \{x \in Z \mid (x - p)(a - p) = 0\} = p + \mathbb{R}(a - p).$$

Proof. $a \in C(p)$ implies $l(p, a) = 0$, i.e. $(a - p)^2 = 0$. Hence we get for all $\alpha \in \mathbb{R}$,

$$p + \alpha(a - p) \in C(p) \cap \{x \in Z \mid (x - p)(a - p) = 0\}.$$

Vice versa, assume $b \in C(p) \backslash \{p\}$ and $(b - p)(a - p) = 0$. From

$$(a - p)^2 = 0, \ (b - p)^2 = 0, \ (a - p)(b - p) = 0$$

and Lemma 60, (β), chapter 3, we obtain that $a - p$, $b - p$ are linearly dependent. Hence

$$b - p = \alpha(a - p),$$

i.e. $b \in p + \mathbb{R}(a - p)$.   □

Proposition 11.   a) *Every tangential hyperplane of a light cone is a lightlike hyperplane.*

   b) *If $a \in E \in LH(Z)$ and if l is a (in fact existing) lightlike line contained in E, and passing through a, then E is a tangential hyperplane of $C(p)$ in a where $p \in l \backslash \{a\}$.*

   c) *All lightlike lines contained in $E = \{x \in Z \mid vx = \alpha\}$, $v \neq 0 = v^2$, are given by $a + \mathbb{R}v$, $a \in E$. All other lines contained in E are spacelike.*

Proof. a) Follows from $(a - p)^2 = 0$.

b) Assume $E = \{x \in Z \mid vx = \alpha\}$, $v^2 = 0 \neq v$. If $a \in E$, then

$$a + \mathbb{R}v \ni a$$

is a lightlike line contained in E. Let $a + \mathbb{R}w$, $w \neq 0 = w^2$, be any lightlike line l contained in E, and take $p = a + \beta w$, $\beta \neq 0$. Hence

$$v^2 = 0, \ w^2 = 0, \ vw = 0,$$

in view of $\alpha = vp = v(a + \beta w) = \alpha + \beta vw$. Lemma 60, (β), chapter 3, yields $w = \gamma v$ with a suitable real γ which must be unequal to 0, since $w \neq 0$. Hence

$$l = a + \mathbb{R}v.$$

Observe $p = a + \beta \gamma v$, $\beta \gamma \neq 0$. Thus

$$0 = (x - p)(a - p = (x - a - \beta \gamma v) \cdot (-\beta \gamma v),$$

i.e. $vx = va = \alpha$.

c) We already realized in step b) that there is one and only one lightlike line contained in E and passing through $a \in E$. Assume now there would exist a timelike line $a + \mathbb{R}w$, $w^2 < 0$, contained in E. Hence

$$v^2 = 0, \ w^2 < 0, \ vw = 0,$$

i.e. $\bar{v}^2 = v_0^2$, $\bar{w}^2 < w_0^2$, $\bar{v}\bar{w} = v_0 w_0$. This implies, by $\bar{v} \neq 0$, since $v \neq 0$, the contradiction

$$(v_0 w_0)^2 = (\bar{v}\bar{w})^2 \leq \bar{v}^2 \bar{w}^2 < v_0^2 w_0^2.$$ □

4.6   Characterization of some hyperplanes

A hyperplane $E = \{x \in Z \mid ax = \alpha\}$, $a \neq 0$, of Z will be called *timelike, spacelike* provided $a^2 > 0$, $a^2 < 0$, respectively, holds true. As in Proposition 9 concerning the lightlike hyperplanes, we also get here invariant notions of $(Z, \mathbb{L}(Z))$ with respect to timelike, spacelike hyperplanes.

Theorem 12. *A hyperplane Z is lightlike if, and only if, it contains a lightlike line, but no timelike lines.*

Proof. If E is a lightlike hyperplane, it contains a lightlike line, but otherwise only lightlike or spacelike lines (see Proposition 11, c)). Assume, vice versa, that the hyperplane

$$E = \{x \in Z \mid ax = \alpha\}, \, a \neq 0,$$

of Z contains at least one lightlike line,

$$l = p + \mathbb{R}v, \, v^2 = 0 \neq v,$$

but otherwise only lightlike or spacelike lines. Hence $ap = \alpha$ and $av = 0$. We will prove that a and v are linearly dependent. This then implies that E is lightlike. Observe $\bar{a} \neq 0$, since otherwise

$$0 = \bar{a}\,\bar{v} = a_0 v_0$$

from $av = 0$, contradicting $a_0 v_0 \neq 0$, by $a \neq 0 \neq v$ and $\bar{v}^2 = v_0^2$.

Case 1. $a^2 = 0$.

$a^2 = 0$, $v^2 = 0$, $va = 0$ imply, by Lemma 60, (β), chapter 3, that a, v are linearly dependent. Thus $a = \gamma v$.

Case 2.1. $a^2 \neq 0$ and $a_0 = 0$.

Hence $at = 0$ with $t = (0, 1)$. Observe $v + \beta t \neq 0$ for all $\beta \in \mathbb{R}$, since otherwise $\bar{v} + \beta \cdot 0 = 0$, i.e. $v = 0$ from $\bar{v}^2 = v_0^2$. Hence the line

$$p + \mathbb{R}\left(v - \frac{v_0}{2}\right)$$

must be contained in E, and must thus be lightlike or spacelike, i.e., by $v_0 \neq 0$,

$$0 \leq \left(v + \frac{v_0}{2}\,t\right)^2 = v_0 v t + \frac{v_0^2}{4}\,t^2 = -\frac{3}{4}\,v_0^2 < 0.$$

Hence, Case 2.1 does not occur.

Case 2.2. $a^2 \neq 0$ and $a_0 \neq 0$.

Observe $b := \left(\bar{a}, \frac{1}{a_0}\,\bar{a}^2\right) \neq 0$, because of $\bar{a} \neq 0$. Obviously,

$$a\,(b + \alpha v) = 0$$

for all $\alpha \in \mathbb{R}$. Moreover, by $\bar{a}\,\bar{v} = a_0 v_0$,

$$vb = \bar{v}\,\bar{a} - v_0 \cdot \frac{\bar{a}^2}{a_0} = -\frac{v_0}{a_0}\, a^2 \neq 0.$$

We also have $b + \alpha v \neq 0$ for all $\alpha \in \mathbb{R}$, since $b + \alpha_0 v = 0$ would imply

$$0 = (b + \alpha_0 v)\, v = bv \neq 0.$$

Hence the line

$$p + \mathbb{R}\,(b + \alpha_1 v)$$

with $\alpha_1 \cdot 2vb := -1 - b^2$, must be contained in E, and must thus be lightlike or spacelike, i.e.

$$0 \leq (b + \alpha_1 v)^2 = b^2 + 2\alpha_1 bv = -1,$$

a contradiction. Thus, Case 2.2 also does not occur. $\qquad\qquad\qquad\square$

Remark. The hyperplane $\{x \in Z \mid tx = 0\}$ contains only spacelike lines $p + \mathbb{R}b$, $tp = 0$, $tb = 0$, $b \neq 0$, since $b = (\bar{b}, 0)$, i.e. $b^2 > 0$.

If $a = (e, 0)$ with $0 \neq e \in X$, then $E = \{x \in Z \mid ax = 0\}$ contains lightlike, timelike and spacelike lines. Take $j \in X$ with $j^2 = 1$ and $ej = 0$. Then the line $\mathbb{R}\,(j, \alpha)$ is contained in E and timelike for $\alpha^2 > 1$, lightlike for $\alpha^2 = 1$, spacelike for $\alpha^2 < 1$.

Theorem 13.   a) $E = \{x \in Z \mid ax = \alpha\}$, $a \neq 0$, *contains only spacelike lines if, and only if, $a^2 < 0$.*

b) E *contains timelike, lightlike and spacelike lines if, and only if, $a^2 > 0$.*

Proof. a) Assume $a^2 < 0$ and $p + \mathbb{R}b \subset E$, $b \neq 0$. Hence $ab = 0$. Observe $a_0 \neq 0$, since $\bar{a}^2 < a_0^2$. Also $\bar{b} \neq 0$ holds true, because otherwise $b = 0$ from $\bar{a}\,\bar{b} = a_0 b_0$. Hence

$$(a_0 b_0)^2 = (\bar{a}\,\bar{b})^2 \leq \bar{a}^2 \bar{b}^2 < a_0^2 \bar{b}^2,$$

i.e. $b_0^2 < \bar{b}^2$, i.e. $b^2 > 0$. Thus $p + \mathbb{R}b$ is spacelike. Assume, vice versa, that E contains only spacelike lines. We hence get $b^2 > 0$ for all $b \neq 0$ in Z which satisfy $ab = 0$. If a_0 were 0, then with $b := t = (0, 1)$,

$$b \neq 0, \ ab = 0, \ b^2 < 0$$

would be the consequence. Hence $a_0 \neq 0$. Put

$$b := \left(\bar{a}, \frac{1}{a_0}\,\bar{a}^2\right)$$

for $\bar{a} \neq 0$. This implies $b \neq 0$, $ab = 0$, whence

$$0 < b^2 = \bar{a}^2 - \frac{1}{a_0^2}\,(\bar{a}^2)^2 = \frac{\bar{a}^2}{a_0^2}\,(a_0^2 - \bar{a}^2),$$

i.e. $a^2 < 0$. If, finally, $a = (0, a_0)$, we get $a^2 < 0$.

b) Assume $a^2 > 0$. Hence $\bar{a}^2 - a_0^2 > 0$, i.e. $\bar{a} \neq 0$. Take $j \in X$ with $j^2 = 1$ and $j\bar{a} = 0$. If $a_0 = 0$, then

$$\frac{\alpha a}{a^2} + \mathbb{R}(j, \gamma) \subset E$$

for all $\gamma \in \mathbb{R}$. This line is lightlike for $\gamma = 1$, timelike for $\gamma = 2$ and spacelike for $\gamma = 1/2$. If $a \neq 0$, then

$$\frac{\alpha a}{a^2} + \mathbb{R}\left(\bar{a} + \gamma j, \frac{1}{a_0}\bar{a}^2\right) \subset E$$

for all $\gamma \in \mathbb{R}$. Since

$$(\bar{a} + \gamma j)^2 - \left(\frac{\bar{a}^2}{a_0}\right)^2 = \gamma^2 - \frac{\bar{a}^2}{a_0^2} a^2 =: \gamma^2 - k, \ k > 0,$$

there are values γ with $\gamma^2 - k > 0, = 0, < 0$. Assume, vice versa, that E contains timelike, lightlike and spacelike lines. This implies $a^2 > 0$, because E does not contain timelike lines for $a^2 \leq 0$. $\qquad \square$

4.7 $\mathbb{L}(Z)$ as subgroup of Lie (X)

Surjective Lorentz transformations ((4.4), here $\omega \in O(X)$) are special Laguerre transformations, (3.124), namely those preserving tangential distances. They are hence Lie transformations (section 12, chapter 3). Let, vice versa, λ be a Lie transformation. If it fixes the Lie cycle ∞, it can be described in terms of Lorentz–Minkowski geometry as a bijection of the set Z of events with

$$\forall_{x,y \in Z} \ l(x, y) = 0 \Rightarrow l(\lambda(x), \lambda(y))$$

(see Theorem 2). In the classical case this means exactly that λ, $\lambda(\infty) = \infty$, as a bijection of $\mathbb{R}^3 \oplus \mathbb{R}$ transforms light signals into light signals. The further assumption then that λ preserves tangential distances leads, as mentioned before, to the Lorentz transformations.

We will collect some other correspondences between Lorentz–Minkowski geometry and Laguerre geometry.

An event as an element of $Z = X \oplus \mathbb{R} = \Gamma$ (see section 10, chapter 3) corresponds to a Laguerre cycle. If x, y are events, then $l(x, y) \geq 0$ (see (3.85)) is the square of the tangential distance of the Laguerre cycles x, y. The Laguerre cycle x touches y if, and only if, $l(x, y) = 0$. In Lorentz–Minkowski geometry this equation means that there is a lightlike line containing the events x and y. The situation that for the Laguerre cycles x, y there is no spear touching both, is characterized by $l(x, y) < 0$: this means in Lorentz–Minkowski geometry that the line through x, y is timelike. In section 4 we already mentioned correspondences between the pencils of Laguerre geometry and specific lines of Z.

Of course, we would like to know what the meaning of spears is in Lorentz–Minkowski geometry. This is answered by

Proposition 14. *The spears s as sets $\{c \in \Gamma \mid c - s\}$ of Laguerre cycles are exactly the ligthlike hyperplanes.*

Proof. Let (a, ξ, α) be coordinates of s, then (see (3.77))

$$
\begin{aligned}
\{c \in \Gamma \mid c - s\} &= \{(m, \tau) \in Z \mid am + \xi\tau = \alpha\} \\
&= \{(\overline{x}, x_0) \in Z \mid (a, -\xi) \cdot x = \alpha\}
\end{aligned}
$$

holds true with $(a, -\xi)^2 = a^2 - (-\xi)^2 = 0$. If, on the other hand, $v \in Z$ is given with $v \neq 0 = v^2$, then

$$
\{(\overline{x}, x_0) \in Z \mid vx = \alpha\}
$$

is the spear with coordinates $(\overline{v}, -v_0, \alpha)$. $\square$

4.8   A characterization of LM-distances

A function $d : Z \times Z \to \mathbb{R}$ will be called a *general Lorentz–Minkowski distance* (LM-distance) of Z provided

$$
d(x, y) = d(\lambda(x), \lambda(y)) \tag{4.29}
$$

holds true for all $x, y \in Z$ and for all bijective Lorentz transformations λ of Z.

Theorem 15. *All general Lorentz–Minkowski distances d of Z are given as follows. Let ϱ be a fixed real number and let $g : \mathbb{R} \to \mathbb{R}$ be an arbitrary function. Then*

$$
d(x, y) = g(l(x, y)) \tag{4.30}
$$

for all elements $x \neq y$ of Z, and $d(x, x) = \varrho$ for all $x \in Z$.

Proof. Obviously, every such d defines a general Lorentz–Minkowski distance. Assume now that $d : Z \times Z \to \mathbb{R}$ is a general Lorentz–Minkowski distance.

a) $d(x, y) = d(x - y, 0)$ holds true for all $x, y \in Z$.

Let $x, y \in Z$ be fixed elements. Then

$$
\lambda(z) := z - y
$$

is a bijective Lorentz transformation of Z. Hence, by (4.29),

$$
d(x, y) = d(\lambda(x), \lambda(y)) = d(x - y, 0).
$$

Define $\varrho := d(0, 0)$. Hence, by a), $d(x, x) = d(0, 0) = \varrho$ for all $x \in Z$.

b) *To $x, y \in Z \backslash \{0\}$ there exists a bijective Lorentz transformation λ with $\lambda(0) = 0$ and $\lambda(x) = y$ if, and only if,*

$$l(x, 0) = \bar{x}^2 - x_0^2 = \bar{y}^2 - y_0^2 = l(y, 0). \tag{4.31}$$

If there exists such a λ, of course, (4.31) holds true. So assume, vice versa, that (4.31) is satisfied.

Case 1. $l(x, 0) < 0$, i.e. $x_0^2 > \bar{x}^2 \geq 0$.

Put $p \cdot x_0 := -\bar{x}$ and observe $p^2 < 1$ and $p \in X$. With $\operatorname{sgn} k := \operatorname{sgn} x_0$ and $k^2(1 - p^2) := 1$, we obtain

$$B_{p,k}(x) = \sqrt{x_0^2 - \bar{x}^2} \, t.$$

Similarly, with suitable q, κ, we get

$$B_{q,\kappa}(y) = \sqrt{y_0^2 - \bar{y}^2} \, t.$$

Hence, by (4.31), $y = B_{-q,\kappa} B_{p,k}(x)$. Also $B_{-q,\kappa} B_{p,k}(0) = 0$.

Case 2. $l(x, 0) = 0$, but $x \neq 0$ and $y \neq 0$.

Obviously, $\bar{x} \neq 0$ and $x_0 \neq 0$. Hence $(\bar{x}/x_0)^2 = 1$. Suppose $x_0 \neq -1$. With

$$p := \frac{x_0^2 - 1}{x_0^2 + 1} \cdot \frac{\bar{x}}{x_0}, \quad \operatorname{sgn} k := \operatorname{sgn} x_0,$$

$k^2(1 - p^2) := 1$, we obtain

$$B_{p,k}\left(\frac{\bar{x}}{x_0}, 1\right) = x. \tag{4.32}$$

Similarly, if $y_0 \neq -1$,

$$B_{q,\kappa}\left(\frac{\bar{y}}{y_0}, 1\right) = y.$$

In the case

$$b := \frac{\bar{x}}{x_0} + \frac{\bar{y}}{y_0} = 0$$

define $\omega(\bar{z}, z_0) = (-\bar{z}, z_0)$ for $z \in Z$, and in the case $b \neq 0$,

$$\omega(\bar{z}, z_0) = \left(2(\bar{z}a)\, a - \bar{z}, \, z_0\right)$$

with $a \cdot \|b\| := b$. Since $\omega : X \to X$ is in both cases an involution, $\omega \in O(X)$. Now

$$y = B_{q,\kappa}\, \omega\, B_{-p,k}(x) \tag{4.33}$$

holds true.

In one of the cases $x_0 = -1$ or $y_0 = -1$, for instance in the case $x_0 = -1$, we replace (4.32) by

$$\lambda \left(\frac{\overline{x}}{x_0}, 1 \right) = (\overline{x}, x_0)$$

with $\lambda(z) := -z$ for $z \in Z$. Then $B_{-p,k}$ in (4.33) needs to be replaced by λ.

Case 3. $l(x, 0) > 0$, i.e. $\overline{x}^2 > x_0^2 \geq 0$.

Observe

$$B_{p,k}(\overline{x}, x_0) = \frac{\overline{x}}{\|x\|} \sqrt{\overline{x}^2 - x_0^2}$$

for $\overline{x}^2 \cdot p := -x_0 \cdot \overline{x}$ and $k > 0$. Take as in Case 2 an $\omega \in O(X)$ such that

$$\omega \left(\frac{\overline{x}}{\|\overline{x}\|} \right) = \frac{\overline{y}}{\|\overline{y}\|}.$$

With

$$B_{q,\kappa}(y) = \frac{\overline{y}}{\|y\|} \sqrt{\overline{y}^2 - y_0^2},$$

$\overline{y}^2 \cdot q := -y_0 \cdot \overline{y}$ and $\kappa > 0$, we obtain

$$y = B_{-q,\kappa} \, \omega \, B_{p,k}(x).$$

c) Suppose that $e \in X$ satisfies $e^2 = 1$ and that ξ is a real number. Define $x = \sqrt{\xi} \cdot e$ for $\xi > 0$, $x = (e, 1)$ for $\xi = 0$ and $x = (0, \sqrt{|\xi|})$ for $\xi < 0$. Then $x \neq 0$ and $l(x, 0) = \xi$ hold true. Define

$$g(\xi) := d(x, 0)$$

for all $\xi \in \mathbb{R}$. If also $\xi = l(y, 0)$ is satisfied for $y \neq 0$, then, by b), there exists $\lambda \in L(Z)$ with $\lambda(0) = 0$ and $\lambda(x) = y$. Hence $d(x, 0) = d(y, 0)$. Thus $d(z, 0) = g\big(l(z, 0)\big)$ holds true for every $z \neq 0$ in Z. For $v \neq w$ in Z we hence obtain, by a),

$$d(v, w) = d(v - w, 0) = g\big(l(v - w), 0\big) = g\big(l(v, w)\big). \qquad \square$$

Define $D(x, y) := \sqrt{|d(x, y)|}$ for $x, y \in Z$. We will call a general Lorentz–Minkowski distance d *additive* provided there exists $e \in X$ with $e^2 = 1$ and

(1) $d(e, t) = d(e, 0) + d(0, t) = 0$, $t = (0, 1)$,

(2) $d(\alpha e, 0) \geq 0$ and $d(\alpha t, 0) \leq 0$ for all real $\alpha > 0$,

(3) $D(0, \beta j) = D(0, \alpha j) + D(\alpha j, \beta j)$ for $j \in \{e, t\}$ and all $\alpha, \beta \in \mathbb{R}$ with $0 \leq \alpha \leq \beta$.

Theorem 16. *Let d be a general Lorentz–Minkowski distance of Z. Then d is additive if, and only if, there exists a fixed non-negative number γ with*

$$d(x, y) = \gamma \cdot l(x, y) \tag{4.34}$$

for all $x, y \in Z$.

Proof. Of course, d of (4.34) is additive. Let now d be an additive general Lorentz–Minkowski distance. (3) implies $D(0,0) = 0$ for $\alpha = \beta = 0$. Hence

$$d(x,x) = d(0,0) = 0$$

for all $x \in Z$. Thus (4.34) holds true for $x = y$. Because of

$$D(\alpha j, \beta j) = D(0, (\beta - \alpha) j)$$

and of (3) there exist constants γ_1 and γ_2 with

$$D(0, \xi e) = \gamma_1 \xi \text{ and } D(0, \xi t) = \gamma_2 \xi$$

for all $\xi \geq 0$: put $\varphi(\xi) := D(0, \xi j)$ for $\xi \geq 0$, observe $\varphi(\xi) \geq 0$ for non-negative ξ and, moreover,

$$\begin{aligned} \varphi(\beta - \alpha) &= D(0, (\beta - \alpha) j) = D(\alpha j, \beta j) \\ &= D(0, \beta j) - D(0, \alpha j) = \varphi(\beta) - \varphi(\alpha) \end{aligned}$$

for $0 \leq \alpha \leq \beta$, whence $\varphi(\xi + \eta) = \varphi(\xi) + \varphi(\eta)$ for $\xi \geq 0, \eta \geq 0$, by defining $\alpha := \xi, \beta := \xi + \eta$ and noticing $0 \leq \alpha \leq \beta$. Now applying the Remark between steps C and D of the proof of Theorem 7 of chapter 1, we obtain $\varphi(\xi) = \gamma \xi, \gamma \geq 0$, i.e.

$$d(0, \xi e) = \gamma_1^2 \xi^2 \text{ and } d(0, \xi t) = -\gamma_2^2 \xi^2,$$

in view of (2) and of $d(x,y) = d(y,x)$ from (4.30). The second part of (1) implies $\gamma_1^2 = \gamma_2^2$. By step c) of the proof of Theorem 15, we obtain

$$\begin{aligned} g(\xi) &= d(\sqrt{\xi} e, 0) = \gamma_1^2 \xi \text{ for } \xi > 0, \\ g(\xi) &= d(\sqrt{|\xi|} t, 0) = \gamma_1^2 \xi \text{ for } \xi < 0. \end{aligned}$$

Moreover,

$$g(0) = d(e + t, 0) = d(B_{0,-1}(e + t), 0) = d(e - t, 0) = d(e, t) = 0.$$

Hence $g(\xi) = \gamma \xi$ for all $\xi \in \mathbb{R}$ with $\gamma = \gamma_1^2 \geq 0$. $\qquad\square$

4.9 Einstein's cylindrical world

Define $C(Z) := \{z \in Z \mid \overline{z}^2 = 1\}$ to be the set of *points* of *Einstein's cylindrical world (Einstein's cylinder universe)* over $Z = X \oplus \mathbb{R}$, and call

$$e(x,y) := [\text{arc}\cos(\overline{x}\,\overline{y})]^2 - (x_0 - y_0)^2 \tag{4.35}$$

with arc $\cos(\overline{x}\,\overline{y}) \in [0, \pi]$ the *Einstein distance* of $x, y \in C(Z)$, by observing $-1 \leq \overline{x}\,\overline{y} \leq 1$, in view of

$$(\overline{x}\,\overline{y})^2 \leq \overline{x}^2 \overline{y}^2 = 1$$

(see section 4, chapter 1).

Theorem 17. *All $f : C(Z) \to C(Z)$ satisfying*

$$e(x, y) = e(f(x), f(y)) \qquad (4.36)$$

for all $x, y \in C(Z)$ are injective and they are given by

$$f(x) = \omega(\overline{x}) + (\varepsilon x_0 + a) t, \ x \in C(Z),$$

with $t := (0, 1) \in Z$ where $\omega : X \to X$ is orthogonal and where ε, a are real numbers such that $\varepsilon^2 = 1$.

Proof. For a solution of (4.36) put

$$f(\overline{x} + x_0 t) =: \varphi(\overline{x}, x_0) + \psi(\overline{x}, x_0) t, \ \overline{x}^2 = 1,$$

with $\varphi \in X$, $\varphi^2 = 1$, and $\psi \in \mathbb{R}$. Of course, we identify $h \in X$ with the element $(h, 0)$ of Z.

a) $f(-\overline{x} + x_0 t) = -\varphi(\overline{x}, x_0) + \psi(\overline{x}, x_0) t$ for all $x \in C(Z)$.
In order to prove this equation, apply (4.36) for $y := -\overline{x} + x_0 t$. Then

$$\pi^2 = \left(\arccos[\varphi(\overline{x}, x_0)\,\varphi(-\overline{x}, x_0)]\right)^2 - \left(\psi(\overline{x}, x_0) - \psi(-\overline{x}, x_0)\right)^2. \qquad (4.37)$$

This implies $\arccos[\varphi(\overline{x}, x_0)\,\varphi(-\overline{x}, x_0)] = \pi$ and $\psi(\overline{x}, x_0) = \psi(-\overline{x}, x_0)$, since otherwise the right-hand side of (4.37) would be smaller than π^2. Hence

$$\varphi(\overline{x}, x_0)\,\varphi(-\overline{x}, x_0) = -1,$$

i.e. $\varphi(-\overline{x}, x_0) = -\varphi(\overline{x}, x_0)$.
b) $\overline{x}\,\overline{y} = \varphi(\overline{x}, x_0)\,\varphi(\overline{y}, y_0)$ for all $x, y \in C(Z)$.
Apply (4.36) for x, y and for $-\overline{x} + x_0 t, y$. Then, by step a),

$$
\begin{aligned}
e(x, y) &= \left(\arccos[\varphi(\overline{x}, x_0)\,\varphi(\overline{y}, y_0)]\right)^2 - A, \\
e(-\overline{x} + x_0 t, y) &= \left(\arccos[-\varphi(\overline{x}, x_0)\,\varphi(\overline{y}, y_0)]\right)^2 - A
\end{aligned}
$$

with $A := \left(\psi(\overline{x}, x_0) - \psi(\overline{y}, y_0)\right)^2$. Subtracting the second equation from the first one, and putting

$$\arccos(\overline{x}\,\overline{y}) =: \alpha, \ \arccos[\varphi(\overline{x}, x_0)\,\varphi(\overline{y}, y_0)] =: \beta,$$

we obtain $\alpha^2 - (\pi - \alpha)^2 = \beta^2 - (\pi - \beta)^2$, i.e. $\alpha = \beta$. Hence

$$\overline{x}\,\overline{y} = \cos\alpha = \cos\beta = \varphi(\overline{x}, x_0)\,\varphi(\overline{y}, y_0).$$

c) $\varphi(h, \xi) = \varphi(h, \eta)$ for all $\xi, \eta \in \mathbb{R}$ and $h \in X$ with $h^2 = 1$.
Apply step b) for $x = h + \xi t$ and $y = h + \eta t$. Then

$$1 = h \cdot h = \varphi(h, \xi)\,\varphi(h, \eta),$$

i.e. $\varphi(h, \xi) = \varphi(h, \eta)$, in view of Lemma 1, chapter 1, and of $\varphi^2 = 1$.

Putting $\varphi(\overline{x}) := \varphi(\overline{x}, 0)$, we may write, by step c)

$$f(\overline{x} + x_0 t) = \varphi(\overline{x}) + \psi(\overline{x}, x_0) t. \tag{4.38}$$

d) Define $K := \{h \in X \mid h^2 = 1\}$. We would like to extend $\varphi : K \to K$ to an orthogonal mapping $\omega : X \to X$. Put $\omega(0) := 0$ and

$$\omega(r) := \|r\| \cdot \varphi\left(\frac{r}{\|r\|}\right)$$

for all $r \neq 0$ in X. Obviously,

$$rs = \omega(r)\,\omega(s) \tag{4.39}$$

for $r, s \in X$ and $0 \in \{r, s\}$. For $r \neq 0 \neq s$ we obtain

$$\omega(r)\,\omega(s) = \|r\| \cdot \|s\| \cdot \varphi\left(\frac{r}{\|r\|}\right)\varphi\left(\frac{s}{\|s\|}\right) = rs,$$

in view of steps b) and c). So (4.39) holds true for all $r, s \in X$. Hence

$$\|\omega(r) - \omega(s)\|^2 = [\omega(r)]^2 - 2\omega(r)\,\omega(s) + [\omega(s)]^2 = \|r - s\|^2$$

for all $r, s \in X$. Thus, by Proposition 3, chapter 1, $\omega : X \to X$ must be orthogonal.

e) There exist fixed $\varepsilon, a \in \mathbb{R}$ with $\varepsilon^2 = 1$ and

$$\psi(\overline{x}, x_0) = \varepsilon x_0 + a \tag{4.40}$$

for all $\overline{x} + x_0 t \in C(Z)$.

Apply (4.36) for $x = r + \xi t$, $y = s + \xi t$ with $\xi \in \mathbb{R}$ and $r, s \in K$. Hence, by $f(z) = \omega(\overline{z}) + \psi(\overline{z}, z_0)\,t$ and (4.39),

$$(\xi - \xi)^2 = \left(\psi(r, \xi) - \psi(s, \xi)\right)^2$$

holds true, i.e. $\psi(r, \xi) = \psi(s, \xi) =: \psi(\xi)$. Applying (4.36) for $x = r + \xi t$, $y = r + \eta t$ with $\xi, \eta \in \mathbb{R}$ and $r \in K$ yields

$$(\xi - \eta)^2 = \left(\psi(\xi) - \psi(\eta)\right)^2. \tag{4.41}$$

Put $a := \psi(0)$. Hence $\xi^2 = \left(\psi(\xi) - a\right)^2$, i.e.

$$\psi(\xi) =: \varepsilon(\xi) \cdot \xi + a$$

for $\xi \neq 0$ with $[\varepsilon(\xi)]^2 = 1$. In the case $\xi \cdot \eta \neq 0$ now (4.41) implies

$$\xi\eta = \varepsilon(\xi)\,\varepsilon(\eta)\xi\eta,$$

i.e. $\varepsilon(\xi) = \varepsilon(\eta) =: \varepsilon$. Hence $\psi(\xi) = \varepsilon\xi + a$ holds true for all $\xi \in \mathbb{R}$.

   f) Because of $f(\overline{x} + x_0 t) = \omega(\overline{x}) + \psi(\overline{x}, x_0) t$ we hence get

$$f(\overline{x} + x_0 t) = \omega(\overline{x}) + (\varepsilon x_0 + a)t, \ \overline{x} + x_0 t \in C(Z). \tag{4.42}$$

Since ω is injective, so must be f $\square$

   A *motion* of $C(Z)$ is a surjective solution $f : C(Z) \rightarrow C(Z)$ of (4.36). Motions are hence bijections of $C(Z)$. Their group will be designated by $MC(Z)$, and the geometry $(C(Z), MC(Z))$ is called *geometry of Einstein's cylindrical world*. Observe that $MC(Z)$ is a subgroup of the Lorentz group $\mathbb{L}(Z)$.

Proposition 18. *The mapping* (4.42) *is a motion if, and only if, ω is surjective.*

Proof. Since $f : C(Z) \rightarrow C(Z)$ of (4.42) is injective, we must show that $\omega : X \rightarrow X$ is surjective if, and only if, f is surjective. If ω is surjective, then

$$f(\overline{x} + x_0 t) = \overline{y} + y_0 t$$

has a solution $\overline{x} + x_0 t$ in $C(Z)$ for given $\overline{y} + y_0 t$ in $C(Z)$, namely

$$\omega^{-1}(\overline{y}) + \varepsilon(y_0 - a)t.$$

If f is surjective, then $\omega(r) = s$ has for given $s \in X$ the solution $r = 0$ for $s = 0$, and $r = \|s\| \cdot \overline{v}$ for $s \neq 0$ and $\overline{v} + v_0 t \in C(Z)$ satisfying

$$f(\overline{v} + v_0 t) = \frac{s}{\|s\|} + at. \qquad\qquad \square$$

4.10   Lines, null-lines, subspaces

If $a \in K = \{h \in X \mid h^2 = 1\}$, define $L(a) = a + \mathbb{R}t$. For $a, b \in K$ and $k, \lambda \in \mathbb{R}$ with $ab = 0$ put

$$L(a, b, k, \lambda) := \{a \cos\varphi + b \sin\varphi + (k\varphi + \lambda)t \mid \varphi \in \mathbb{R}\}.$$

The sets $L(a)$, $L(a, b, k, \lambda)$ of points of $C(Z)$ are called the *lines* of $C(Z)$. A *null-line* of $C(Z)$ is a line $L(a, b, k, \lambda)$ with $k^2 = 1$. The lines $L(a)$ are euclidean lines, $L(a, b, 0, \lambda)$ are euclidean circles and every $L(a, b, k, \lambda)$ with $k \neq 0$ is a circular helix on the circular cylinder

$$\{a \cos\varphi + b \sin\varphi + \lambda t \mid \varphi, \lambda \in \mathbb{R}\}.$$

Proposition 19. *The line L of $C(Z)$ is a null-line if, and only if, there exist to every $p \in L$ and to every real $\varepsilon > 0$ a point $q \neq p$ on L satisfying $e(p, q) = 0$ and*

$$(\overline{p} - \overline{q})^2 + (p_0 - q_0)^2 < \varepsilon.$$

Proof. a) For two distinct points $a + \alpha t$, $a + \beta t$ of $L(a)$, we obtain

$$e(a + \alpha t, a + \beta t) = -(\alpha - \beta)^2 < 0,$$

and for two distinct points

$$v := a \cos \alpha + b \sin \alpha + \lambda t, \; w := a \cos \beta + b \sin \beta + \lambda t$$

of $L(a, b, 0, \lambda)$, $0 \le \alpha < \beta < 2\pi$, obviously,

$$
\begin{aligned}
e(v, w) &= (\beta - \alpha)^2 && \text{for } \beta - \alpha \le \pi, \\
e(v, w) &= [2\pi - (\beta - \alpha)]^2 && \text{for } \beta - \alpha \ge \pi,
\end{aligned}
$$

i.e. $e(v, w) > 0$.

b) Let now L be the line $L(a, b, k, \lambda)$ with $k \ne 0$. Assume $p \in L$ with

$$p = a \cos \varphi + b \sin \varphi + (k\varphi + \lambda) t. \tag{4.43}$$

If $k^2 = 1$ and if $\varepsilon > 0$ is given, choose $\xi \in]0, 1[$ with $2\xi^2 < \varepsilon$. Put

$$q = a \cos(\varphi + \xi) + b \sin(\varphi + \xi) + (k[\varphi + \xi] + \lambda) t \tag{4.44}$$

and observe $p \ne q \in L$, $e(p, q) = 0$ and

$$(\bar{p} - \bar{q})^2 + (p_0 - q_0)^2 = 2(1 - \cos \xi) + \xi^2 < \varepsilon,$$

since $2(1 - \cos \xi) < \xi^2$ for $0 < \xi < 1$, and since $2\xi^2 < \varepsilon$.

If $0 \ne k^2 \ne 1$, put $\varepsilon = \left(\frac{k}{10} \right)^2$. We are interested in all points (4.44) such that

$$(\bar{p} - \bar{q})^2 + (p_0 - q_0)^2 = 2(1 - \cos \xi) + k^2 \xi^2 < \left(\frac{k}{10} \right)^2 \tag{4.45}$$

holds true. (4.45) implies $|\xi| < \frac{1}{10}$, since $2(1 - \cos \xi) \ge 0$. But for p, q of (4.43), (4.44) with $|\xi| < \frac{1}{10}$, we obtain

$$e(p, q) = [\text{arc} \cos(\cos \xi)]^2 - k^2 \xi^2,$$

and we must choose arc $\cos(\cos \xi)$ in $[0, \pi]$. Since

$$\text{arc} \cos(\cos \xi) = \text{arc} \cos(\cos|\xi|)$$

and $|\xi| < \frac{1}{10}$, we get $e(p, q) = \xi^2 - k^2 \xi^2 \ne 0$ for $\xi \ne 0$. Hence L can not be a null-line. $\qquad \square$

Points and lines are called *subspaces* of $C(Z)$, also $\emptyset$. The other subspaces are defined by

$$C(V) := \{v + \lambda t \mid v \in V, v^2 = 1, \lambda \in \mathbb{R}\}$$

where V is any subspace of dimension ≥ 2 of the vector space X.

Theorem 20. *If p is a point and $L \ni p$ a line of $C(Z)$ which is not a null-line, then*

$$A(p, L) := \lim_{L \ni x \to p} \frac{e(x, p)}{l(x, p)} = 1,$$

where $l(x, p)$ designates the Lorentz–Minkowski distance of x, p.

Proof. Observe $d(x, p) = (\bar{x} - \bar{p})^2 - (x_0 - p_0)^2 = 2(1 - \bar{x}\,\bar{p}) - (x_0 - p_0)^2$.

Case 1. $L = a + \mathbb{R}t$.

$p \in L$ implies $L = p + \mathbb{R}t$. For $x = p + \xi t$ we define $x \to p$ by $\xi \to 0$. More generally, $x \to p$ is defined by

$$(\bar{x} - \bar{p})^2 + (x_0 - p_0)^2 \to 0. \tag{4.46}$$

Observe

$$\frac{e(x, p)}{l(x, p)} = \frac{e(p + \xi t, p)}{l(p + \xi t, p)} = \frac{-\xi^2}{-\xi^2} = 1.$$

Case 2. $L = \{a \cos \varphi + b \sin \varphi + [k\varphi + \lambda] t \mid \varphi \in \mathbb{R}\}$, $k^2 \neq 1$. Put $p = a \cos \varphi + b \sin \varphi + [k\varphi + \lambda] t$ and

$$x = a \cos(\varphi + \xi) + b \sin(\varphi + \xi) + [k(\varphi + \xi) + \lambda] t.$$

Because of $x \to p$ we obtain, by (4.46),

$$2(1 - \cos \xi) + k^2 \xi^2 \to 0,$$

i.e. $\xi \to 0$. Assume that $|\xi| \neq 0$ is small enough. Then

$$\frac{e(x, p)}{l(x, p)} = \frac{\xi^2 - k^2 \xi^2}{2(1 - \cos \xi) - k^2 \xi^2}, \tag{4.47}$$

is well-defined by observing

$$\lim_{\xi \to 0} \frac{2(1 - \cos \xi)}{\xi^2} = 1,$$

i.e. $2(1 - \cos \xi) \neq k^2 \xi^2$ since $k^2 \neq 1$ and since $|\xi|$ is small enough. Hence

$$\frac{e(x, p)}{l(x, p)} = \frac{1 - k^2}{\frac{2(1 - \cos \xi)}{\xi^2} - k^2} \to 1. \qquad \square$$

4.11   2-point invariants of $(C(Z), MC(Z))$

Suppose that $W \neq \emptyset$ is a set. We would like to determine all 2-point invariants of Einstein's cylinder universe, i.e. all

$$d : C(Z) \times C(Z) \to W \tag{4.48}$$

such that

$$d\left(x,y\right) = d\left(f\left(x\right),\, f\left(y\right)\right) \tag{4.49}$$

holds true for all $x, y \in C\left(Z\right)$ and all $f \in MC\left(Z\right)$. Define the cartesian product

$$W_0 := [-1, 1] \times \mathbb{R}_{\geq 0}$$

with $\mathbb{R}_{\geq 0} = \{r \in \mathbb{R} \mid r \geq 0\}$.

Theorem 21. *Let g be a mapping from W_0 into W. Then*

$$d\left(x,y\right) = g\left(\overline{x}\,\overline{y},\, |x_0 - y_0|\right) \tag{4.50}$$

is a solution of the functional equation (4.49). If, on the other hand, (4.48) is a solution of (4.49), there exists a function $g : W_0 \to W$ such that (4.50) holds true.

Proof. a) Suppose that $f\left(x\right) = \omega\left(\overline{x}\right) + \left(\varepsilon x_0 + a\right) t$ is in $MC\left(Z\right)$ and put

$$v = f\left(x\right),\ \ w = f\left(y\right)$$

for given $x, y \in C\left(Z\right)$. We obtain

$$\overline{v}\,\overline{w} = \omega\left(\overline{x}\right)\omega\left(\overline{y}\right) = \overline{x}\,\overline{y}$$

and $|v_0 - w_0| = |\left(\varepsilon x_0 + a\right) - \left(\varepsilon y_0 + a\right)| = |x_0 - y_0|$. This implies that (4.50) is a 2-point invariant of $MC\left(Z\right)$.

b) Assume that (4.48) solves (4.49). If $\left(\gamma, \delta\right) \in W_0$, define

$$a := i,\ \ b := \gamma i + \sqrt{1 - \gamma^2}\, j + \delta t,$$

where $i, j \in X$ satisfy $ij = 0$ and $i^2 = 1 = j^2$. Then put $g\left(\gamma, \delta\right) := d\left(a, b\right)$. We then have to show

$$d\left(x,y\right) = g\left(\gamma, \delta\right) \tag{4.51}$$

for all $x, y \in C\left(Z\right)$ with $\overline{x}\,\overline{y} = \gamma$ and $|x_0 - y_0| = \delta$. If we are able to find a mapping $f \in MC\left(Z\right)$ for such a special pair x, y with $x = f\left(a\right)$ and $y = f\left(b\right)$, then

$$d\left(x,y\right) = d\left(f\left(a\right),\, f\left(b\right)\right) = d\left(a, b\right) = g\left(\gamma, \delta\right),$$

and (4.51) is proved for that special pair.

If $\omega \in O\left(X\right)$ satisfies

$$\omega\left(i\right) = \overline{x}\ \text{and}\ \omega\left(\gamma i + \sqrt{1 - \gamma^2}\, j\right) = \overline{y}, \tag{4.52}$$

then

$$f\left(z\right) := \omega\left(\overline{z}\right) + \left(\varepsilon z_0 + x_0\right) t$$

with $\varepsilon = 1$ for $x_0 = y_0$ and $\delta\varepsilon := y_0 - x_0$ for $x_0 \neq y_0$ (observe $\delta = |x_0 - y_0|$) is such a mapping, i.e. satisfies $x = f(a)$ and $y = f(b)$.

c) So what we actually have to find is an $\omega \in O(X)$ such that (4.52) holds true for a pair $x, y \in C(Z)$ satisfying $\overline{x}\,\overline{y} = \gamma$ and $|x_0 - y_0| = \delta$. Observe $i^2 = 1$ and $\overline{x}^2 = 1$. Because of step A of the proof of Theorem 7 of chapter 1 there exists $\omega_1 \in O(X)$ with $\omega_1(i) = \overline{x}$. Define $\overline{y} =: \omega_1(v)$ and observe

$$iv = \omega_1(i)\,\omega_1(v) = \overline{x}\,\overline{y} = \gamma. \tag{4.53}$$

d) If we are able to find $\alpha \in O(X)$ such that

$$\alpha(i) = i \text{ and } \alpha\left(\gamma i + \sqrt{1 - \gamma^2}\, j\right) = v \tag{4.54}$$

holds true, $\omega := \omega_1 \alpha$ solves (4.52). Define

$$X_0 := \{h \in X \mid hi = 0\}$$

and note $X_0 \ni j, v - \gamma i$, because of (4.53). Since X is a real inner product space of dimension at least 2, X_0 must be a real inner product space with $\dim X_0 \geq 1$. Observe

$$(v - \gamma i)^2 = v^2 - 2\gamma vi + \gamma^2 i^2 = 1 - \gamma^2, \tag{4.55}$$

in view of (4.53), $i^2 = 1$ and $1 = \overline{y}^2 = \omega_1(v)\,\omega_1(v) = v \cdot v$ with $y \in C(Z)$.

e) There exists $\beta \in O(X_0)$ satisfying

$$\sqrt{1 - \gamma^2}\, \beta(j) = v - \gamma i.$$

In the case $\gamma^2 = 1$ put $\beta = \mathrm{id}$, in view of (4.55). If $\gamma^2 < 1$ and $\dim X_0 \geq 2$, apply again step A of the proof of Theorem 7, chapter 1, by observing

$$\left(\frac{v - \gamma i}{\sqrt{1 - \gamma^2}}\right)^2 = 1.$$

If $\gamma^2 < 1$ and $\dim X_0 = 1$, then $j, v - \gamma i$ are linearly dependent, and the existence of β is trivial.

Define now $\alpha(h + \xi i) := \beta(h) + \xi i$ for all $h \in X_0$ and note $\alpha \in O(X)$, since $\beta \in O(X_0)$, and since (4.54) holds true. $\qquad\square$

Remark. A characterization theorem for Einstein's distance function $e(x, y)$ is proved in W. Benz [5].

4.12   De Sitter's world

Define $\Sigma(Z) := \{z \in Z \mid z^2 = 1\}$ to be the set of *points* of *de Sitter's world* over $Z = X \oplus \mathbb{R}$, and call every restriction $\lambda \mid \Sigma(Z)$ a *motion* of $\Sigma(Z)$ where λ is a surjective Lorentz transformation of Z satisfying $\lambda(0) = 0$. We thus get the group $M\Sigma(Z)$ of motions of de Sitter's world $\Sigma(Z)$, by observing $\lambda(z) \in \Sigma(Z)$ for $z \in \Sigma(Z)$ for a Lorentz transformation λ of Z with $\lambda(0) = 0$:

$$1 = z^2 = l(z, 0) = l\big(\lambda(z),\, \lambda(0)\big) = [\lambda(z)]^2.$$

The geometry $\big(\Sigma(Z),\, M\Sigma(Z)\big)$ is called *geometry of de Sitter's world*. The points $a, b \in \Sigma(Z)$ are called *separated* provided $a \neq b \neq -a$. Such a pair must be linearly independent. Otherwise $\alpha a = \beta b$ would hold true with real α, β which are not both 0. But $(\alpha a)^2 = (\beta b)^2$, i.e. $\beta \in \{\alpha, -\alpha\}$ is impossible.

   If $a, b \in \Sigma(Z)$ are separated, every ellipse, every euclidean line, every branch of a hyperbola in

$$\{\xi a + \eta b \mid \xi, \eta \in \mathbb{R}\} \cap \Sigma(Z) \tag{4.56}$$

is called a *line* of $\Sigma(Z)$. All $\xi a + \eta b$ in (4.56) are characterized by the equation

$$1 = (\xi a + \eta b)^2 = \big(\xi + s(a,b)\eta\big)^2 + \big(1 - [s(a,b)]^2\big)\eta^2 \tag{4.57}$$

with

$$s(a,b) := ab. \tag{4.58}$$

We will call $s(a,b) = ab$ the *de Sitter distance* of $a, b \in \Sigma(Z)$, also in the cases $b = a$ or $b = -a$. Observe that (4.58) is a 2-point invariant: for $\lambda \in M\Sigma(Z)$,

$$l(a, b) = l\big(\lambda(a),\, \lambda(b)\big)$$

implies $(a-b)^2 = [\lambda(a)-\lambda(b)]^2$, i.e. $ab = \lambda(a)\lambda(b)$. In the cases $(ab)^2 < 1$, $(ab)^2 = 1$, $(ab)^2 > 1$, respectively, we obtain in (4.57) an ellipse (a *closed line*), two euclidean lines (two *null-lines*), two branches of a hyperbola (two *open lines*), respectively, of $\Sigma(Z)$. The lines of $\Sigma(Z)$ are also called its *geodesics*.

4.13   2-point invariants of $\big(\Sigma(Z),\, M\Sigma(Z)\big)$

Theorem 22. *$M\Sigma(Z)$ acts transitively on $\Sigma(Z)$. If $\dim X \geq 3$ and if a, b and c, e are pairs of separated points, there exists $\delta \in M\Sigma(Z)$ with $\delta(a) = c$ and $\delta(b) = e$ if, and only if, $ab = ce$ holds true.*

Proof. a) In step b) of the proof of Theorem 15 we showed that to $x, y \in Z \backslash \{0\}$ there exists a bijective Lorentz transformation λ with $\lambda(0)$ and $\lambda(x) = y$ if, and only if, $l(x, 0) = l(y, 0)$. Suppose that x, y are points of $\Sigma(Z)$. Then

$$l(x, 0) = 1 = l(y, 0)$$

holds true. There hence exists a motion δ with $\delta(x) = y$.

b) If a, b and c, e are pairs of separated points, and if $\delta \in M\Sigma(Z)$ satisfies $\delta(a) = c$ and $\delta(b) = e$, then $ab = \delta(a)\delta(b) = ce$, as we already know.

c) Let a, b and c, e be pairs of separated points with $ab = ce$. Because of Theorem 1, Lorentz transformations of Z fixing 0, must be linear. Separated points x, y must hence be transformed into separated points under motions. In view of step a) we thus may assume $a = c$ without loss of generality. If $h \in X$ satisfies $h^2 = 1$, which especially implies $h \in \Sigma(Z)$, we even may assume $a = h$, in view of step a). Then $ab = ae$ reads $hb = he$ or $h\bar{b} = h\bar{e}$, since

$$h = \bar{h} + h_0 t = \bar{h} + 0 \cdot t.$$

Put $Z_0 := \{z \in Z \mid zh = 0\}$. Obviously, $t \in Z_0$. Again, we would like to apply step b) of the proof of Theorem 15, but this time for Z_0 instead of Z for the points

$$\xi := b - (bh)h \text{ and } \eta := e - (eh)h,$$

which both belong to Z_0. This can be done, since

$$\dim X_0 \geq 2 \text{ with } X_0 := \{z \in Z_0 \mid zt = 0\},$$

because of $\dim X \geq 3$. Observe

$$l(\xi, 0) = \xi^2 = b^2 - (bh)^2 = 1 - (bh)^2 = 1 - (eh)^2 = l(\eta, 0).$$

There hence exists a bijective Lorentz transformation λ_0 of Z_0 satisfying

$$\lambda_0(0) = 0 \text{ and } \lambda_0\big(b - (bh)h\big) = e - (eh)h. \tag{4.59}$$

The problem now is to extend λ_0 to a bijective Lorentz transformation λ of Z. This will be accomplished by putting

$$\lambda(x) := \lambda_0\big(x - (xh)h\big) + (xh)h \tag{4.60}$$

for all $x \in Z$. That λ is an extension of λ_0 follows from

$$xh = 0 \text{ for all } x \in Z_0.$$

With $x_h := x - (xh)h$ for $x \in Z$, we obtain

$$l\big(\lambda(x), \lambda(y)\big) = \big(\lambda(x) - \lambda(y)\big)^2 = \big([\lambda_0(x_h) - \lambda_0(y_h)] + [xh - yh]h\big)^2,$$

i.e., because of $x_h \cdot h = 0$, $\lambda_0 : Z_0 \to Z_0$ and of $zh = 0$ for $z \in Z_0$,

$$l\big(\lambda(x), \lambda(y)\big) = l\big(\lambda_0(x_h), \lambda_0(y_h)\big) + (xh - yh)^2.$$

Similarly,

$$l(x, y) = (x - y)^2 = \big([x_h - y_h] + [xh - yh]h\big)^2,$$

i.e. $l\left(x,y\right)=l\left(x_h,y_h\right)+\left(xh-yh\right)^2$. Hence

$$l\big(\lambda\left(x\right),\,\lambda\left(y\right)\big)=l\left(x,y\right),$$

since λ_0 is a Lorentz transformation of Z_0, i.e. satisfies

$$l\big(\lambda_0(x_h),\,\lambda_0(y_h)\big)=l\left(x_h,y_h\right).$$

We finally show $\lambda\left(h\right)=h$ and $\lambda\left(b\right)=e$. In fact, by (4.60),

$$\lambda\left(h\right)=\lambda_0(h-h^2\cdot h)+h^2h=\lambda_0(0)+h=h,$$

and, by (4.59) and $bh=eh$,

$$\lambda\left(b\right)=\lambda_0\big(b-(bh)\,h\big)+(bh)\,h=[e-(eh)\,h]+(bh)\,h=e. \qquad \square$$

Theorem 23. *Let W be a set and $g:\mathbb{R}\to W$ be a function, and let w_0,w_1 be fixed elements of W. Then*

$$d\left(x,y\right)=\begin{cases} g\left(xy\right) & \text{for}\quad x,y \text{ separated}\\ w_0 & \text{for}\quad x=y\\ w_1 & \text{for}\quad x=-y \end{cases} \tag{4.61}$$

with $x,y\in\Sigma\left(Z\right)$ is a solution

$$d:\Sigma\left(Z\right)\times\Sigma\left(Z\right)\to W \tag{4.62}$$

of the functional equation

$$\forall_{x,y\in\Sigma\left(Z\right)}\ \forall_{f\in M\Sigma\left(Z\right)}\ d\left(x,y\right)=d\big(f\left(x\right),\,f\left(y\right)\big). \tag{4.63}$$

If, on the other hand, (4.62) solves (4.63), then there exists a function $g:\mathbb{R}\to W$ and elements $w_0,w_1\in W$ such that (4.61) holds true.

Proof. a) Obviously, (4.61) solves (4.63) for all motions f and all $x,y\in\Sigma\left(Z\right)$ as was shown almost at the end of section 12 by means of the formula

$$ab=\lambda\left(a\right)\lambda\left(b\right)$$

for $a,b\in\Sigma\left(Z\right)$ and $\lambda\in\Sigma\left(Z\right)$. (Here we only need this formula for a,b separated.)
b) Assume now that $d:\Sigma\left(Z\right)\times\Sigma\left(Z\right)\to W$ is a solution of (4.63). Take elements $i,j\in X$ with $i^2=1=j^2$ and $ij=0$. For $k\in\mathbb{R}$ define $g\left(k\right)$ by means of

$$g\left(k\right):=d\left(i,ki+j+kt\right). \tag{4.64}$$

Observe here $i\in\Sigma\left(Z\right)$, $ki+j+kt\in\Sigma\left(Z\right)$ and

$$k=i\left(ki+j+kt\right). \tag{4.65}$$

Moreover, put $w_0 := d(i, i)$ and $w_1 := d(i, -i)$. If $x \in \Sigma(Z)$, there exists $f \in MS(Z)$ with $f(i) = x$ on account of Theorem 22. Hence

$$d(x, x) = d(f(i), f(i)) = d(i, i) = w_0.$$

Since f is linear, we also get

$$d(x, -x) = d(f(i), f(-i)) = d(i, -i) = w_1.$$

Suppose now that $x, y \in \Sigma(Z)$ are separated. If $xy =: k$, then, according to (4.65), there exists $f \in M\Sigma(Z)$ satisfying

$$f(i) = x \text{ and } f(ki + j + kt) = y.$$

Hence $d(x, y) = d(f(i), f(ki + j + kt)) = d(i, ki + j + kt) = g(k)$, because of (4.64). Thus

$$d(x, y) = g(k) = g(xy). \qquad \square$$

Remark. Concerning the spacetimes of Einstein, de Sitter and others see also W.-L. Huang [1], [2], J. Lester [3]–[8] where finite-dimensional cases are treated.

4.14   Elliptic and spherical distances

The basis of the remaining part of this book is again a real inner product space X of dimension at least 2. As in the sections before we do not exclude the case that the dimension of X is infinite.

We define the *elliptic distance* $\varepsilon(x, y)$ of $x, y \in X_0 := X \backslash \{0\}$ by means of $\varepsilon(x, y) \in \left[0, \frac{\pi}{2}\right]$ and

$$\cos \varepsilon(x, y) = \frac{|xy|}{\|x\| \cdot \|y\|} \tag{4.66}$$

with $\|x\| = \sqrt{x^2}$. The *spherical distance* $\sigma(x, y)$ of $x, y \in X_0$ is given by $\sigma(x, y) \in [0, \pi]$ and

$$\cos \sigma(x, y) = \frac{xy}{\|x\| \cdot \|y\|}. \tag{4.67}$$

In view of the inequality of Cauchy–Schwarz (see section 4 of chapter 1), the right-hand side of (4.66) must be in $[0, 1]$, and that of (4.67) in $[-1, 1]$. Observe that

$$\varepsilon(\lambda x, \mu y) = \varepsilon(x, y) \tag{4.68}$$

holds true for all $x, y \in X_0$ and all $\lambda, \mu \in \mathbb{R} \backslash \{0\}$. Moreover,

$$\sigma(\lambda x, \mu y) = \sigma(x, y) \tag{4.69}$$

is satisfied for all $x, y \in X_0$ and all real λ, μ with $\lambda \cdot \mu > 0$. If $\omega : X \to X$ is an orthogonal mapping, obviously, by (4.66),

$$\cos \varepsilon (x, y) = \cos \varepsilon \left(\omega (x), \omega (y) \right)$$

holds true, and hence $\varepsilon (x, y) = \varepsilon \left(\omega (x), \omega (y) \right)$ for $x, y \in X_0$, since

$$\varepsilon (x, y), \ \varepsilon \left(\omega (x), \omega (y) \right) \in \left[0, \frac{\pi}{2} \right].$$

Similarly, $\sigma (x, y) = \sigma \left(\omega (x), \omega (y) \right)$. This implies that orthogonal mappings of X preserve elliptic and spherical distances. Notice that the orthogonal mappings $\omega : X \to X$ of X are Lorentz transformations of $Z = X \oplus \mathbb{R}$ of the form $(x, \varrho) \to \left(\omega (x) \varrho \right)$ for $(x, \varrho) \in Z$ (see Theorem 1 of chapter 4 and the discussion following this theorem).

Proposition 24. *Suppose* $x, y, z \in X_0$. *Then the following statements hold true.*

(a) $\varepsilon (x, y) = 0$ *if, and only if,* x, y *are linearly dependent.*

(b) $\varepsilon (x, y) = \varepsilon (y, x)$ *and*

$$\varepsilon (x, z) \leq \varepsilon (x, y) + \varepsilon (y, z). \tag{4.70}$$

Proof. Observe that $\varepsilon (x, y) + \varepsilon (y, z) \leq \pi$ holds true for all $x, y, z \in X_0$. The inequality (4.70) must hence be equivalent with

$$\cos \varepsilon (x, z) \geq \cos [\varepsilon (x, y) + \varepsilon (y, z)].$$

In view of (4.68) we may assume $x^2 = y^2 = z^2 = 1$. So we have to show

$$|\beta| \geq |\gamma| \cdot |\alpha| - \sqrt{1 - \gamma^2} \sqrt{1 - \alpha^2} \tag{4.71}$$

where we put $\alpha := yz$, $\beta := zx$, $\gamma := xy$. With

$$p := x - \gamma y, \ q := z - \alpha y$$

we get

$$\sqrt{1 - \gamma^2} \sqrt{1 - \alpha^2} = \sqrt{p^2} \sqrt{q^2} \geq |pq| = |\beta - \alpha \gamma| \geq |\alpha \gamma| - |\beta|,$$

i.e. (4.71). $\qquad\square$

Proposition 25. *Suppose that* x, y, z *are in* X_0. *Then the following statements hold true.*

(a) $\sigma (x, y) = 0$ *if, and only if,* $y = \lambda x$ *with* $\lambda > 0$.

(b) $\sigma (x, y) = \sigma (y, x)$ *and*

$$\sigma (x, z) \leq \sigma (x, y) + \sigma (y, z). \tag{4.72}$$

Proof. $\sigma\left(x, y\right) = 0$ implies

$$xy = \|x\| \cdot \|y\|,$$

i.e., by Lemma 1, chapter 1, $y = \lambda x$, i.e. $\lambda x^2 = \|\lambda\| x^2$. Hence $\lambda > 0$. On the other hand, we get from $y = \lambda x$, obviously, $\sigma\left(x, y\right) = 0$.

In order to prove (4.72), we may assume

$$\sigma\left(x, y\right) + \sigma\left(y, z\right) \leq \pi,$$

without loss of generality since $\sigma\left(x, z\right) \in [0, \pi]$. In this case, (4.72) is equivalent with

$$\cos\sigma\left(x, z\right) \geq \cos\left(\sigma\left(x, y\right) + \sigma\left(y, z\right)\right).$$

Applying the notations $\alpha, \beta, \gamma, p, q$ of the proof of Proposition 24, we have to show that

$$\beta \geq \gamma\alpha - \sqrt{1 - \gamma^2}\,\sqrt{1 - \alpha^2}$$

holds true, again assuming $x^2 = y^2 = z^2 = 1$. But here

$$\sqrt{1 - \gamma^2}\,\sqrt{1 - \alpha^2} = \sqrt{p^2}\,\sqrt{q^2} \geq |pq| = |\beta - \alpha\gamma| \geq \alpha\gamma - \beta,$$

since $\varrho, -\varrho \leq |\varrho|$ for all $\varrho \in \mathbb{R}$. $\square$

4.15   Points

Suppose that $Q \neq \emptyset$ is a set and that d is a mapping from $Q \times Q$ into $\mathbb{R}$. We assume that the structure (Q, d) satisfies

  (i) $d\left(x, x\right)\ \ = 0$ for all $x \in Q$,

  (ii) $d\left(x, y\right)\ \ = d\left(y, x\right)$ and

$$d\left(x, z\right) \leq d\left(x, y\right) + d\left(y, z\right)$$

for all $x, y, z \in Q$.

We will call such a structure (Q, d) an ES-space, since only the cases (X_0, ε), (X_0, σ) will be of interest for us. Since

$$0 = d\left(x, x\right) \leq d\left(x, y\right) + d\left(y, x\right) = 2d\left(x, y\right)$$

holds true, distances $d\left(x, y\right)$ are always non-negative. We will call $x, y \in Q$ equivalent, $x \sim y$, provided $d\left(x, y\right) = 0$. Because of (ii),

$$x \sim y \Rightarrow y \sim x$$

holds true, and, moreover,

$$x \sim y \text{ and } y \sim z \text{ imply } x \sim z.$$

The last statement follows from

$$0 \le d\left(x, z\right) \le d\left(x, y\right) + d\left(y, z\right) = 0.$$

We shall call the equivalence classes

$$[x] := \{y \in Q \mid y \sim x\}$$

points. If p, q are points, then we define

$$d\left(p, q\right) := d\left(x, y\right) \tag{4.73}$$

in the case that $x \in p$ and $y \in q$. In order to prove that $d\left(p, q\right)$ is well-defined, we must show

$$d\left(x, y\right) = d\left(x', y'\right)$$

for all $x, y, x', y' \in Q$ with $x \sim x'$ and $y \sim y'$. But

$$d\left(x, y\right) \le d\left(x, x'\right) + d\left(x', y'\right) + d\left(y', y\right) = d\left(x', y'\right),$$

and, of course, also $d\left(x', y'\right) \le d\left(x, y\right)$. It is now trivial to check that the set of points of (Q, d) is a metric space with respect to the distance notion (4.73).

The points of (X_0, ε) are called the *elliptic points* of X, the points of (X_0, σ) the *spherical points* of X. By $E\left(X\right)$, $S\left(X\right)$ we designate the set of elliptic points, of spherical points of X, respectively. In view of our considerations before, $\left(E\left(X\right), \varepsilon\right)$ and $\left(S\left(X\right), \sigma\right)$ are metric spaces.

We would like to describe other representations of the equivalence classes of (X_0, ε), and of those of (X_0, σ).

If $x \in X_0$, then, obviously, $[x] = \mathbb{R}x \backslash \{0\}$ in the case (X_0, ε). It is hence possible to identify the points of $\left(E\left(X\right), \varepsilon\right)$ with the euclidean lines $\mathbb{R}x$ of X through the *origin* 0. The distance $\varepsilon\left(x, y\right)$ then measures the smaller angle between the lines $\mathbb{R}x$ and $\mathbb{R}y$. The class $[x]$, $x \ne 0$, of $\left(E\left(X\right), \varepsilon\right)$ can also be identified with the following pair of points,

$$\left\{\frac{x}{\|x\|}, -\frac{x}{\|x\|}\right\}, \tag{4.74}$$

which are on the euclidean ball $B\left(0, 1\right) = \{y \in X \mid y^2 = 1\}$. Of course, they are the points of intersection of the line $\mathbb{R}x$ and the ball $B\left(0, 1\right)$.

For $x \in X_0$ we obtain $[x] = \mathbb{R}_{>0}x := \{\lambda x \mid 0 < \lambda \in \mathbb{R}\}$ in the case (X_0, σ). We hence may identify $[x]$ with the half-line $\mathbb{R}_{\ge 0}x$, but also with the point

$$\frac{x}{\|x\|}$$

of $B\left(0, 1\right)$. The distance $\sigma\left(x, y\right)$ measures the angle $\in [0, \pi]$ between the half-lines $\mathbb{R}_{\ge 0}x$ and $R_{\ge 0}y$. In the case $x^2 = 1 = y^2$ and $x \ne y \ne -x$ take the circle through x and y with center 0. Then $\sigma\left(x, y\right)$ measures the smaller distance along the circle from x to y. Of course $\sigma\left(x, x\right) = 0$ and $\sigma\left(x, -x\right) = \pi$.

4.16   Isometries

Suppose that (Q, d) is an ES-space. An *isometry* of (Q, d) is a mapping

$$f : Q \to Q$$

satisfying $d(x, y) = d(f(x), f(y))$ for all $x, y \in Q$. Denote by $\Pi(Q)$ the set of all points of (Q, d).

Proposition 26. *If $f : Q \to Q$ is an isometry, then*

$$F([x]) = [f(x)], \quad x \in Q, \tag{4.75}$$

is an isometry of the metric space $\big(\Pi(Q), d\big)$.

Proof. F is well-defined. In fact, $x \sim x'$ implies $d(x, x') = 0$ and hence

$$d(f(x), f(x')) = d(x, x') = 0,$$

i.e. $f(x) \sim f(x')$ and hence $[f(x)] = [f(x')]$. Moreover,

$$d\big([f(x), [f(y)]\big) = d\big(f(x), f(y)\big) = d(x, y) = d([x], [y]). \qquad \square$$

Proposition 27. *All isometries f of an ES-space (Q, d) are given as follows. If F is any isometry of the metric space $\big(\Pi(Q), d\big)$,*

$$F : \Pi(Q) \to \Pi(Q),$$

then define $f : Q \to Q$ arbitrarily up to the restriction that

$$f(x) \in F([x]) \tag{4.76}$$

is satisfied for all $x \in Q$.

Proof. Since $d([x], [y]) = d(F[x], F[y])$ holds true, and, moreover,

$$d([x], [y]) = d(x, y)$$

and $d(F[x], F[y]) = d\big(f(x), f(y)\big)$, by (4.76), we obtain

$$d(x, y) = d\big(f(x), f(y)\big). \qquad \square$$

We already know (see before Proposition 24) that orthogonal mappings of X preserve elliptic and spherical distances. We now prove a theorem which even holds in suitable non-real situations as was shown, based on other methods, by A. Alpers and E.M. Schröder [1].

Theorem 28. *All isometries of $E(X)$ or $S(X)$ are given as follows. Take arbitrarily an orthogonal mapping of X, then its restriction f on X_0, and, finally, the corresponding mapping (4.75) of f.*

Proof. a) Suppose that F is an isometry of $E(X)$ or $S(X)$, respectively. This mapping must be injective since $F([x]) = F([y])$ implies

$$d([x], [y]) = d(F[x]), F([y])) = 0,$$

i.e. $[x] = [y]$. Here d designates the distance notion ε or σ, respectively.

b) If $\dim X < \infty$, Theorem 28 is a consequence of the more general Theorems A.8.2 and A.9.2 of [2] (W. Benz), chapter 2.

c) Assume now $\dim X \geq 3$. In view of step a), F induces an injective mapping $\gamma : \Sigma \to \Sigma$ of the set Σ of all 1-dimensional subspaces

$$\langle x \rangle := [x] \cup [-x] \cup \{0\}, \ x \in X_0, \tag{4.77}$$

of the vector space X. In the case of $E(X)$, of course, $[x] = [-x]$ holds true. In the case of $S(X)$, the image of (4.77) must also be a 1-dimensional subspace of X since

$$\sigma(F([x]), F([-x])) = \sigma([x], [-x]) = \pi,$$

i.e. since $F([x]) =: [\xi]$ implies $F([-x]) = [-\xi]$. We would like to show that γ maps 2-dimensional subspaces of X, considered as sets of 1-dimensional subspaces, *onto* 2-dimensional subspaces. Let φ be such a 2-dimensional subspace of the vector space X and let x, a, b be elements of φ satisfying $a \perp b$ and $x^2 = a^2 = b^2 = 1$. Define

$$F([x]) =: [x'], \ F([a]) =: [a'], \ F([b]) =: [b'].$$

Without loss of generality we may assume $x'^2 = a'^2 = b'^2 = 1$. Now we get

$$x =: \alpha a + \beta b$$

with suitable $\alpha, \beta \in \mathbb{R}$ and

$$xa = \varepsilon_1 x'a', \ xb = \varepsilon_2 x'b', \ ab = \varepsilon_3 a'b' = 0$$

with $\varepsilon_i^2 = 1$, $i = 1, 2, 3$, since equations like $d([x], [a]) = d([x'], [a'])$ hold true. If x' were not an element of

$$\text{span} \{a', b'\} := \{\delta_1 a' + \delta_2 b' \mid \delta_1, \delta_2 \in \mathbb{R}\},$$

then we would have

$$x' := \lambda a' + \mu b' + \tau c, \ \tau \neq 0,$$

with a suitable $c \in \text{span} \{a', b', x'\}$ satisfying $c \perp a', b'$ and $c^2 = 1$. But this contradicts

$$\alpha = \varepsilon_1 \lambda, \ \beta = \varepsilon_2 \mu, \ 1 = \alpha^2 + \beta^2, \ 1 = \lambda^2 + \mu^2 + \tau^2,$$

since $\lambda^2 + \mu^2 = \alpha^2 + \beta^2 = 1 = \lambda^2 + \mu^2 + \tau^2$. Hence $\gamma(\varphi) \subseteq \text{span} \{a', b'\}$. In order to show that $\text{span} \{a', b'\}$ is also a subset of $\gamma(\varphi)$, we consider any $y = \lambda a' + \mu b'$ with $\lambda \mu \neq 0$ and $\lambda^2 + \mu^2 = 1$, and we define

$$z_1 := \lambda a + \mu b \text{ and } z_2 := \lambda a - \mu b.$$

The 1-dimensional subspace $\langle y \rangle$ must be the image of $\langle z_1 \rangle$ or $\langle z_2 \rangle$ under γ, since

$$\{\gamma(\langle z_1 \rangle), \gamma(\langle z_2 \rangle)\} \subseteq \{\langle \lambda a' + \mu b' \rangle, \langle \lambda a' - \mu b' \rangle\}$$

and $\langle z_1 \rangle \neq \langle z_2 \rangle$ hold true.

d) We now define Y to be the set of all elements x of X such that there exists $\xi \in X_0$ with $x \in \gamma(\langle [\xi] \rangle)$. This implies that Y is a subspace of X and that γ is an isomorphism between the projective spaces over X and Y. The proof of Theorem 31 of [3], W. Benz, 219 ff, with obvious modifications, implies that there exists an injective linear mapping δ of X into itself satisfying

$$\gamma(\text{ span } \{x\}) = \text{ span } \{\delta(x)\} \tag{4.78}$$

for all $x \in X_0$. Since

$$\frac{(xy)^2}{x^2 y^2} = \frac{[\delta(x) \delta(y)]^2}{[\delta(x)]^2 [\delta(y)]^2}$$

holds true for all $x, y \in X_0$, we obtain that $x \perp y$ implies $\delta(x) \perp \delta(y)$. Moreover, $x^2 = y^2$ implies $(x - y)(x + y) = 0$, i.e. $[\delta(x)]^2 = [\delta(y)]^2$.

e) Let j be a fixed element of X with $j^2 = 1$. Define $\delta(j) =: k$. For $x \in X_0$ we hence have

$$\left(\frac{x}{\|x\|} \right)^2 = 1 = j^2,$$

and thus

$$\left[\delta\left(\frac{x}{\|x\|} \right) \right]^2 = k^2,$$

i.e. $\|\delta(x)\| = \|k\| \cdot \|x\|$. If we replace δ in (4.78) by the orthogonal mapping Δ,

$$\Delta(x) := \frac{1}{\|k\|} \delta(x), \; x \in X,$$

then (4.78) remains true. F is hence induced by Δ in the form of (4.75) with

$$f(x) := \Delta(x)$$

for all $x \in X_0$.   □

The surjective isometries of $E(X)$, $S(X)$ are called *motions* of $E(X)$, $S(X)$, respectively. Their groups $ME(X)$, $MS(X)$ are called *elliptic, spherical group* of X, the geometries

$$\big(E(X), \, ME(X) \big), \, \big(S(X), \, MS(X) \big)$$

elliptic, spherical geometry over X, respectively. Observe that all these motions are induced (see Theorem 28) by orthogonal mappings, i.e. by Lorentz transformations.

4.17 Distance functions of X_0

A function $d : X_0 \times X_0 \to \mathbb{R}_{\geq 0}$ will be called a *distance function* of X_0. Observe that the distance functions we were interested in in sections 10 and 11 of chapter 1 were functions from $X \times X$ into $\mathbb{R}_{\geq 0}$. This time the element 0 of X will be excluded. Therefore step B of the proof of Theorem 7 of chapter 1 can not be applied for our present purposes without modifications:

Proposition 29. *Define*

$$L := \{(\xi_1, \xi_2, \xi_3) \in \mathbb{R}^3 \mid \xi_1, \xi_2 \in \mathbb{R}_{>0} \ and \ \xi_3^2 \leq \xi_1 \xi_2\}$$

with $\mathbb{R}_{>0} = \mathbb{R}_{\geq 0} \backslash \{0\}$. Take $f : L \to \mathbb{R}_{\geq 0}$ arbitrarily. Then

$$d(x, y) = f(x^2, y^2, xy) \tag{4.79}$$

is a distance function of X_0 satisfying

$$(*) \qquad d(x, y) = d(\omega(x), \omega(y)) \ \text{for all } \omega \in O(X) \ \text{and all } x, y \in X_0.$$

If, vice versa, d is a distance function of X_0 such that $(*)$ holds true, then there exists $f : L \to \mathbb{R}_{\geq 0}$ with (4.79) for all $x, y \in X_0$.

Proof. Obviously, (4.79) satisfies $(*)$. So assume that d is a distance function of X_0 with $(*)$. Suppose that (ξ_1, ξ_2, ξ_3) is in L and that j, k are fixed elements of X with $j^2 = 1 = k^2$ and $jk = 0$. Put

$$x_0 := j \sqrt{\xi_1} \ \text{and} \ y_0 \sqrt{\xi_1} := j\xi_3 + k \sqrt{\xi_1 \xi_2 - \xi_3^2},$$

and observe $x_0 \neq 0 \neq y_0$. Define

$$f(\xi_1, \xi_2, \xi_3) := d(x_0, y_0).$$

The function $f : L \to \mathbb{R}_{\geq 0}$ is hence defined for all elements of L. We now have to prove that (4.79) holds true. Let x, y be elements of X_0, and put

$$\xi_1 := x^2, \ \xi_2 := y^2, \ \xi_3 := xy.$$

$x, y \in X_0$ implies $\xi_1 > 0$, $\xi_2 > 0$. Moreover, (ξ_1, ξ_2, ξ_3) must be in L, in view of the Cauchy–Schwarz inequality. If we are able to prove the existence of an $\omega \in O(X)$ with

$$\omega(x_0) = x \ \text{and} \ \omega(y_0) = y,$$

where x_0, y_0 are the already defined elements with respect to (ξ_1, ξ_2, ξ_3), then, by $(*)$,

$$d(x, y) = d(x_0, y_0) = f(\xi_1, \xi_2, \xi_3) = f(x^2, y^2, xy)$$

holds true and (4.79) is established. Without loss of generality we may assume $x = x_0$, in view of the fact that $x^2 = x_0^2$ (see step A of the proof of Theorem 7 of chapter 1). Suppose that $z := y - y_0$ is unequal to 0 and define

$$M := \{m \in X \mid m \perp z\}.$$

Then M is a maximal subspace of X because $p \in X \backslash M$ implies

$$pz^2 - (pz)\, z \in M$$

and hence $p \in \mathbb{R}z \oplus M$. Since $x = x_0$, we get $x \in M$ from $xy = x_0 y_0$. Define

$$w\,(\alpha z + m) = -\alpha z + m$$

for all $\alpha \in \mathbb{R}$ and $m \in M$. Notice $w \in O\,(X)$, $w^2 = \mathrm{id}$, and $w\,(x) = x$, in view of $x \in M$. Finally observe $w\,(y_0) = y$ because of

$$y_0 = -\frac{1}{2}\,z + \frac{1}{2}\,(y + y_0),\ y + y_0 \perp z. \qquad \Box$$

Theorem 30. *Let d be a distance function of X_0 satisfying $(*)$ and*

(S) $d\,(x, y) = d\,(\lambda x, \mu y)$ *for all $\lambda, \mu \in \mathbb{R}_{>0}$ and all $x, y \in X_0$.*

Then there exists a function

$$g : [0, \pi] \rightarrow \mathbb{R}_{\geq 0}$$

with $d\,(x, y) = g\left(\sigma\,(x, y)\right)$ for all $x, y \in X_0$.

Proof. Because of Proposition 29 there exists $f : L \rightarrow \mathbb{R}_{\geq 0}$ with (4.79). Define

$$g\,(\xi) := f\,(1, 1, \cos \xi)$$

for $\xi \in [0, \pi]$. Hence, by (S) and (4.79),

$$f\,(x^2, y^2, xy) = f\,(\lambda^2 x^2, \mu^2 y^2, \lambda \mu x y)$$

for $\lambda \cdot \sqrt{x^2} := 1 =: \mu\,\sqrt{y^2}$, i.e.

$$d\,(x, y) = f\left(1, 1, \frac{xy}{\sqrt{x^2}\,\sqrt{y^2}}\right) = f\,(1, 1, \cos[\sigma\,(x, y)]) = g\left(\sigma\,(x, y)\right)$$

for all $x, y \in X_0$. $\Box$

Theorem 31. *Let d be a distance function of X_0 satisfying $(*)$, (S) and*

(E) $d\,(x, y) = d\,(-x, y)$ *for all $x, y \in X_0$.*

Then there exists a function

$$h : \left[0, \frac{\pi}{2}\right] \rightarrow \mathbb{R}_{\geq 0}$$

with $d\,(x, y) = h\left(\varepsilon\,(x, y)\right)$ for all $x, y \in X_0$.

Proof. As already in the proof of Theorem 30, we obtain

$$d\left(x,y\right) = f\left(1,1,\frac{xy}{\sqrt{x^2}\sqrt{y^2}}\right).$$

Hence, by (E),

$$f\left(1,1,\frac{xy}{\sqrt{x^2}\sqrt{y^2}}\right) = f\left(1,1,-\frac{xy}{\sqrt{x^2}\sqrt{y^2}}\right),$$

and thus

$$d\left(x,y\right) = f\left(1,1,\frac{|xy|}{\sqrt{x^2}\sqrt{y^2}}\right).$$

This implies

$$d\left(x,y\right) = f\left(1,1,\cos[\varepsilon\left(x,y\right)]\right) = h\left(\varepsilon\left(x,y\right)\right)$$

in view of $h\left(\xi\right) := f\left(1,1,\cos\xi\right)$ for $\xi \in [0,\frac{\pi}{2}]$. $\qquad\qquad\square$

4.18   Subspaces, balls

If V is a subspace of dimension $r \geq 1$ of the vector space X, then the set $\{[x] \mid 0 \neq x \in V\}$ is called a *subspace* of $E\left(X\right)$, or also of $S\left(X\right)$, depending on the definition of points. If V is of dimension 2, then the corresponding subspace is called a *line*. Obviously, isometries transform subspaces onto subspaces of the same dimension. *Balls (hyperspheres)* are defined by

$$B\left([m],\varrho\right) := \{[x] \mid x \in X_0 \text{ and } d\left([m],[x]\right) = \varrho\}$$

for

1. $$d = \varepsilon \text{ and } \varrho \in \left[0,\frac{\pi}{2}\right]$$

or

2. $$d = \sigma \text{ and } \varrho \in \left(0,\pi\right]$$

where $m \in X_0$.

In the elliptic case the following Proposition holds true.

Proposition 32. *If $m^2 = 1$ and $\varrho \in \left[0,\frac{\pi}{2}\right]$, then*

$$B\left([m],\varrho\right) = \{[p \cdot \sin\varrho + m \cdot \cos\varrho] \mid p \in m^\perp \text{ and } p^2 = 1\}$$

where $m^\perp := \{x \in X \mid x \perp m\}$.

Proof. $[p \cdot \sin \varrho + m \cdot \cos \varrho]$ belongs to $B\left([m], \varrho\right)$ for $p \in m^{\perp}$ and $p^2 = 1$, since

$$\cos \varepsilon \left([m], [p \sin \varrho + m \cos \varrho]\right) = |m \cdot (p \sin \varrho + m \cos \varrho)| = \cos \varrho.$$

If, on the other hand,

$$\cos \varrho = \frac{|mx|}{\|x\|}$$

holds true for $x \in X_0$, then, since there exists $p \in m^{\perp}$ with $p^2 = 1$ and

$$x = \alpha p + \beta m \tag{4.80}$$

with suitable $\alpha, \beta \in \mathbb{R}$,

$$|\beta| = \sqrt{\alpha^2 + \beta^2} \, \cos \varrho \tag{4.81}$$

holds true, i.e. $|\beta| \sin \varrho = |\alpha| \cos \varrho$. We may assume $\beta \geq 0$, because otherwise we would work with $-x$ instead of x in (4.80). We also may assume $\alpha \geq 0$, since otherwise we would replace p by $-p$ in (4.80). Hence there exists $k \in \mathbb{R}$ with

$$(\alpha, \beta) = k \cdot (\sin \varrho, \cos \varrho). \tag{4.82}$$

Now $x \neq 0$ in (4.80) implies $k \neq 0$. $\qquad\square$

In the spherical case we again assume $m^2 = 1$, without loss of generality. We obtain

Proposition 33. *If $\varrho \in [0, \pi]$, then*

$$B\left([m], \varrho\right) = \{[p \cdot \sin \varrho + m \cdot \cos \varrho] \mid p \in m^{\perp} \text{ and } p^2 = 1\}.$$

In this situation (4.81) reads as $\beta = \sqrt{\alpha^2 + \beta^2} \, \cos \varrho$, so that $\cos \varrho$ and β have the same sign. Hence

$$\beta \sin \varrho = |\alpha| \cos \varrho.$$

We may assume $\alpha \geq 0$, because otherwise we would replace p by $-p$ in (4.80). Notice, moreover, that in the new situation, $k > 0$ must hold true in (4.82).

In the spherical case

$$B\left([m], \varrho\right) = B\left([-m], \pi - \varrho\right)$$

holds true.

4.19   Periodic lines

Suppose that ϱ is a positive real number and that

$$x : [0, \varrho[\rightarrow M \tag{4.83}$$

with $[0, \varrho[:= \{\xi \in \mathbb{R} \mid 0 \le \xi < \varrho\}$, is a mapping into the set M of a metric space (M, d) satisfying

$$(*) \qquad d\left(x\left(\xi\right), x\left(\eta\right)\right) = \begin{cases} |\xi - \eta| & \text{if} \quad |\xi - \eta| \le \frac{\varrho}{2} \\ \varrho - |\xi - \eta| & \text{if} \quad |\xi - \eta| > \frac{\varrho}{2} \end{cases}$$

for all $\xi, \eta \in [0, \varrho[$. Then

$$\{x\left(\xi\right) \mid 0 \le \xi < \varrho\} \qquad (4.84)$$

is called (W. Benz [15]) a *ϱ-periodic line* of (M, d). The mapping (4.83) of a ϱ-periodic line (4.84) must be injective. Assume $x\left(\xi\right) = x\left(\eta\right)$ for $\xi, \eta \in [0, \varrho[$. If $|\xi - \eta| \le \frac{\varrho}{2}$, then

$$0 = d\left(x\left(\xi\right), x\left(\eta\right)\right) = |\xi - \eta|$$

implies $\xi = \eta$, and if $|\xi - \eta| > \frac{\varrho}{2}$, then

$$0 = d\left(x\left(\xi\right), x\left(\eta\right)\right) = \varrho - |\xi - \eta|$$

is impossible, since $\xi, \eta \in [0, \varrho[$ yields $|\xi - \eta| < \varrho$. According to the beginning of section 18, a line l of $S\left(X\right)$ can be defined as follows: take $V = \text{span } \{p, q\}$ with $p^2 = 1 = q^2$ and $pq = 0$, and define

$$l = \{[\alpha p + \beta q] \mid \alpha^2 + \beta^2 = 1\}.$$

Identifying the set of points of (X_0, σ) with $\{x \in X \mid x^2 = 1\}$, we may write

$$l = \{p \cos \xi + q \sin \xi \mid 0 \le \xi < 2\pi\}. \qquad (4.85)$$

We will call these lines the *classical lines* of $S\left(X\right)$.

Theorem 34. *The 2π-periodic lines of the metric space $\left(S\left(X\right), \sigma\right)$ are exactly the classical lines of $S\left(X\right)$.*

Proof. a) Define $x\left(\xi\right) := p \cos \xi + q \sin \xi$ on the basis of (4.85) and observe, by definition of σ,

$$\cos \sigma \left(x\left(\xi\right), x\left(\eta\right)\right) = x\left(\xi\right) x\left(\eta\right) = \cos|\xi - \eta|. \qquad (4.86)$$

Hence $\sigma \left(x\left(\xi\right), x\left(\eta\right)\right) = |\xi - \eta|$ for $|\xi - \eta| \le \pi$, since $\sigma \left(x\left(\xi\right), x\left(\eta\right)\right) \in [0, \pi]$. In the case $|\xi - \eta| > \pi$ with $\xi, \eta \in [0, 2\pi[$ we obtain

$$\cos \sigma \left(x\left(\xi\right), x\left(\eta\right)\right) = \cos(2\pi - |\xi - \eta|)$$

from (4.86), and hence $\sigma \left(x\left(\xi\right), x\left(\eta\right)\right) = 2\pi - |\xi - \eta|$. Thus (4.85) is a 2π-periodic line.

b) Suppose that $x : [0, 2\pi[\rightarrow \{x \in X \mid x^2 = 1\}$ satisfies the functional equation
(∗) in the case $\varrho = 2\pi$ for all $\xi, \eta \in [0, 2\pi[$. Hence

$$x (\xi)\, x (\eta) = \cos \sigma \left(x (\xi),\, x (\eta) \right) = \cos(\xi - \eta). \tag{4.87}$$

Thus $p, q \in S(X)$, $pq = 0$, for $p := x(0)$, $q =: x\left(\tfrac{\pi}{2}\right)$. Observe, by (4.87),

$$x (\xi) \cdot p \;\; = x (\xi)\, x (0) \;\;\;\; = \cos \xi,$$
$$x (\xi) \cdot q \;\; = x (\xi)\, x \left(\tfrac{\pi}{2}\right) = \sin \xi$$

for $\xi \in [0, 2\pi[$. Hence

$$x (\xi)(p \cos \xi + q \sin \xi) = 1, \tag{4.88}$$

i.e. $[x (\xi)(p \cos \xi + q \sin \xi)]^2 = [x (\xi)]^2 \cdot [p \cos \xi + q \sin \xi]^2$. Thus, by Lemma 1, chapter 1,

$$x (\xi) = \lambda (\xi) \cdot (p \cos \xi + q \sin \xi),$$

with $\lambda (\xi) \in \{1, -1\}$. But $\lambda (\xi) = -1$ is not possible, because otherwise, by (4.88),

$$1 = x (\xi)(p \cos \xi + q \sin \xi) = -(p \cos \xi + q \sin \xi)^2 = -1.$$

Hence $x (\xi) = p \cos \xi + q \sin \xi$, $\xi \in [0, 2\pi[$, is a classical line. $\qquad\square$

Now we will work with the metric space $\left(E (X),\, \varepsilon \right)$ by identifying $E (X)$, as already described earlier, with

$$\left\{\{x, -x\} \subset X \mid x^2 = 1\right\}$$

and by writing

$$\sigma \left(\{x, -x\}, \{y, -y\}\right) \;\; \in \;\; \left[0, \frac{\pi}{2}\right],$$
$$\cos \sigma \left(\{x, -x\}, \{y, -y\}\right) \;\; = \;\; |xy|$$

for all $\{x, -x\}, \{y, -y\} \in E (X)$.

We shall write $R = \{1, -1\}$ and $\{x, -x\} = R \cdot x = Rx$. Observe $Rx = R \cdot (-x)$.

Again, according to the beginning of section 18, a line l of $E (X)$ can be defined by

$$l = \{[\alpha p + \beta q] \mid \alpha^2 + \beta^2 = 1\}$$

with $p, q \in X$ such that $p^2 = q^2 = 1$, $pq = 0$, i.e. by

$$l = \{R (p \cos \xi + q \sin \xi) \mid \xi \in [0, \pi[\}. \tag{4.89}$$

These lines will be called the *classical lines* of $E (X)$.

Theorem 35. *The π-periodic lines of the metric space $\left(E (X),\, \varepsilon \right)$ are exactly the classical lines of $E (X)$.*

Proof. a) For $p, q \in X$ with $p^2 = q^2 = 1$, $pq = 0$, define

$$x(\xi) := p \cos \xi + q \sin \xi, \xi \in [0, \pi[.$$

We are now interested in the mapping

$$\xi \to Rx(\xi), \xi \in [0, \pi[,$$

from $[0, \pi[$ into $E(X)$. Observe

$$\cos \varepsilon \left(Rx(\xi), Rx(\eta) \right) = |x(\xi) x(\eta)| = |\cos|\xi - \eta||.$$

If $|\xi - \eta| \leq \frac{\pi}{2}$, then $\cos|\xi - \eta| \geq 0$, and hence

$$\varepsilon \left(Rx(\xi), Rx(\eta) \right) = |\xi - \eta|.$$

In the other case, $\frac{\pi}{2} < |\xi - \eta| < \pi$, we obtain

$$0 < \pi - |\xi - \eta| < \frac{\pi}{2}$$

and $\cos \varepsilon \left(Rx(\xi), Rx(\eta) \right) = |\cos|\xi - \eta|| = |\cos(\pi - |\xi - \eta|)|$, i.e.

$$d \left(Rx(\xi), Rx(\eta) \right) = \pi - |\xi - \eta|.$$

Hence (4.89) is a π-periodic line.

b) Suppose the mapping $\varphi : [0, \pi[\to E(X)$ solves the functional equation $(*)$, x there replaced by φ now, in the case $\varrho = \pi$ for all $\xi, \eta \in [0, \pi[$. We shall write

$$\varphi(\xi) =: Rx(\xi)$$

with a suitable function $x : [0, \pi[\to \{x \in X \mid x^2 = 1\}$. Hence

$$|x(\xi) x(\eta)| = \cos \varepsilon \left(Rx(\xi), Rx(\eta) \right) = \begin{cases} \cos|\xi - \eta| & \text{if } |\xi - \eta| \leq \frac{\pi}{2}, \\ \cos(\pi - |\xi - \eta|) & \text{if } |\xi - \eta| > \frac{\pi}{2} \end{cases}$$

for all $\xi, \eta [0, \pi[$, i.e.

$$|x(\xi) x(\eta)| = |\cos(\xi - \eta)|. \tag{4.90}$$

Put $p := x(0)$, $q := x\left(\frac{\pi}{2}\right)$. Hence $p^2 = q^2 = 1$, $pq = 0$. Equation (4.90) yields

$$|x(\xi) p| = |\cos \xi|, |x(\xi) q| = \sin \xi \tag{4.91}$$

for $\xi \in [0, \pi[$. Put $D := [0, \pi[\setminus \{0, \frac{\pi}{2}\}$ and

$$\alpha(\xi) := \frac{x(\xi) p}{\cos \xi}, \beta(\xi) := \frac{x(\xi) q}{\sin \xi} \tag{4.92}$$

for $\xi \in D$. Hence, by (4.91), $[\alpha(\xi)]^2 = 1 = [\beta(\xi)]^2$. Observe, by (4.92), for all $\xi \in D$,

$$x(\xi)(p\alpha(\xi)\cos\xi + q\beta(\xi)\sin\xi) = 1.$$

Thus, for all $\xi \in D$,

$$[x(\xi)(p\alpha(\xi)\cos\xi + q\beta(\xi)\sin\xi)]^2 = [x(\xi)]^2[p\alpha(\xi)\cos\xi + q\beta(\xi)\sin\xi]^2.$$

Hence, by Lemma 1, chapter 1, for all $\xi \in D$,

$$Rx(\xi) = R\left(p\cos\xi + q\gamma(\xi)\sin\xi\right), \ \gamma(\xi) := \frac{\beta(\xi)}{\alpha(\xi)},$$

with $[\gamma(\xi)]^2 = 1$.

Case 1. $\gamma(\xi) = 1$ for all $\xi \in D$. Then $Rx(\xi) = R(p\cos\xi + q\sin\xi)$, which also holds true for $\xi = 0$ or $\xi = \frac{\pi}{2}$. We hence get a classical line.

Case 2. $\gamma(\xi) = -1$ for all $\xi \in D$. Put $q' = -q$ and observe $p^2 = (q')^2 = 1$, $pq' = 0$. Thus

$$Rx(\xi) = R(p\cos\xi + q'\sin\xi),$$

which also holds true for $\xi = 0$ or $\xi = \frac{\pi}{2}$. We again get a classical line.

Case 3. There exists $\xi_1, \xi_2 \in D$ with $\gamma(\xi_1) = 1 = -\gamma(\xi_2)$. Here (4.90) implies

$$\begin{aligned}
|\cos(\xi_1 - \xi_2)| &= |x(\xi_1)x(\xi_2)| \\
&= |(p\cos\xi_1 + q\sin\xi_1)(p\cos\xi_2 - q\sin\xi_2)| = |\cos(\xi_1 + \xi_2)|,
\end{aligned}$$

i.e. $[\cos(\xi_1 - \xi_2)]^2 = [\cos(\xi_1 + \xi_2)]^2$. Hence

$$\cos\xi_1\cos\xi_2\sin\xi_1\sin\xi_2 = 0.$$

This is a contradiction, since $\xi_1, \xi_2 \in D$. Hence Case 3 does not occur. $\qquad\square$

Concerning metric (periodic) lines in Lorentz–Minkowski geometry see R. Höfer [1], [2].

4.20   Hyperbolic geometry revisited

Let X be a real inner product space containing two linearly independent elements. As in earlier sections define $Z = X \oplus \mathbb{R}$. Put

$$H(Z) := \{z = (\bar{z}, z_0) \in Z \mid z^2 = -1 \text{ and } z_0 \geq 0\}$$

where $z_1 \cdot z_2$, so especially $z^2 = z \cdot z$, designates the product (3.88). The mapping

$$\mu : X \to H(Z) \tag{4.93}$$

with $\mu(x) := (x, \sqrt{1 + x^2})$ for $x \in X$ turns out to be a bijection. In fact, if

$$z = (\bar{z}, z_0) \in H(Z)$$

is given, $\bar{z}^2 - z_0^2 = -1$ and $z_0 \geq 0$ hold true, i.e. $z_0 = \sqrt{1 + \bar{z}^2}$. Hence $\mu(\bar{z}) = z$.

Theorem 36. *Let f be a hyperbolic motion of (X, hyp). Then there exists a bijective Lorentz transformation $\lambda : Z \to Z$ with $\lambda(0) = 0$ and $\lambda(H(Z)) = H(Z)$ such that*

$$\mu(f(x)) = \lambda\mu(x) \tag{4.94}$$

holds true for all $x \in X$. We will call λ an induced Lorentz transformation of the hyperbolic motion f.

Proof. Because of step I of the proof of Theorem 7 of chapter 1, it is sufficient to prove Theorem 36 for $f \in O(X)$ and for $f = T_t$ where T_t is a translation with axis $e \in X$, $e^2 = 1$.

Case 1. $f := \omega \in O(X)$.

If $z = (\bar{z}, z_0) \in Z$, define $\lambda(z) := (\omega(\bar{z}), z_0)$. This is a Lorentz transformation of Z (see section 4.1) with $\lambda(0) = 0$ and $\lambda(H(Z)) = H(Z)$, since

$$\left(\omega(x), \sqrt{1 + [\omega(x)]^2}\right) = \left(\omega(x), \sqrt{1 + x^2}\right) \tag{4.95}$$

implies $\lambda(H(Z)) \subseteq H(Z)$, and $\omega^{-1} \in O(X)$, obviously,

$$\lambda^{-1}(H(Z)) \subseteq H(Z).$$

In order to prove (4.94), observe

$$\lambda\mu(x) = \lambda\left(x, \sqrt{1 + x^2}\right) = \left(\omega(x), \sqrt{1 + x^2}\right) = \mu\left(\omega(x)\right),$$

by (4.95).

Case 2. $f = T_t$.

Put $c := \cosh t$ and $s := \sinh t$. Because of (1.8), we obtain for $x \in X$,

$$1 + [T_t(x)]^2 = \left(c\sqrt{1 + x^2} + (xe)s\right)^2. \tag{4.96}$$

If $(xe)s \geq 0$, then $A := c\sqrt{1 + x^2} + (xe)s \geq 0$, since $c \geq 0$. In the case $(xe)s < 0$ we get, by $(xe)^2 \leq x^2e^2 = x^2$,

$$(1 + x^2)c^2 - (xe)^2s^2 \geq (1 + x^2)c^2 - x^2s^2 = c^2 + x^2 \geq 0,$$

i.e. $c\sqrt{1 + x^2} \geq |(xe)s| = -(xe)s$, i.e. again $A \geq 0$. Hence, by (4.96),

$$\sqrt{1 + [T_t(x)]^2} = c\sqrt{1 + x^2} + (xe)s. \tag{4.97}$$

Still applying the abbreviations $c := \cosh t$, $s := \sinh t$, we define

$$\lambda_t(z) := z + (\bar{z}e)\left((c - 1)e, s\right) + z_0(se, c - 1) \tag{4.98}$$

for $z = (\bar{z}, z_0) \in Z$. Hence λ_t is a linear mapping from Z into Z. Observe $\cosh(-t) = c$ and $\sinh(-t) = -s$ and thus

$$\lambda_{-t}(z) = z + (\bar{z}e)\big((c-1)\,e, -s\big) + z_0(-se,\, c-1).$$

This implies for all $z \in Z$,

$$\lambda_{-t}\lambda_t(z) = z = \lambda_t\lambda_{-t}(z).$$

Hence, by applying the first Remark of section 1,7, $\lambda_t : Z \to Z$ must be bijective. In order to prove that λ_t is a Lorentz transformation, we will show (compare (3.141))

$$l(z_1, z_2) = l\big(\lambda_t(z_1),\, \lambda_t(z_2)\big)$$

for all $z_1, z_2 \in Z$. Since λ_t is linear, the last equation is equivalent with

$$(z_1 - z_2)^2 = [\lambda_t(z_1 - z_2)]^2,$$

i.e. with $z^2 = [\lambda_t(z)]^2$ for all $z \in Z$. Put

$$a := \big((c-1)\,e,\, s\big),\ b := (se, c-1)$$

and observe, by (4.98),

$$\lambda_t(z) = z + (\bar{z}e)\,a + z_0 b.$$

In view of $a^2 = 2(1-c)$, $ab = 0$, $b^2 = 2(c-1) = -a^2$, we obtain

$$\Delta := [\lambda_t(z)]^2 - z^2 = \big((\bar{z}e)^2 - z_0^2\big)\,a^2 + 2(\bar{z}e)(za) + 2z_0(zb),$$

i.e. $\Delta = 0$, because of $za = (c-1)(\bar{z}e) - sz_0$, $zb = s(\bar{z}e) - (c-1)z_0$. So we know that λ_t is a bijective and linear Lorentz transformation of Z. Obviously, $\lambda_t(0) = 0$. That $\lambda_t\big(H(Z)\big) \subseteq H(Z)$ holds true, follows from (4.98), (1.8), (4.97) by

$$\lambda_t(x, \sqrt{1+x^2}) \ = \big(x + [(\bar{x}e)(c-1) + \sqrt{1+x^2}s]\,e,\ (xe)\,s + \sqrt{1+x^2}c\big)$$

$$= \big(T_t(x),\ \sqrt{1 + [T_t(x)]^2}\big)$$

for all $x \in X$. Replacing t by $-t$, we also get $\lambda_t^{-1}\big(H(Z)\big) \subseteq H(Z)$. We finally must prove (4.94), i.e.

$$\mu\big(T_t(x)\big) = \lambda\mu(x) = \lambda(x, \sqrt{1+x^2}).$$

But this is already clear, since $\mu\big(T_t(x)\big) = \big(T_t(x), \sqrt{1 + [T_t(x)]^2}\big)$. $\square$

More precisely, we will denote the Lorentz transformation $\lambda_t : Z \to Z$ of (4.98) also by $L_{e,t}$. Observe

$$L_{e,t} = L_{-e,-t} \tag{4.99}$$

for all $t \in \mathbb{R}$ and all $e \in X$ satisfying $e^2 = 1$.

Proposition 37. *A Lorentz transformation σ of Z with*

$$\sigma\big(H\,(Z)\big) = H\,(Z)$$

satisfies $\sigma\,(0) = 0$.

Proof. Put $\sigma\,(0) =: d = (\bar{d}, d_0)$. Suppose that $j \in X$ satisfies $j^2 = 1$ and $\bar{d}j = 0$. Let ϱ be an arbitrary real number and define $x_\varrho \in X$ by

$$\sigma\,(x_\varrho, \sqrt{1 + x_\varrho^2}) = (\varrho j, \sqrt{1 + \varrho^2}),$$

by applying $\sigma\,\big(H\,(Z)\big) = H\,(Z)$. Hence, by (4.3),

$$-1 = l\big(0, (x_\varrho, \sqrt{1 + x_\varrho^2})\big) = l\,\big(d, (\varrho j, \sqrt{1 + \varrho^2})\big),$$

i.e. $0 = d^2 - 2d \cdot (\varrho j, \sqrt{1 + \varrho^2})$, i.e., by $\bar{d}j = 0$,

$$\vec{d}^2 - d_0^2 + 2d_0 \sqrt{1 + \varrho^2} = 0 \tag{4.100}$$

for all $\varrho \in \mathbb{R}$. Hence $d_0 = 0$ and $\vec{d}^2 = 0$, i.e. $d = 0$, by applying (4.100) for $\varrho = 0$ and $\varrho = 1$. $\qquad\square$

Proposition 38. *A Lorentz transformation φ of Z satisfies*

$$\varphi\,\big(H\,(Z)\big) = H\,(Z) \tag{4.101}$$

if, and only if, φ is linear and orthochronous.

Proof. Assume that φ is linear and orthochronous. If $z \in H\,(Z)$, we obtain

$$-1 = l\,(0, z) = l\,\big(\varphi\,(0), \varphi\,(z)\big) = l\,\big(0, \varphi\,(z)\big),$$

i.e. $[\varphi\,(z)]^2 = -1$. Moreover, $0 \le z$ implies $0 \le \varphi\,(z)$, i.e. $\varphi\,(z) \in H\,(Z)$. Hence the left-hand side of (4.101) is contained in $H\,(Z)$. Since φ^{-1} is linear and orthochronous as well, we obtain (4.101).— Assume now that $\varphi \in \mathbb{L}\,(Z)$ satisfies the equation (4.101). From Proposition 37 we get $\varphi\,(0) = 0$ and from Theorem 1 that φ is of the form

$$\varphi\,(z) = B_{p,k}\omega\,(z).$$

Since $(0, 1)$ is in $H\,(Z)$, so must be $\varphi\,(0, 1)$. Hence $0 \le \varphi\,(0, 1)$ and

$$\varphi\,(0, 1) = B_{p,k}\omega\,(0, 1) = B_{p,k}(0, 1).$$

Here k must be unequal to -1, since otherwise $\varphi\,(0, 1) = (0, -1)$, i.e. $0 \not\le \varphi\,(0, 1)$. Hence (see section 3.14)

$$\varphi\,(0, 1) = B_{p,k}(0, 1) = (0, 1) + k\,(p, 0) + \frac{k^2}{k + 1}\,(0, p^2),$$

i.e. $\varphi\,(0,1) = (0,1) + (kp, k-1)$ with $k^2(1-p^2) = 1$. Now $0 \leq \varphi\,(0,1)$ implies, by definition, that the last component of $\varphi\,(0,1)$, namely k, must be non-negative. Hence the Lorentz boost $B_{p,k}$ must be proper, i.e. $B_{p,k}\omega$ is orthochronous (see Theorem 5 and observe that $k \geq 0$ and $k^2(1-p^2) = 1$ imply $k \geq 1$.   □

Proposition 39. *Let φ_1, φ_2 be linear and orthochronous Lorentz transformations of Z with $\varphi_1(z) = \varphi_2(z)$ for all $z \in H\,(Z)$. Then $\varphi_1 = \varphi_2$.*

Proof. Since also $\varphi := \varphi_2^{-1}\varphi_1$ is linear and orthochronous, it is sufficient to consider only the case $\varphi_2 = \mathrm{id}$. Put $\varphi_1 =: \varphi$ and

$$\varphi\,(z) = B_{p,k}\omega\,(z) \tag{4.102}$$

with $k \geq 0$. From $\varphi\,(z) = z$ for all $z \in H\,(Z)$ and (3.120) we get

$$B_{-p,k}(\bar{z}, z_0) = \big(\omega\,(\bar{z}), z_0\big)$$

for all $z = (\bar{z}, z_0) \in H\,(Z)$. Applying this for $z = (0,1)$, we obtain (see section 3.14 and observe $k^2p^2 = k^2 - 1$ from $k^2(1-p^2) = 1$),

$$(0,1) + (-kp, k-1) = B_{-p,k}(0,1) = \big(\omega\,(0), 1\big) = (0,1),$$

i.e. $p = 0$, $k = 1$, i.e. $B_{p,k} = \mathrm{id}$. Hence $z = \varphi\,(z) = \omega\,(z)$ from (4.102) for all $z \in H\,(Z)$, i.e.

$$\big(\bar{z}, \sqrt{1+\bar{z}^2}\big) = \big(\omega\,(\bar{z}), \sqrt{1+\bar{z}^2}\big)$$

for all $\bar{z} \in X$, i.e. $\omega = \mathrm{id}$. Thus φ is the identity mapping of Z.   □

From Theorem 36 and Propositions 38 and 39 we know that there exists exactly one induced Lorentz transformation $\tau\,(f)$ of a given hyperbolic motion f and, moreover, that $\tau\,(f)$ is linear and orthochronous. We will designate by $\mathbb{L}_{\mathrm{orth}}(Z)$ the group of all orthochronous Lorentz transformations of (Z) leaving fixed $0 \in Z$. After a while we will see that every element of $\mathbb{L}_{\mathrm{orth}}(Z)$ is an induced Lorentz transformation of a certain hyperbolic motion and, moreover, that hyperbolic geometry over X and

$$\big(H\,(Z), \mathbb{L}_{\mathrm{orth}}(Z)\big) \tag{4.103}$$

are isomorphic, where $\mathbb{L}_{\mathrm{orth}}$, though acting on Z, is considered here as acting on $H\,(Z)$ only, namely via the restrictions of all $\varphi \in \mathbb{L}_{\mathrm{orth}}(Z)$ on $H\,(Z)$.

The induced Lorentz transformation $\tau\,(\omega)$ of the hyperbolic motion $x \rightarrow \omega\,(x)$, $\omega \in O\,(X)$, is given by (see Case 1 of the proof of Theorem 36)

$$(\bar{z}, z_0) \rightarrow \big(\omega\,(\bar{z}), z_0\big), \tag{4.104}$$

and that one of $x \rightarrow T_t(x)$ by

$$L_{e,t}(z) = z + (\bar{z}e)\big((c-1)\,e, s\big) + z_0(se, c-1), \tag{4.105}$$

in view of (4.98), where e is the axis of T_t and where we put $c := \cosh t$, $s := \sinh t$.

A *geometrical interpretation for proper Lorentz boosts* yields the following statement.

Theorem 40. *Suppose that $p \in X$ and $k \in \mathbb{R}_{\geq 0}$ satisfy $0 < p^2 < 1$ and $k^2(1 - p^2) = 1$. Then*

$$B_{p,k} = L_{e,t} \tag{4.106}$$

holds true with

$$k = \cosh t, \, t > 0, \, and \, p =: e \tanh t. \tag{4.107}$$

Remark. Exactly one proper Lorentz boost is missing in (4.106), namely that one with $p = 0$ and, consequently, with $k = 1$, because of $k^2(1 - p^2) = 1$ and $k \geq 0$.— On the other hand, to T_t with $t > 0$ and axis e there belongs, (by (4.106), the induced transformation $B_{p,k}$ with (4.107). If $t < 0$, we may apply that the translations T_t with axis e, and T_{-t} with axis $(-e)$ coincide.

Proof of Theorem 40. Since $k \geq 0$, we obtain, by section (3.14) and by $z = (\bar{z}, z_0) \in Z$,

$$B_{p,k}(z) = z + k\,(z_0 p, \bar{z}p) + \frac{k^2}{k+1}\left((\bar{z}p)\,p, z_0 p^2\right). \tag{4.108}$$

$k^2(1 - p^2) = 1$ implies $k > 1$ and, by $\cosh t := k, \, t > 0$,

$$\|p\| = \tanh t.$$

Define $e \in X$ by $p =: e \cdot \tanh t$ and put $c := \cosh t$, $s := \sinh t$. Hence, by (4.108),

$$B_{p,k}(z) = z + \left((\bar{z}e)(c - 1)\,e + sz_0 e, \, s\,(\bar{z}e) + z_0(c - 1)\right) = L_{e,t}(z). \qquad \square$$

Proposition 41. *Suppose that t is a real number and that $e \in X$ satisfies $e^2 = 1$. If e is the axis of the hyperbolic translation T_t, then the induced Lorentz transformation of T_t is given by*

$$\tau\,(T_t) = B_{e \tanh t, \, \cosh t}. \tag{4.109}$$

Proof. This follows for $t > 0$ from (4.106) and (4.107), since $L_{e,t} = \tau\,(T_t)$. In the case $t = 0$ we get $\tau\,(T_t) = \mathrm{id}$ and also $B_{e \tanh t, \, \cosh t} = B_{0,1} = \mathrm{id}$. Assume, finally, $t < 0$, and put $r := -t > 0$. If T'_r is a translation with axis $-e$, we obtain

$$T_t = T'_{-t} = T'_r,$$

i.e., by the first part of this proof, since $r > 0$,

$$\tau\,(T_t) = \tau\,(T'_r) = B_{(-e) \tanh r, \, \cosh r} = B_{e \tanh t, \, \cosh t}. \qquad \square$$

We now would like to show that every $\varphi \in \mathbb{L}_{\text{orth}}$ is induced by a motion. Assume

$$\varphi(z) = B_{p,k}\omega(z) = B_{p,k}\big(\omega(\bar{z}), z_0\big)$$

with $k \geq 1$. There is nothing to prove for $k = 1$, since then $\varphi(z) = \omega(z)$. We hence may assume $k > 1$ and $p \neq 0$. Put $k = \cosh t$, $t > 0$, and $p =: e\tanh t$. The translation T_t with axis e then induces $B_{p,k}$, by Theorem 40. So we would like to verify that the motion

$$x \rightarrow T_t\omega(x)$$

induces φ. If λ is induced by $x \rightarrow \omega(x)$ (see Case 1 of the proof of Theorem 36), we obtain, by (4.94),

$$\omega(x) = \mu^{-1}\lambda\mu(x)$$

and $T_t(x) = \mu^{-1}B_{p,k}\mu(x)$, i.e.

$$T_t\omega(x) = \mu^{-1}B_{p,k}\mu\big(\omega(x)\big) = \mu^{-1}B_{p,k}\lambda\mu(x) = \mu^{-1}\varphi\mu(x).$$

Hence φ is induced by $T_t\omega$.

Proposition 42. $\big(X, M(X, hyp)\big) \cong \big(H(Z), \mathbb{L}_{orth}(Z)\big)$.

Proof. Observe that $\mu : X \rightarrow H(Z)$ with $\mu(x) = (x, \sqrt{1+x^2})$ is a bijection. Moreover, associate to the hyperbolic motion f the restriction on $H(Z)$,

$$\tau(f)(z) = \mu f\mu^{-1}(z), \ z \in H(Z),$$

of the induced Lorentz transformation $\tau(f)$ of f. Hence

$$\tau : M(X, \text{hyp}) \rightarrow \mathbb{L}_{orth}(Z)$$

is an isomorphism satisfying

$$\mu\big(f(x)\big) = \mu f\mu^{-1}\big(\mu(x)\big) = \tau(f)\big(\mu(x)\big),$$

i.e. we obtain (1.15). $\qquad\square$

Remark. If $k \in \mathbb{R}$ and $p \in X$ satisfy $k > 1$ and $k^2(1-p^2) = 1$, Theorem 7 (chapter 4), case 2, implies

$$B_{0,-1}B_{p,k}(\bar{z}, z_0) = B_{-p,-k}\big(\omega(\bar{z}), z_0\big)$$

with

$$\omega(\bar{z}) = \bar{z} - \frac{2p\bar{z}}{p^2}\,p = \omega^{-1}(\bar{z}),$$

i.e. with $\omega \in O(X)$. Hence $B_{-p,-k} = B_{0,-1}B_{p,k}\omega$ represents a geometrical interpretation for improper Lorentz boosts $B_{-p,-k} \neq B_{0,-1}$, since, on the one hand, $B_{p,k}$ is induced by a hyperbolic translation and since, on the other hand, $B_{0,-1}$ and ω are simple geometrical mappings of Z (compare 3.6).

Expressing the hyperbolic distance in terms of $\left(H\left(Z\right), \mathbb{L}_{orth}(Z)\right)$ yields

$$\cosh \; \mathrm{hyp}\,(z_1, z_2) = -z_1 \cdot z_2 = 1 + \frac{1}{2}(z_1 - z_2)^2$$

for $z_1 =: (x, \sqrt{1 + x^2})$, $z_2 =: (y, \sqrt{1 + y^2})$. Because of (4.2) we also may write

$$2 \sinh\left(\frac{1}{2}\; \mathrm{hyp}\,(z_1, z_2)\right) = \sqrt{l\,(z_1, z_2)}, \qquad\qquad (4.110)$$

since $\sqrt{1 + x^2}\,\sqrt{1 + y^2} - xy \geq 1$ (see section 1.10).

Remark. $(x, \sqrt{1 + x^2})$ is said to be the *Weierstrass coordinates* of the hyperbolic point $x \in X$. Generally speaking, let

$$\psi : X \to \mathbb{R}$$

be an arbitrary function, for instance $\psi\,(x) = 0$, $\psi\,(x) = \sqrt{1 + x^2}$ or $\psi\,(x) = \|x\|$. Define

$$H_\psi(Z) := \left\{(x, \psi\,(x)) \in Z \mid x \in X\right\}$$

and the trivial bijection $m : X \to H_\psi(Z)$ by $m\,(x) = (x, \psi\,(x))$. Furthermore, let Γ_ψ be the group

$$\Gamma_\psi := \{\tau\,(f) := mfm^{-1} \mid f \in M\,(X, \mathrm{hyp})\}$$

which, of course, is a subgroup of Perm $H_\psi(Z)$. Obviously, the geometries

$$\left(X, M\,(X, \mathrm{hyp})\right),\; \left(H_\psi(Z), \Gamma_\psi\right)$$

are isomorphic, since $\tau\,:\, M\,(X, \mathrm{hyp}) \to \Gamma_\psi$ is a group isomorphism satisfying $\tau\,(f)(\mu\,(x)) = \mu\,(f\,(x))$ for all $x \in X$.

Appendix A

Notations and symbols

Theorems, propositions, and lemmata are numbered consecutively in each chapter, so that Lemma 1 may be followed by Proposition 2 and that by Theorem 3. Chapters are subdivided into sections but numbering of formulas is within chapters, not sections. The end of a proof is indicated by $\square$. The symbols $:=$ or $=:$ mean that the side of the equation, where the colon is, is defined by the other side. Sometimes *provided* is used as an abbreviation for *if and only if*.

$$A \Rightarrow B \quad \text{means} \quad A \text{ implies } B,$$

$$A \Leftrightarrow B \quad \text{is defined by} \quad (A \Rightarrow B) \text{ and } (B \Rightarrow A),$$

$$\forall \quad \text{abbreviates} \quad \textit{for all}.$$

Moreover,

$$\forall_{x \in S} \, A(x) \Rightarrow B(x) \quad \text{means} \quad A(x) \text{ implies } B(x) \text{ for all } x \in S,$$

$$\exists, \quad\quad\quad\quad\quad\quad\quad\quad \text{there exist(s)}$$

and

$$f : A \rightarrow B \text{ that } f \text{ is a mapping from } A \text{ into } B.$$

If f is a mapping from B into C and g a mapping from A into B, then $fg : A \rightarrow C$ is defined by $(fg)(x) := f[g(x)]$ for all $x \in A$.

If f is a mapping from A into B and if H is a subset of M, then $f \mid H$ (the so–called *restriction* of f on H) denotes the mapping $\varphi : H \rightarrow B$ with $\varphi(x) := f(x)$ for all $x \in H$.

If S is a set, then $\mathrm{id} : S \rightarrow S$ designates the mapping defined by $\mathrm{id}(x) = x$ for all $x \in S$.

If S is a set, then $\{x \in S \mid P(x)\}$ denotes the set of all x in S which satisfy property P.

If A, B are sets, then $A\backslash B := \{x \in A \mid x \notin B\}$.

$\mathbb{R}$ denotes the set of all real numbers, furthermore,

$$\mathbb{R}_{\geq 0} := \{x \in \mathbb{R} \mid x \geq 0\},$$
$$\mathbb{R}_{> 0} := \mathbb{R}_{\geq 0}\backslash\{0\}.$$

If $A_1, \ldots, A_n$ are sets, their *cartesian product* is

$$A_1 \times A_2 \times \cdots \times A_n := \{(x_1, \ldots, x_n) \mid x_i \in A_i \text{ for } i = 1, \ldots, n\}.$$

If M is a set, $\#M$ designates its cardinality.

If a is a non–negative real number, $\sqrt{a}$ denotes the real number $b \geq 0$ satisfying $b^2 = a$.

Appendix B

Bibliography

ACZÉL, J.:

  [1] Lectures on functional equations and their applications. Academic Press, New York – London, 1966.

ACZÉL, J. and DHOMBRES, J.:

  [1] Functional equations in several variables. Cambridge University Press. Cambridge, New York, 1989.

AL-DHAHIR, M.W., BENZ, W. AND GHALIEH, K.:

  [1] A Groupoid of the Ternary Ring of a Projective Plane. Journ. Geom. 42 (1991) 3–16.

ALEXANDROV, A.D.:

  [1] Seminar Report. Uspehi Mat. Nauk. 5 (1950), no. 3 (37), 187.

  [2] A contribution to chronogeometry. Canad. J. Math. 19 (1967) 1119–1128.

  [3] Mappings of Spaces with Families of Cones and Space-Time-Transformations. Annali di Matematica 103 (1975) 229–257.

ALEXANDROV, A.D. and OVCHINNIKOVA, V.V.:

  [1] Note on the foundations of relativity theory. Vestnik Leningrad. Univ. 11, 95 (1953).

ALPERS, A. and SCHRÖDER, E.M.:

  [1] On mappings preserving orthogonality of non-singular vectors. Journ. Geom. 41 (1991) 3–15.

ARTZY, R.:

  [1] Linear Geometry. Addison-Wesley, New York, 1965.

  [2] Geometry. BI Wissenschaftsverlag, Mannheim, Leipzig, Wien, Zürich, 1992.

BAER, R.:

  [1] Linear algebra and projective geometry. Academic Press, New York, 1952.

BATEMAN, H.:

  [1] The Transformation of the electrodynamical Equations. Proc. of the London Math. Soc., 2nd Series, 8 (1910) 223–264.

BECKMAN, F.S. and QUARLES JR., D.A.:

  [1] On Isometries of Euclidean Spaces. Proc. Amer. Math. Soc. 4 (1953) 810–815.

BENZ, W.:

[1] Abstandsräume und eine einheitliche Definition verschiedener Geometrien. Math. Sem. ber. 28 (1981) 189–201.

[2] Geometrische Transformationen (unter besonderer Berücksichtigung der Lorentztransformationen). BI Wissenschaftsverlag, Mannheim, Leipzig, Wien, Zürich, 1992.

[3] Real Geometries. BI Wissenschaftsverlag, Mannheim, Leipzig, Wien, Zürich, 1994.

[4] Ebene Geometrie. Einführung in Theorie und Anwendung. Spektrum Akademischer Verlag, Heidelberg – Berlin – Oxford, 1997.

[5] Einstein distances in Hilbert spaces. Result. Math. 36 (1999) 195–207.

[6] Hyperbolic distances in Hilbert spaces. Aequat. Math. 58 (1999) 16–30.

[7] Elliptic distances in Hilbert spaces. Aequat. Math. 59 (2000) 177–190.

[8] Mappings preserving two hyperbolic distances. Journ. Geom. 70 (2001) 8–16.

[9] Lie Sphere Geometry in Hilbert Spaces. Result. Math. 40 (2001) 9–36

[10] A common characterization of classical and relativistic addition. Journ. Geom. 74 (2002) 38–43.

[11] On Lorentz–Minkowski Geometry in Real Inner Product Spaces. Advances in Geometry. Special Issue (2003) 1–12.

[12] Möbius Sphere Geometry in Inner Product Spaces. Aequat. Math. 66 (2003) 284–320.

[13] A common characterization of Euclidean and Hyperbolic Geometry by Functional Equations. Publ. Math. Debrecen 63 (2003) 495–510.

[14] Extensions of distance preserving mappings in euclidean and hyperbolic geometry. Journ. Geom. 79 (2004) 19–26.

[15] Metric and Periodic Lines in Real Inner Product Space Geometries. Monatsh. Math. 141 (2004) 1–10.

[16] De Sitter distances in Hilbert spaces. Rocznik Nauk-Dydakt. Prace Mat. 17 (2000) 49–56.

BENZ, W. and BERENS, H.:

[1] A Contribution to a Theorem of Ulam and Mazur. Aequat. Math. 34 (1987) 61–63.

BEZDEK, K. and CONNELLY, R.:

[1] Two-distance preserving functions from euclidean space. Period. Math. Hung. 39 (1999) 185–200.

BLASCHKE, W.:

[1] Collected Works, Volumes 1 (1982), 2–5 (1985), 6 (1986). Edited by W. Burau – Hamburg, S.S. Chern – Berkeley, K. Leichtweiß – Stuttgart, H.R. Müller – Braunschweig, L.A. Santalo – Buenos Aires, U. Simon – Berlin, K. Strubecker – Karlsruhe. Thales Verlag, Essen.

[2] Vorlesungen über Differentialgeometrie und geometrische Grundlagen von Einsteins Relativitätstheorie III. Differentialgeometrie der Kreise und Kugeln. Bearbeitet von G. Thomsen. Grundlehren der math. Wissenschaften in Einzeldarstellungen. Bd. 29. Berlin, Springer, 1929.

BLUMENTHAL, L.M. and MENGER, K.:

[1] Studies in Geometry. W.H. Freeman and Comp., San Francisco, 1970.

BLUNCK, A.:

[1] Reguli and chains over skew fields. Beiträge Algebra Geom. 41 (2000) 7–21.

BRILL, A.:

[1] Das Relativitätsprinzip. Jber. DMV 21 (1912) 60–87.

BUEKENHOUT, F.:

[1] Handbook of Incidence Geometry. ed. by F. Buekenhout. Elsevier Science B.V., Dordrecht, Boston, 1995.

CACCIAFESTA, F.:

[1] An observation about a theorem of A.D. Alexandrov concerning Lorentz transformations. Journ. Geom. 18 (1982) 5–8.

DARÓCZY, Z.:

[1] Über die stetigen Lösungen der Aczél–Benz'schen Funktionalgleichung. Abh. Math. Sem. Univ. Hamburg 50 (1980) 210–218.

[2] Elementare Lösung einer mehrere unbekannte Funktionen enthaltenden Funktionalgleichung. Publ. Math. Debrecen 8 (1961) 160–168.

FARRAHI, B.:

[1] A characterization of isometries of absolute planes. Result. Math. 4 (1981) 34–38.

HAVLICEK, H.:

[1] On the geometry of field extensions. Aequat. Math. 45 (1993) 232–238.

[2] Spreads of right quadratic skew field extensions. Geom. Dedicata 49 (1994) 239–251.

HERGLOTZ, G.:

[1] Über die Mechanik des deformierbaren Körpers vom Standpunkt der Relativitätstheorie. Ann. d. Physik 36 (1911) 493–533.

HERZER, A.:

[1] Chain Geometries, in: Buekenhout, F., Handbook of Incidence Geometry. Elsevier Science B.V., Dordrecht, Boston, 1995.

HÖFER, R.:

[1] Metric Lines in Lorentz–Minkowski Geometry, to appear Aequat. Math.

[2] Periodic Lines in Lorentz–Minkowski Geometry, to appear.

HUANG, W.-L.:

[1] Transformations of Strongly Causal Space-Times Preserving Null Geodesics. Journ. Math. Phys. 39 (1998) 1637–1641.

[2] Null Line Preserving Bijections of Schwarzschild Spacetime. Comm. Math. Phys. 201 (1999) 471–491.

KUCZMA, M.:

[1] Functional equations in a single variable. Monografie Mat. Vol. 46, P.W.N., Warszawa, 1968.

[2] An introduction to the theory of functional equations and inequalities. Cauchy's equation and Jensen's inequality. Uniw. Slask-P.W.N., Warszawa, 1985.

KUZ'MINYH, A.V.:

[1] Mappings preserving a unit distance. Sibirsk. Mat. Ž. 20 (1979) 597–602.

[2] On the characterization of isometric and similarity mappings. Dokl. Akad. Nauk. SSSR 244 (1979) 526–528.

LESTER, J.A.:

[1] Cone preserving mappings for quadratic cones over arbitrary fields. Canad. J. Math. 29 (1977) 1247–1253.

[2] The Beckman–Quarles theorem in Minkowski space for a spacelike square-distance. Archiv d. Math. 37 (1981) 561–568.

[3] Alexandrov-Type Transformations on Einstein's Cylinder Universe. C.R. Math. Rep. Acad. Sci. Canada IV (1982) 175–178.

[4] Transformations of Robertson–Walker spacetimes preserving separation zero. Aequat. Math. 25 (1982) 216–232.

[5] Separation-Preserving Transformations of de Sitter Spacetime. Abhdlgn. Math. Sem. Hamburg 53 (1983) 217–224.

[6] A Physical Characterization of Conformal Transformations of Minkowski Spacetime. Ann. Discrete Math. 18 (1983) 567–574.

[7] The Causal Automorphisms of de Sitter and Einstein Cylinder Spacetimes. J. Math. Phys. 25 (1984) 113–116.

[8] Transformations Preserving Null Line Sections of a Domain. Result. Math. 9 (1986) 107–118.

[9] Distance-preserving transformations. In Handbook of Geometry. ed. by F. Buekenhout. Elsevier Science B.V., Dordrecht, Boston, 1995.

MENGER, K.:

[1] Selecta Mathematica I. Springer, Wien, New York, 2002.

[2] Untersuchungen über allgemeine Metrik. Math. Ann. 100 (1928) 75–163.

MOSZNER, Z. and TABOR, J.:

[1] L'équation de translation sur une structure avec zéro. Ann. Polon. Math. 31 (1975) 255–264.

PAMBUCCIAN, V.:

[1] Tarskian remarks on symmetric ternary relations, to appear.

PFEFFER, W.F.:

[1] Lorentz transformations of a Hilbert space. Amer. J. of Math. 103 (1981) 691–709.

RADÓ, F., ANDREESCŬ, D. and VÀLCAN, D.:

[1] Mappings of E^n into E^m preserving two distances. Babes–Bolyai University, Fac. of Math., Research Sem., Sem. on Geometry. Preprint No. 10 (1986) 9–22, Cluj–Napoca.

RÄTZ, J.:

[1] Comparison of inner products. Aequat. Math. 57 (1999) 312–321.

[2] Characterization of inner product spaces by means of orthogonally additive mappings. Aequat. Math. 58 (1999) 111–117.

SAMAGA, H.-J.:

[1] Miquel-Sätze in Minkowski-Ebenen. I. Aequat. Math. 49 (1995) 98–114, II. Result. Math. 25 (1994) 341–356, III. Geom. Ded. 56 (1995) 53–73.

SCHAEFFER, H.:

[1] Der Satz von Benz–Radó. Aequat. Math. 31 (1986) 300–309.

[2] Automorphisms of Laguerre geometry and cone-preserving mappings of metric vector spaces. Lecture Notes in Math. 792 (1980) 143–147.

SCHRÖDER, E.M.:

[1] Eine Ergänzung zum Satz von Beckman und Quarles. Aequat. Math. 19 (1979) 89–92.

[2] Vorlesungen über Geometrie I, II, III. BI-Wissenschaftsverlag, Mannheim, Wien, Zürich, 1991, 1991, 1992.

[3] On 0-distance preserving permutations of affine and projective quadrics. Journ. Geom. 46 (1993) 177–185.

[4] Zur Kennzeichnung distanztreuer Abbildungen in nichteuklidischen Räumen. Journ. Geom. 15 (1980) 108–118.

[5] Ein einfacher Beweis des Satzes von Alexandrov–Lester. Journ. Geom. 37 (1990) 153–158.

[6] Zur Kennzeichnung der Lorentztransformationen. Aequat. Math. 19 (1979) 134–144.

SCHWERDTFEGER, H.:

[1] Geometry of Complex Numbers. Circle Geometry, Moebius Transformation, Non-euclidean Geometry. Dover Publications, New York, 1979.

SEXL, R. and URBANTKE, H.K.:

[1] Relativität, Gruppen, Teilchen. Springer-Verlag, Wien, New York, 1976.

TIMERDING, H.E.:

[1] Über ein einfaches geometrisches Bild der Raumzeitwelt Minkowskis. Jber. DMV 21 (1912) 274–285.

ZEEMAN, E.C.:

[1] Causality implies the Lorentz group. Journ. Math. Phys. 5 (1964) 490–493.

Appendix C

Index